AF358721

Material Modeling in Multiphysics Simulation

Editors

Francesco De Bona
Jelena Srnec Novak
Francesco Mocera

Basel • Beijing • Wuhan • Barcelona • Belgrade • Novi Sad • Cluj • Manchester

Editors
Francesco De Bona
University of Udine
Udine
Italy

Jelena Srnec Novak
University of Rijeka
Rijeka
Croatia

Francesco Mocera
Politecnico di Torino
Torino
Italy

Editorial Office
MDPI
St. Alban-Anlage 66
4052 Basel, Switzerland

This is a reprint of articles from the Special Issue published online in the open access journal *Metals* (ISSN 2075-4701) (available at: https://www.mdpi.com/journal/metals/special_issues/Modeling_Multiphysics_Simulation).

For citation purposes, cite each article independently as indicated on the article page online and as indicated below:

Lastname, A.A.; Lastname, B.B. Article Title. *Journal Name* **Year**, *Volume Number*, Page Range.

ISBN 978-3-7258-1085-7 (Hbk)
ISBN 978-3-7258-1086-4 (PDF)
doi.org/10.3390/books978-3-7258-1086-4

Contents

About the Editors

Francesco De Bona

He received the Ph.D degrees in Applied Mechanics from the Polytechnic University of Turin, Italy in 1987. He was head of the Micromechanics laboratory at Sincrotrone Trieste from 1988 to 1995. From 1996 till now he is at Polytechnic Department of Engineering and Architecture (DPIA), University of Udine, currently as full professor of Machine Design. His research activity includes computational and experimental mechanics in microfabrication and microsystems design, multiphysics simulation, low cycle fatigue and cyclic plasticity. He was scientific coordinator of 3 EU research projects. He is author of more than 100 scientific publications and 3 patents.

Jelena Srnec Novak

In 2016, she received her Ph.D in Industrial and Information Engineering at the University of Udine (Italy), where she also worked as a postdoctoral researcher at the Polytechnic Department of Engineering and Architecture (DPIA) until 2020. She is an assistant professor in the Faculty of Engineering at the University of Rijeka (Croatia). Her main research interests are focused on the mechanical behavior of materials, cyclic plasticity, material characterization, low-cycle fatigue prediction, non-linear finite element simulations, and nanoindentation. She is the author and co-author of more than 30 scientific papers in the above areas published in international journals and presented at conferences.

Francesco Mocera

He received his Ph.D in Mechanical Engineering from Politecnico di Torino, Italy, in 2019. He has been an assistant professor at Politecnico di Torino since 2020. His research interests span all of the mechanical and electrical components of electric and hybrid non-road mobile machineries. His expertise covers the design and numerical modeling of electro-mechanical powertrain components, the development and optimization of control strategies aimed at machine productivity and efficiency and the use of field data to characterize performance and environmental footprint. He also conducts studies on the electrothermo–mechanical behavior of lithium-ion cells, currently the most widespread technology for on-board energy storage. At Politecnico di Torino, he is currently in charge of two courses of machine design and a new course entitled "Mechanics for Agriculture" covering the basic principles and innovation in powertrain design in agricultural machineries.

Editorial

Material Modeling in Multiphysics Simulation

Francesco De Bona [1,*]**, Francesco Mocera** [2] **and Jelena Srnec Novak** [3]

[1] Department Polytechnic of Engineering and Architecture, University of Udine, Via delle Scienze 208, 33100 Udine, Italy

[2] Department of Mechanical and Aerospace Engineering (DIMEAS), Politecnico di Torino, Corso Duca degli Abruzzi 24, 10129 Torino, Italy; francesco.mocera@polito.it

[3] Faculty of Engineering & Centre for Micro- and Nanosciences and Technologies, University of Rijeka, Vukovarska 58, 51000 Rijeka, Croatia; jsrnecnovak@uniri.hr

* Correspondence: debona@uniud.it; Tel.: +39-0432-558269

1. Introduction

Virtual prototyping techniques, generally based on numerical methods, are widely used in the process of designing an industrial product [1,2]. In recent decades, the demand for strong improvements in terms of productivity, reliability, and cost reduction have been fundamental considerations in this form of design, often requiring more than one simultaneously occurring physical field (thermal, mechanical, electrical, metallurgical, etc.) to be taken into account. At present, a huge amount of commercial code and a huge number of new algorithms have been developed for performing multiphysics simulations [3,4]; nevertheless, the availability of a suitable material model often presents a bottleneck in obtaining reliable results. For example, see [5,6]. The Special Issue is thus aimed at investigating metallic material modeling techniques for virtual prototypes, with an emphasis on both the theoretical aspects and experimental identification and verification. The simulation of additive manufacturing techniques is paid special attention as an emerging field [7]. Nevertheless, more traditional metal-forming processes, continuous casting in particular [8], still require new approaches given increases in the casting speed and in the dimensions of the final products. A wealth of other topics could benefit from multiphysics simulations, in particular metal forming [9,10], joining techniques [11], the thermal treatment of metals [12,13], and manufacturing processes [14]. This Special Issue collates papers that provide state-of-the-art knowledge on material modeling for multiphysics simulations and in which the above-mentioned topics are developed and applied to relevant engineering case studies.

2. An Overview of the Published Articles

Additive manufacturing (AM) techniques for metals constitute a fascinating challenge in materials science and engineering. Firstly, AM techniques offer the possibility of manufacturing unique geometries, which is not possible in traditional metallurgical processing. Even more promisingly, AM processes enable local control over the microstructure and properties. It must be pointed out that it is possible that manufacturing particular geometries could cause a clash with the accuracy of the techniques used, and this requires accurate calibration of the process parameters. In (contribution 1), the distortion of thin-walled structures obtained using the laser power bed fusion (LPBF) process is thermo-mechanically analyzed. In particular, a variety of thin-walled components are printed using LPBF with different wall thicknesses and building-heights. Different open and closed shapes are compared, and a 3D scanner is used to measure their actual warpage. This experimental scheme facilitates the calibration of an FE model of the AM process. In particular, the adopted numerical strategy is based on the inherent strain method. The inherent strain tensor is assumed as a user-defined material parameter depending on the material and process characteristics. Its value was obtained experimentally. This numerical tool was also

Citation: De Bona, F.; Mocera, F.; Srnec Novak, J. Material Modeling in Multiphysics Simulation. *Metals* **2024**, *14*, 296. https://doi.org/10.3390/met14030296

Received: 29 November 2023
Accepted: 27 February 2024
Published: 1 March 2024

used to define a structural optimization strategy to mitigate the warpage of the thin-walled parts printed using LPBF. Another promising possibility offered by the AM process is the possibility of attaining local control over the microstructure and properties; nevertheless, such an approach is quite difficult in practice due to the complex multiscale relationships between the material parameters and processing conditions. This aspect is considered in (contribution 2), where coupled modeling of the process, microstructure, and properties was investigated using three different numerical strategies. The time–temperature history of the AM raster patterns from a computational fluid dynamics module serves as the input to multilayer simulations of the grain structure using cellular-automata-based code. Finally, a crystal plasticity finite element model is used to simulate the micromechanical response and properties. The aim is to assess the grain size and texture as a function of the number of layers, as well as the influence of heterogeneous nucleation.

While AM techniques are still in the development phase, nevertheless, other processes have largely been consolidated, and their set-up seems to be well defined in most cases, for instance, the continuous casting process of steels. However, even in this case, since the production requirements are becoming increasingly demanding, new strategies for overcoming the complexity of process simulation must be adopted. In particular, solidifying steel follows highly nonlinear thermo-mechanical behavior depending on the loading history, temperature, and metallurgical phase fraction calculations (liquid, ferrite, and austenite). Because most process events (e.g., solidification, segregation, defect production) are temperature-driven, formulating a tool for analyzing the thermal field is a fundamental preliminary objective in the development of an accurate material model. Theoretically, to describe such a process, a three-dimensional approach would be required. In this case, the computational cost of obtaining results is often unreasonably resource- and time-intensive. To overcome these issues, a faster bidimensional approach known as the traveling slice approach was developed and is now frequently utilized. Although this modeling strategy is frequently encountered in literature reviews, an assessment of its limitations remains lacking, especially within the current context of increases in casted product sizes, which may call into question the validity of this modeling technique. In (contribution 3), the traveling slice model is compared to a non-approximated analysis, notably using large-dimension products. The traveling slice approach can be considered a proven modeling technique for describing continuous casting processes. In (contribution 4), a numerical model with a computationally challenging multiphysics approach is used in high-performance computing to generate sufficient training and testing data for subsequent deep learning. It has thus been demonstrated how innovative sequence deep learning methods can learn from multiphysics modeling data on a solidifying slice traveling in a continuous caster and correctly and instantly capture the complex history and temperature-dependent phenomena in test data samples. The use of machine learning techniques to support multiphysics and multi-phase simulations of continuous casting processes is also proposed in (contribution 5), where an industrial contactless vertical casting process has been modeled using evolving domain and dynamic mesh techniques. In particular, an augmented genetic algorithm machine learning approach enables the implementation of an accurate material. It was thus possible to increase the accuracy and flexibility of the process simulations while limiting the required computational time and resources. The technique is sufficiently flexible to implement using mainstream commercial solvers for material processes.

Many other technological processes in metal production can benefit from material modeling, as is the case for ultrasonic treatment. Cavitation, caused by high-intensity ultrasonic treatment, is used in a wide range of industrial applications and becomes more and more pertinent to metallurgy and foundry processes. It can be used as an effective method for modifying a material's microstructure and improving its mechanical properties, especially in the context of the treatment of aluminum alloys. In (contribution 6), the capabilities of computational fluid dynamics to model the formation and dynamics of acoustic cavitation in an aluminium alloy are investigated. Among the different metal-forming techniques, hot metal extrusion is generally preferred when the final product is characterized by a high aspect ratio—for example, turbine blades. This process induces higher plastic deformation in comparison to forging. Thus, finer process parameter tuning is required to avoid unwanted microstructural effects. A typical failure occurring during hot metal extrusion is the formation of surface cracks. Numerical modeling could thus be a useful tool to more deeply investigate the influence of the process parameters on the final characteristics of the product. In (contribution 7), an FE-based multiphysics numerical model of the extrusion process for a superalloy component was devised. In parallel, a series of experimental tests were designed and carried out to examine the extrusion of pre-heated Inconel 718 billets, thus allowing for comprehensive validation of the numerical model and its respective results. The validated model was then used to perform parametric analyses in order to pinpoint the ranges of the processing conditions that avoid the formation of unwanted features. In particular, the microstructure evolution can be simulated using a semi-empirical model, taking into account both the recrystallization (dynamic and static) and grain growth, providing the average grain size and the fraction of recrystallized grains.

Numerical simulation can also profitably support the fine-tuning of thermal treatment. In (contribution 8), a simulation procedure for predicting the influence of the carbon content and quenching process parameters on the phase composition and hardness distribution after heat treatment is proposed. Experiments applying quenching in the form of high-pressure gas quenching and quenching oil were employed to validate the computational results. In particular, careful set-up of a carbon diffusivity model incorporating the influence of the alloying elements was undertaken in this research. Thermal data such as thermal conductivity and heat capacity also required adjustments; therefore, experimental data were used. The material model must also take into account phase transformations, which, depending on the cooling rate, can be classified as diffusion-controlled or diffusionless. The Johnson–Mehl–Avrami–Kolmogorov model was adopted to describe the kinetics of the isothermal phase transformation, during which diffusion is the governing phenomenon, according to the nucleation and growth of the new phase. Martensitic transformation is a diffusionless transformation that occurs upon rapid quenching of the austenite phase. As martensitic transformation is athermal, that is, is not controlled by the thermal history of the material, the volume fraction of the transformed phase is calculated based on an equation incorporating the degree of undercooling of the material. To describe the transformation kinetics, a Koistinen–Marburger model was used. Finally, the Maynier model was adopted for hardness calculations. In (contribution 9), the isothermal decomposition of austenite in steel is mathematically modeled and computer-simulated. This research has significant implications in the field of the thermal treatment of steels. In fact, the isothermal decomposition of austenite implies the steel is quenched from the austenite range to the temperature of isothermal transformation, where all of the austenite decomposes at a constant temperature. The microstructure composition and mechanical properties of the steel considered can thus be optimized and controlled. The proposed mathematical model was verified experimentally, confirming that the characteristic parameters included in the model of ferrite, pearlite, and bainite transformation can be successfully evaluated.

Furthermore, joining techniques often require a multiphysics approach. In (contribution 10), a mathematical model of induction soldering for waveguide assembly components was developed, which could help with testing and calibrating the induction soldering process for the thin-walled aluminum waveguides found in spacecraft. To verify the developed models, simulations were compared with experiments, confirming the prediction accuracy.

Finally, in (contribution 11), fundamental research on metals was shown to potentially better our understanding of how a material model can be more accurately predicted and interpreted. In particular, the temperature dependence of resistivity over a very wide temperature range is explained according to free randomly moving electrons scattering due to electronic defects, accounting for the thermal energy exchange between phonons and free randomly moving electrons.

Acknowledgments: The Guest Editors appreciate the contributions of the distinguished authors and the time and effort of the reviewers, editors, and the editorial staff of *Metals*.

Conflicts of Interest: The authors declare no conflicts of interest.

List of Contributions:

1. Lu, X.; Chiumenti, M.; Cervera, M.; Tan, H.; Lin, X.; Wang, S. Warpage Analysis and Control of Thin-Walled Structures Manufactured by Laser Powder Bed Fusion. *Metals* **2021**, *11*, 686. https://doi.org/10.3390/met11050686.
2. Lchigo, M.; Carson, R.; Belak, J. Understanding Uncertainty in Microstructure Evolution and Constitutive Properties in Additive Process Modeling. *Metals* **2022**, *12*, 324. https://doi.org/10.3390/met12020324.
3. Bazzaro, G.; De Bona, F. Effectiveness of Travelling Slice Modeling in Representing the Continuous Casting Process of Large Product Sections. *Metals* **2023**, *13*, 1505. https://doi.org/10.3390/met13091505.
4. Koric, S.; Abueidda, D.W. Deep Learning Sequence Methods in Multiphysics Modeling of Steel Solidification. *Metals* **2021**, *11*, 494. https://doi.org/10.3390/met11030494.
5. Horr, A.M.; Kronsteiner, J. Dynamic Simulations of Manufacturing Processes: Hybrid-Evolving Technique. *Metals* **2021**, *11*, 1884. https://doi.org/10.3390/met11121884.
6. Riedel, E.; Bergedieck, N.; Scharf, S. CFD Simulation Based Investigation of Cavitation Dynamics during High Intensity Ultrasonic Treatment of A356. *Metals* **2020**, *10*, 1529. https://doi.org/10.3390/met10111529.
7. Bacchetti, S.; Coppola, M.A.; De Bona, F.; Lanzutti, A.; Miotti, P.; Salvati, E.; Sordetti, F. Experimental and Numerical Investigation of Hot Extruded Inconel 718. *Metals* **2023**, *13*, 1129. https://doi.org/10.3390/met13061129.
8. Iżowski, B.; Wojtyczka, A.; Motyka, M. Numerical Simulation of Low-Pressure Carburizing and Gas Quenching for Pyrowear 53 Steel. *Metals* **2023**, *13*, 371. https://doi.org/10.3390/met13020371.
9. Smoljan, B.; Iljkić, D.; Smokvina Hanza, S.; Hajdek, K. Mathematical Modelling of Isothermal Decomposition of Austenite in Steel. *Metals* **2021**, *11*, 1292. https://doi.org/10.3390/met11081292.
10. Tynchenko, V.; Kurashkin, S.; Tynchenko, V.; Bukhtoyarov, V.; Kukartsev, V.; Sergienko, R.; Kukartsev, V.; Bashmur, K. Mathematical Modeling of Induction Heating of Waveguide Path Assemblies during Induction Soldering. *Metals* **2021**, *11*, 697. https://doi.org/10.3390/met11050697.
11. Palenskis, V.; Jonkus, V. Study of the Free Randomly Moving Electron Transport Peculiarities in Metals. *Metals* **2023**, *13*, 1551. https://doi.org/10.3390/met13091551.

References

1. Moatamedi, M.; Rahulan, T.; Khawaja, H. *Multiphysics Simulations in Automotive and Aerospace Applications*, 1st ed.; Academic Press: London, UK, 2022; ISBN 978-0-12-817899-7.
2. Zhang, Q.; Cen, S. *Multiphysics Modeling: Numerical Methods and Engineering Applications*; Elsevier: London, UK, 2016.
3. Zimmerman, W.B.J. *Multiphysics Modeling with Finite Element Methods*; World Scientific Ed.: Singapore, 2006.
4. Yang, Z. *Multiphysics Modeling with Application to Biomedical Engineering*; CRC Press: Boca Raton, FL, USA, 2020.
5. Moro, L.; Benasciutti, D.; De Bona, F. Simplified numerical approach for the thermo-mechanical analysis of steelmaking components under cyclic loading: An anode for electric arc furnace. *Ironmak. Steelmak.* **2019**, *46*, 56–65. [CrossRef]
6. Moro, L.; Srnec Novak, J.; Benasciutti, D.; De Bona, F. Thermal distortion in copper moulds for continuous casting of steel: Numerical study on creep and plasticity effect. *Ironmak. Steelmak.* **2019**, *46*, 97–103. [CrossRef]
7. Francois, M.M.; Sun, A.; King, W.E.; Henson, N.J.; Tourret, D.; Bronkhorst, C.A.; Abdeljawad, F. Modeling of additive manufacturing processes for metals: Challenges and opportunities. *Curr. Opin. Solid State Mater. Sci.* **2017**, *21*, 198–206. [CrossRef]
8. Thomas, B.G. Review on modeling and simulation of continuous casting. *Steel Res. Int.* **2018**, *89*, 1700312. [CrossRef]
9. Nielsen, C.V.; Martin, P.A.F. *Metal Forming: Formability, Simulation, and Tool Design*, 1st ed.; Academic Press: London, UK, 2021.
10. Dal, M.; Fabbro, R. An overview of the state of art in laser welding simulation. *Opt. Laser Technol.* **2016**, *78*, 2–14. [CrossRef]
11. Yanagimoto, J.; Banabic, D.; Banu, M.; Madej, L. Simulation of metal forming–Visualization of invisible phenomena in the digital era. *CIRP Ann.* **2022**, *71*, 599–622. [CrossRef]
12. Simsir, C.; Hakan Gür, C. Simulation of Quenching. In *Handbook of Thermal Process Modelling of Steels*; Hakan Gür, C., Pan, J., Eds.; CRC Press: Boca Raton, FL, USA, 2009; pp. 341–426.

13. Rohde, J.; Jeppsson, A. Literature review of heat treatment simulations with respect to phase transformation, residual stresses and distortion. *Scand. J. Metall.* **2000**, *29*, 47–62. [CrossRef]
14. Jabbari, M.; Baran, I.; Mohanty, S.; Comminal, R.; Sonne, M.R.; Nielsen, M.W.; Spangenberg, J.; Hattel, J.H. Multiphysics modelling of manufacturing processes: A review. *Adv. Mech. Eng.* **2018**, *10*, 1687814018766188. [CrossRef]

metals

Article

Study of the Free Randomly Moving Electron Transport Peculiarities in Metals

Vilius Palenskis and Vytautas Jonkus *

Faculty of Physics, Vilnius University, Sauletekio av. 9, III bld., Vilnius LT-10222, Lithuania; vilius.palenskis@ff.vu.lt
* Correspondence: vytautas.jonkus@ff.vu.lt

Abstract: In this study, we review some aspects of the application of free randomly moving (RM) electron density and its probability density function distribution to the main free electron transport characteristics of elemental metals. It is shown that metal atom thermal vibrations not only produce free RM electrons, but also produce the same number of electronic defects (weakly shielded ions). The general expressions for the drift mobility, diffusion coefficient, and mean free path of randomly moving electrons are presented. It is shown that the scattering of free RM electrons is mainly due to electronic defects, which cause the distortion of the periodic potential (or the charge density) distribution in the periodic lattice. The resistivity of the elemental metal is caused by electronic defect scattering, taking into account the exchange in the thermal energies between phonons and free RM electrons. Special attention is paid to the analysis of the Hall effect measurement data: the Hall coefficient is presented for two types of RM electrons and holes, taking into account electron-like and hole-like densities of states. The paramagnetism and diamagnetism of the free RM electrons are simply explained using the definition of free RM electron density.

Keywords: density of free randomly moving (RM) electrons; probability density function; density of states (DOS); mean free path of electrons; resistivity of metals; diffusion coefficient; drift mobility; Hall effect; Hall mobility

Citation: Palenskis, V.; Jonkus, V. Study of the Free Randomly Moving Electron Transport Peculiarities in Metals. *Metals* **2023**, *13*, 1551. https://doi.org/10.3390/met13091551

Academic Editors: Francesco De Bona and Asit Kumar Gain

Received: 3 July 2023
Revised: 8 August 2023
Accepted: 29 August 2023
Published: 3 September 2023

1. Introduction

The foundations of electron transport were laid at the beginning of the last century by the Drude–Lorentz–Sommerfeld free electron theory [1–3]. The main achievement of the Drude–Lorentz theory was the prediction of the Wiedemann–Franz law. Sommerfeld solved the electronic heat capacity problem by considering the Fermi–Dirac statistics of electrons in metals. The main limitations of the Sommerfeld model of free electrons are related to the suggestion that all of the valence electrons in the metal are free, as a result of which the estimates of the free electron density, their mobility, the Hall coefficient, and the Fermi-level energy are not correct. Later, very important works on the electron theory of metals were published [4–17], but they did not solve the mentioned limitations. Even recently [18], explanations of electrical conductivity in metals have been proposed based on the Drude–Sommerfeld model in which all valence electrons are free, which is completely inapplicable to metals.

According to quantum mechanics, in the ideal periodic lattice of a metal with periodic potential energy distribution, free electrons can move without any scattering by lattice atoms as Bloch waves [5,12]. The scattering of the free electrons can take place in spots of chemical and structural imperfection. These defects cause the metal's residual resistivity at very low temperatures. It is a well-known fact that at temperatures above the Debye temperature, the resistivity of elemental metals changes linearly with temperature, while it usually changes as T^{-5} below it [11]. This has been thought to be caused by lattice atom vibrations [6,8,9], but this cannot explain why the real electron mean free path is many orders of magnitude greater than the interatomic distance. As shown in [19–23], the lattice vibrations play another role. In this study, the estimation of the density of free randomly moving (RM) electrons and their probability density distribution as a function of electron energy and the applications of these characteristics to free electron transport in elemental metals are presented.

Section 2 deals with free electron effective density and the probability density distribution of free electrons on energy; Section 3 deals with the mean free path of free RM electrons and resistivity temperature dependences of elemental metals; Section 4 deals with the Hall effect in elemental metals; Section 5 deals with the magnetic properties caused by free RM electrons in elemental metals; and Section 6 deals with the plasma frequency of free RM electrons.

2. The Effective Density of Free RM Electrons and Their Probability Density Distribution as a Function of Energy

As shown in [19–23], the effective density n_{eff} of the free RM electrons can be expressed as

$$n_{\text{eff}} = \int_0^\infty g(E)f(E)f_1(E)\mathrm{d}E = kT\int_0^\infty g(E)[-\partial f(E)/\partial E]\mathrm{d}E, \tag{1}$$

where $g(E)$ is the density of states (DOS) in the conduction band; $f(E)$ is the Fermi distribution function; and $f_1(E) = 1 - f(E)$ is the probability that at a given temperature T, an electron can be thermally scattered or change its energy state due to the external field.

Such a description of free RM electrons is valid for any degree of degeneracy of the electron gas in materials, and it is valid for semiconductors, metals and superconductors in the normal state. From Equation (1), it follows that the function

$$p(E) = -\partial f(E)/\partial E = f(E)[1 - f(E)/kT], \tag{2}$$

is the probability density function of the free RM electron distribution as a function of energy, which fulfils all the requirements of probability theory [24,25] and is consistent with the Pauli exclusion principle. It should be noted that the probability density function allows us to determine the average value of any random function $x(E)$ as a function of energy:

$$< x(E) >= \int_0^\infty x(E)p(E)g(E)\mathrm{d}E. \tag{3}$$

For materials that have a non-degenerate electron gas, the probability $f_1(E) = 1 - f(E) \approx 1$, and Equation (1) adopts the classical view. Thus, all electrons in the conduction band n are free and participate in conduction:

$$n_{\text{eff}} = n = \int_0^\infty g(E)f(E)\mathrm{d}E. \tag{4}$$

Equation (4) is not appropriate for describing free RM electrons in metals. For metals, the probability density function $p(E)$ has a sharp maximum at the Fermi energy $E = E_F$. Under these conditions, Equation (1) becomes

$$n_{\text{eff}} = g(E_F)kT \ll n, \tag{5}$$

where $g(E_F) = g(E)$ at $E = E_F$. The DOS at Fermi-level energy $g(E_F)$ can be obtained from the electronic heat capacity data for elemental metals [7]. The DOS $g(E_F)$ values of the elemental metal distribution in the periodic table are shown in Figure 1. As can be seen from Figure 1, the smallest $g(E_F)$ values are those for Be and alkali metals, while the largest ones are those for V, Ni, Pd, and Nb.

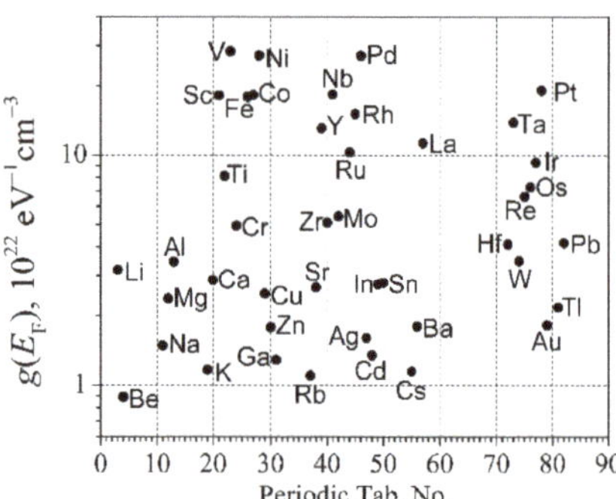

Figure 1. The DOS $g(E_F)$ values of elemental metal distribution in the periodic table.

Figure 2 provides a schematic hypothetical illustration of the DOS for noble- and transition-group metals at room temperature, as described in [26]. In the case of noble metals (Figure 2a), the area beneath the curve $g_s(E)f(E)$ in yellow is obtained using Equation (4) and represents the total density of valence electrons in the conduction band; the red area corresponds to the effective density of free RM electrons evaluated using Equation (1). It can be seen that the d-band is not created in noble metals, because this band is lower by 2 eV than the Fermi-level energy. For transition metals, the d-band is not entirely filled, and the total effective density of the free RM electrons can be described as $n_{eff} = g_\Sigma(E_F)kT = [g_s(E_F) + g_d(E_F)]kT$, where $g_\Sigma(E_F)$ is obtained from the electronic heat measurements. As can be seen from Figure 1, the DOS $g(E_F)$ at the Fermi surface for transition-group metals is much larger than that for alkali and noble metals.

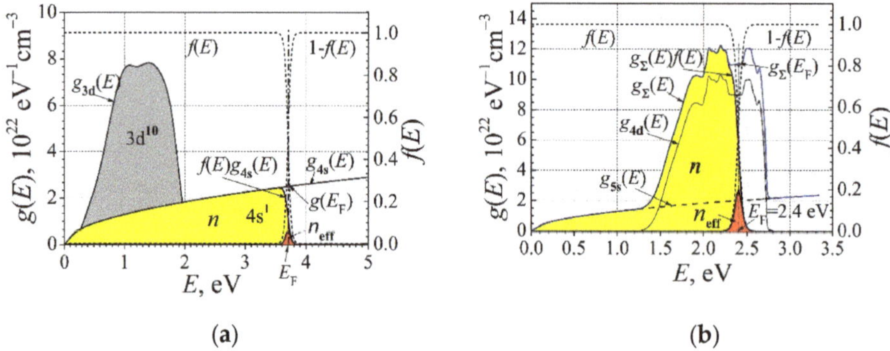

(a)

(b)

Figure 2. Schematic illustration of the energy dependence of the functions $g(E)$, $g(E)f(E)$, and $g(E)f(E)$ $[1 - f(E)]$ for noble (**a**) and transition group (**b**) metals at room temperature. The yellow area represents the total density n of valence electrons in the conduction band and the red area represents the effective density n_{eff} of free RM electrons; $g_\Sigma(E) = g_{5s}(E) + g_{4d}(E)$.

Since only free RM electrons n_{eff} electrons participate in the conduction of metals, the other parts of the valence electrons ($n - n_{eff}$) cannot change their energy because all neighbor energy levels are occupied, and also because of the Pauli exclusion principle. Therefore, those parts of the electrons have a constant energy and can only move around the native ions. It is essential to emphasize that free RM electrons in equilibrium conditions do not interact with these ($n - n_{eff}$) electrons as they cannot change their energy.

The valence electrons of a given elemental metal are distributed in the conduction band from 0 to the Fermi level energy E_F. Those atoms whose valence electrons have energies close to the Fermi level energy due to lattice thermal vibrations can be excited and leave the native atoms, becoming free and able to move randomly. The atoms that lose their valence electrons partially are shielded by the valence electrons of neighboring atoms. Such partially shielded ions, which are referred to as electronic defects, cause a distortion of the periodic potential in the periodic lattice. Therefore, thermal vibrations of lattice atoms produce free RM electrons and the same number of electronic defects, as well as the local distortion spots in the potential periodicity. In Figure 3, there is presented a schematic illustration of a two-dimensional lattice pattern of metal atoms. The waves of valence electrons partially overlap with those of neighboring atoms, and they move in the field of a central force of the native ions [27]. When a particular atom, due to its thermal vibration, excites a free RM electron, as shown in Figure 3A in the row (b), it also produces a distortion of the potential $U(x)$ periodicity, as shown schematically in Figure 3B in case (b). Therefore, atoms that produce free RM electrons produce the same number of ionic spots (electronic defects), which are not completely screened by the neighbor valence electrons. Considering that the density of free RM electrons is equal to $n_{\text{eff}} = g(E_F)kT$, this means that the average density of electronic defects can be estimated as [23]:

$$N_{\text{eff}} = n_{\text{eff}} = g(E_F)kT, \tag{6}$$

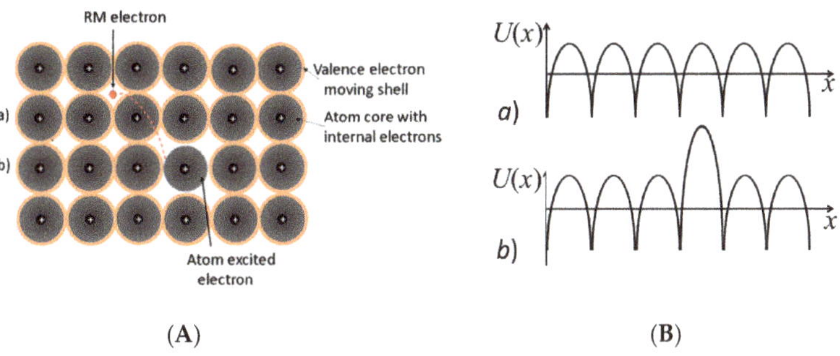

Figure 3. (**A**) Schematic representation of a two-dimensional pattern of metal atoms, and (**B**) a schematic potential $U(x)$ distribution. In row (b) one atom excites the RM electron, causing the distortion of the potential periodicity in row (b) in Figure 3B.

The other parts of the atoms $(N - N_{\text{eff}})$ (where N is the total atom density) do not have enough vibrational (phonon) energy to produce free RM electrons close to the Fermi level energy. It is worth pointing out that the thermal vibrations of the lattice atoms play another role, as described in [4,6–12].

As shown in [19–23], the conductivity σ of a homogeneous material can be expressed as

$$\sigma = q^2 n_{\text{eff}} D / kT = q n_{\text{eff}} \mu_{\text{drift}}, \tag{7}$$

where q is the electron charge, D is the diffusion coefficient, and μ_{drift} is the drift mobility of free RM electrons. From this expression follows the Einstein relation $D / \mu_{\text{drift}} = kT / q$, which is valid for materials with any degree of the degeneration of electron gas with one type of free charge carriers (electrons and holes). The relationship between the electrical conductivity and both the diffusion coefficient and the drift mobility is described by Equation (7) and is shown in Figure 4. Considering that values of the electrical conductivity and the electronic heat capacity are well known and are available from the Handbook [28], values of the diffusion coefficient and the drift mobility of free RM electrons were calculated using Equations (6) and (7).

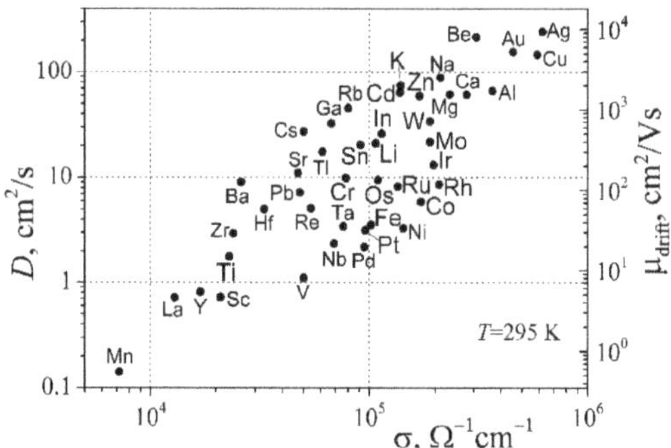

Figure 4. The diffusion coefficient and the drift mobility relation with the conductivity of elemental metals at room temperature (the right scale has been chosen to correspond to the left scale, so that the positions of the points in the pattern are the same for both scales).

Accounting for the total effective density of free RM electrons being equal to $n_{\text{eff}} = g_\Sigma(E_F)kT = [g_s(E_F) + g_d(E_F)]kT$, and free charge carriers at the Fermi surface having the same Fermi energy, the same diffusion coefficients and the same velocities and relaxation times, thus, each quasi-particle (electron or hole) carries in the same contribution to the electronic heat capacity and conductivity. Moreover, each quasi-particle makes the same contribution to the energy fluctuation variance:

$$< \Delta \varepsilon_1^2 > = \frac{< (E - < E >)^2 >}{n_{\text{eff}}} = \frac{\int_0^\infty (E - E_F)^2 g(E) f(E) [1 - f(E)] \mathrm{d}E}{n_{\text{eff}}}, \tag{8}$$

where

$$< (E - E_F)^2 > = g(E_F)(kT)^3 \int_0^\infty (\varepsilon - \varepsilon_F)^2 f(\varepsilon)[1 - f(\varepsilon)] \mathrm{d}\varepsilon = (\pi^2/3) g(E_F)(kT)^3, \tag{9}$$

and here, $\varepsilon = E/kT$. Then,

$$< \Delta \varepsilon_1^2 > = (\pi^2/3)(kT)^2. \tag{10}$$

This result is consistent with the theory of energy fluctuations of free randomly moving particles [29]. It also shows that quasi-particle scattering in metals is an inelastic process. Therefore, the term free electron can be used here as a quasi-particle.

3. The Mean Free Path of RM Electrons and the Temperature Dependence of the Resistivity of the Elemental Metals

The mean free path is the most important parameter characterizing the scattering mechanism of randomly moving charge carriers [30].

The mean free path of free RM electrons in metals can be estimated as

$$l_F = v_F \tau_F = \frac{1}{\sigma_{\text{eff}} N_{\text{eff}}} = \frac{1}{\sigma_{\text{eff}} g(E_F) kT}, \tag{11}$$

and the average relaxation time as

$$\tau_F = \frac{1}{\sigma_{\text{eff}} v_F N_{\text{eff}}} = \frac{1}{\sigma_{\text{eff}} g(E_F) v_F kT}, \tag{12}$$

where σ_{eff} is the effective scattering cross-section of free RM electrons by electronic defects. It should be noted that above the Debye temperature, the effective electron scattering cross-section σ_{eff} is temperature independent. Thus, the statement that the scattering cross-section is proportional to temperature in the mentioned temperature range [4–7,11,12] contradicts the probability density function of free RM electrons. Equation (11) directly shows that the electron mean free path at temperatures above the Debye temperature is inversely proportional to both the temperature and the DOS at the Fermi surface.

The electron-scattering cross-section of elemental metal distribution on DOS at the Fermi level is shown in Figure 5a at room temperature. The product $l_F \sigma_{\text{eff}} = 1/n_{\text{eff}} = V_{1\text{el}}$, where $V_{1\text{el}}$ is the volume for one free RM electron. The dependence of the volume $V_{1\text{el}}$ on the periodic table is shown in Figure 5b at room temperature. The largest volume $V_{1\text{el}}$ has Be and alkali metals and the smallest one has V, Ni, and Pd.

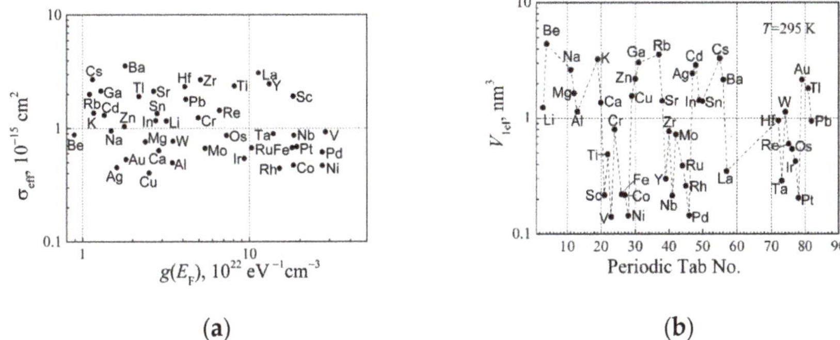

(a)   (b)

Figure 5. (**a**) The effective scattering cross-section σ_{eff} of free RM electrons caused by electronic defects and the relation with DOS at the Fermi surface at room temperature; (**b**) the volume for one free RM electron distribution on the periodic table at room temperature.

Figure 6 shows the electric resistivity of Al, Cu, and Pd as a function of temperature from 1 K to 1000 K. As has been mentioned, the resistivity at temperatures above the Debye temperature is proportional to temperature T, and it is not related to the scattering cross-section dependence on temperature. What are the effects that cause a steep resistivity decrease with temperatures below the Debye temperature Θ? The electric resistivity below 10 K is independent and caused by the scattering of free RM electrons by chemical and structural imperfections. Since each free electron absorbs or excites one phonon during scattering, the scattering cross-section depends on the ratio of the thermal energy exchange between free RM electrons and electronic defects.

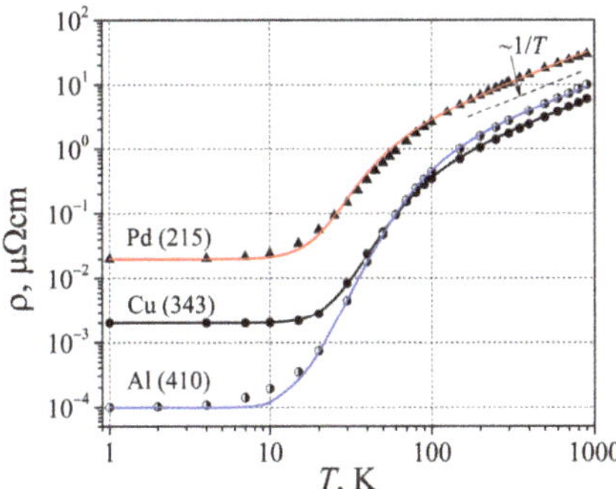

Figure 6. The resistivity dependence on temperature for Pd, Cu, and Al. The points are experimental data taken from [28], the number in brackets near the chemical symbol is the Debye temperature Θ in K used to calculate resistivity, and the solid lines are calculated with Equation (22).

The average thermal energy [5] of the free electrons estimated for one free RM electron is:

$$E_{1\text{el}} = (\pi^2/6)g(E_F)(kT)^2/n_{\text{eff}} \approx 1.64\, kT. \tag{13}$$

The average phonon thermal energy at $T > \Theta$ is $3kT$ because all the lattice vibration waves are excited, but at $T < \Theta$ only the low-frequency phonons become excited. The average thermal energy of a single phonon accounting for the excitation and annihilation of the phonon can be described as [12,23,31,32]:

$$E_{1ph} = 3kT(T/\Theta)^4 \int_0^{\Theta/T} 4x^5 / \left[(e^x - 1)(1 - e^{-x}) \right] \mathrm{d}x. \tag{14}$$

Then,

$$E_{1ph} / E_{1el} 1.83 \, \eta_{ph}(T/\Theta), \tag{15}$$

where

$$\eta_{ph}(T/\Theta) = (T/\Theta)^4 \int_0^{\Theta/T} 4x^5 / \left[(e^x - 1)(1 - e^{-x}) \right] \mathrm{d}x, \tag{16}$$

is the phonon mediation factor for free RM electron scattering. Then, the resultant scattering cross-section σ_{res} of free RM electrons can be described as

$$\sigma_{res} = \sigma_{eff} \eta_{ph}(T/\Theta), \tag{17}$$

where σ_{eff} can be obtained from Equation (11) at room temperature.

At temperatures above the Debye temperature, the resistivity of the elemental metal is given by

$$\rho = 1/\sigma = 1/[q^2 g(E_F) D\,(T) = \rho(T_0)(T/T_0) \tag{18}$$

where $T_0 = 300$ K, and $\rho(T_0) = 1/[q^2 g(E_F) D\,(T_0)]$. The resultant relaxation time τ_{res} of the free RM electrons now can be expressed as

$$1/\tau_{res} = (1/\tau_{eff}) + (1/\tau_{def}), \tag{19}$$

where

$$1/\tau_{eff} = \sigma_{res} N_{eff} v_F = \sigma_{res} g(E_F) v_F kT, \tag{20}$$

and

$$1/\tau_{def} = \sigma_{def} N_{def} v_F, \tag{21}$$

where σ_{def} is the average scattering cross-section of the residual defects (impurities), and N_{def} is their density. Thus, the resistivity of elemental metals in the temperature range from 1 K to 1000 K can be expressed as

$$\rho = \rho_0 + \rho(T_0)(T/T_0) \eta_{ph}(T/\Theta). \tag{22}$$

The calculated resistivity dependences on temperature by Equation (22) are shown in Figure 6 as solid lines with $T_0 = 300$ K. Here, the calculations have been performed using the constant Debye temperature values Θ. Considering that the Debye temperature is not completely constant [29], there may be a small difference between calculated and experimental data in some temperature ranges. It can be seen that Equation (22) describes the resistivity temperature dependence well enough, though the metals presented have very different Fermi surfaces and different DOS at the Fermi surface.

4. Hall Effect of Metals

The Hall coefficient R_H, when the direct current flows in the x-direction and the magnetic field is directed along the z-axis, is given by

$$R_H = \frac{E_y}{j_x B_z} = \frac{E_H}{\sigma E_x B_z} \tag{23}$$

where j_x is the direct current density in the x-direction, B_z is the magnetic flux density in the z-direction, $E_y = E_H = U_H/w$, and here U_H is the Hall voltage in a material plate with dimensions $d \times w \times l$ (thickness $\times$ width $\times$ length). $E_x = U_x/l$ is the applied electric field strength in the x-direction, and σ is the conductivity of the material wafer. The Hall effect for the small direct current and weak magnetic fields can also be characterized by the Hall angle φ:

$$\varphi = E_y/E_x = E_H/E_x. \tag{24}$$

On the other hand, the Hall angle can be expressed in terms of the electron cyclotron angular frequency $\omega_c = qB_z/m^*$ [33]:

$$\varphi = \omega_c < \tau^2 > / < \tau > = (qB_z/m^*) < \tau > r_H, \tag{25}$$

where $r_H = < \tau^2 > / < \tau >^2$ is the Hall factor, which depends on the free charge carrier scattering mechanism. For highly degenerate electron gas $r_H = 1$, it can be noted that ω_c is independent of the kinetic energy of the particle, and it does not depend on the radius of motion of the electrons and on their velocity.

Such a general expression for materials with one type of the free charge carriers follows from Equations (23)–(25):

$$R_H \sigma = (q < \tau > /m^*)r_H = \mu_{Hall}. \tag{26}$$

Considering that the quantity μ_H has the same dimension as the drift mobility of the free charge carriers, it is called Hall mobility. Equation (26) is valid for both non-degenerate and degenerate electron gas materials with one type of the free RM charge carriers.

In the general case, the absolute value of the drift mobility μ_{drift} for one type of charge carriers can be described as

$$\mu_{drift} = \frac{qD}{kT} = \frac{q < \tau >}{m^*} \cdot \frac{< E >}{(3/2)kT} = \frac{q < \tau >}{m^*} \cdot \alpha_\varepsilon, \tag{27}$$

where the coefficient α_ε indicates by how many times the average kinetic energy $<E>$ of the free RM electron is higher than the classical particle thermal energy $(3/2)kT$. Accounting that the conductivity can be described by such a general expression:

$$\sigma = qn_{eff} \frac{q < \tau >}{m^*} \cdot \alpha_\varepsilon \tag{28}$$

then the Hall coefficient is described by the following general relationship:

$$R_H = r_H/(qn_{eff}\alpha_\varepsilon), \tag{29}$$

which is valid for any degree of degeneration of the electron gas with one type of free charge carrier. In the case of a material with non-degenerate electron gas, $\alpha_\varepsilon = 1$ and the Hall coefficient takes the classical form:

$$R_H = r_H/(qn_{eff}) = r_H/(qn). \tag{30}$$

Considering that for metals $r_H = 1$, $n_{eff} = g(E_F)kT$ and $\alpha_\varepsilon = E_F/[(3/2)kT]$, the Hall coefficient for metals with one type of free randomly moving charge carrier can be expressed as [20–23]:

$$R_H = 3/[2qg(E_F)E_F]. \tag{31}$$

For an ideal spherical Fermi surface, the product $g(E_F)E_F = 3n/2$ [5], and one can obtain the classical relation $R_H = 1/(qn)$, where n is the total density of electrons in the conduction band, but this does not mean that all the electrons in the conduction band are free, and can move randomly and participate in conduction. Many elemental metals do not have spherical Fermi surfaces; they have very complex Fermi surfaces [5,7,26].

The experimental data of the Hall coefficient distribution in the periodic table for elemental metals at room temperature is shown in Figure 7a.

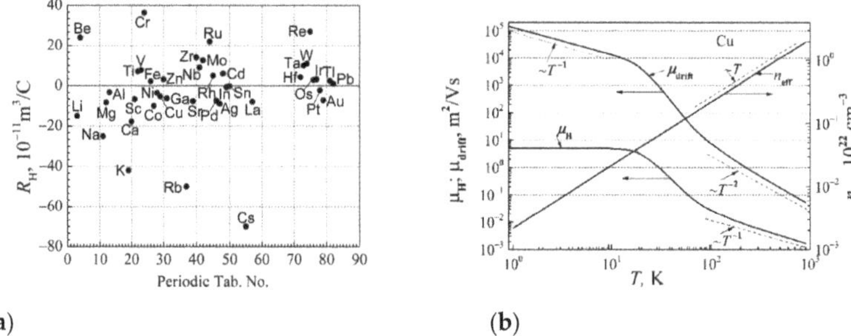

Figure 7. (**a**) The Hall coefficient distribution on location in the periodic table for elemental metals at room temperature. Data are taken from [7,34,35]. (**b**) The free RM electron density n_{eff}, and their drift mobility μ_{drift} and Hall mobility μ_{H} (absolute values) as a function of temperature for copper.

The magnetic flux density during the Hall coefficient measurement was in the range (0.5–1.3) T. As can be seen from Figure 7a, the Hall coefficient for different elemental metals has different signs, i.e., in different elemental metals prevail in electronic-type or hole-type charge carriers.

The general expressions for conductivity and the Hall coefficient for two types of free charge carriers (electrons and holes) can be described as [12,36]:

$$\sigma = \sigma_{\text{e}} + \sigma_{\text{h}}, \tag{32}$$

$$R_{\text{H2}} = \frac{R_{\text{He}}\sigma_e^2 + R_{\text{Hh}}\sigma_h^2}{\sigma^2} = \frac{\mu_{\text{He}}\sigma_e + \mu_{\text{Hh}}\sigma_h}{\sigma^2}. \tag{33}$$

Here, the relation $R_{\text{He,h}}\sigma_{\text{e,h}} = \mu_{\text{He,h}}$ is fulfilled for each type of charge carrier. The total density of DOS obtained from the electronic heat capacity results is equal to $g_{\text{total}}(E_F) = g_{\text{el}}(E_F) + g_{\text{hole}}(E_F)$, where $g_{\text{el}}(E_F)$ and $g_{\text{hole}}(E_F)$ are the DOS components caused by electron-like and hole-like DOS at the Fermi surface, respectively. The electrical conductivity, the electronic heat capacity, and the thermal noise measurements [37] show that they do not depend on the quasi-particle origin (electron or hole); all these particles move randomly with the Fermi velocity.

Considering that the Fermi energy for a given elemental metal is the same for free RM electrons and for free RM holes, this means that the average velocity v_F and the average relaxation time τ_F at the Fermi surface are the same for all free RM charge carriers. Then, the absolute value of the drift mobility for each type of free RM charge carrier can be expressed as

$$\mu_{\text{e,h drift}} = \mu_{\text{drift}} = \frac{qD}{kT} = \frac{qv_F^2\tau_F}{3kT} \tag{34}$$

From this expression, it can be seen that for elemental metals the drift mobility of free RM charge carriers does not depend on the effective mass of the charge carrier, and changes with temperature as $1/T^2$, because $\tau_F \sim 1/T$. Therefore, the absolute value of the Hall mobility for free RM charge carriers can be expressed as

$$\mu_{\text{He,h}} = \mu_{\text{H}} = R_{\text{He,h}}\sigma_{\text{e,h}} = q\tau_F/m. \tag{35}$$

The Hall mobility is determined by the relaxation time and is independent of the energy of free charge carriers. An illustration of the effective density of free RM electrons, and their absolute values of the drift and Hall mobility as a function of temperature for copper, is shown in Figure 7b.

At temperatures higher than room temperature, the free RM electron density increases with temperature as $\sim T$, and the drift mobility decreases as $\sim 1/T^2$, causing the resistivity to increase with temperature as $\sim T$. As can be seen from Figure 7b, the absolute value of the drift mobility in copper at cryogenic temperatures exceeds the Hall mobility by three orders.

It follows from Equations (27) and (35) that the Hall mobility for a single type of free charge carriers can also be expressed as $\mu_H = \mu_{\text{drift}}/\alpha_\varepsilon$. The absolute values of the Hall mobilities of the free RM charge carriers of elemental metals' distribution in the periodic table are shown in Figure 8a at room temperature. The Hall mobility values are distributed in the range 10–70 cm^2/Vs. The highest Hall mobilities have Ag, Be, and Au, and the lowest ones have La, Sc, Y, and V. The absolute values of the drift mobilities of free RM charge carriers of the elemental metal distribution in the periodic table are shown in Figure 8b at room temperature. The drift mobility values for different elemental metals are distributed in a very wide range, from 30 cm^2/Vs to 10^4 cm^2/Vs, and this causes the very wide distribution of the conductivity values of elemental metals. The highest values of drift mobility have Ag, Be, Au, and Cu, and the lowest ones have Sc, Y, La, and V.

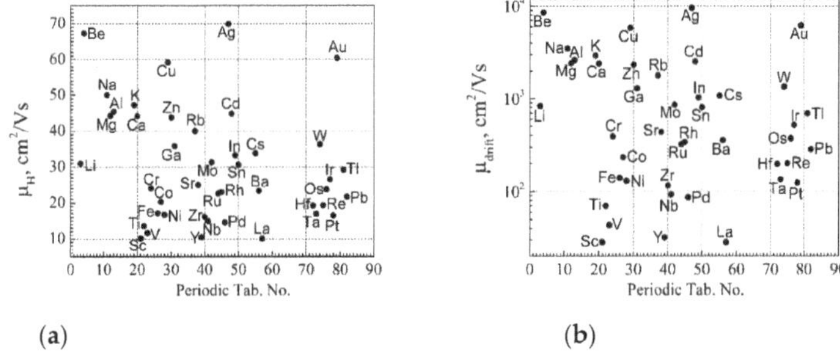

(a) (b)

Figure 8. (a) The absolute value of the Hall mobilities for a single type of free RM charge carrier of elemental metal distribution in the periodic table at room temperature; (b) the absolute value of the drift mobilities for a single type of free RM charge carriers of elemental metal distribution in the periodic table at room temperature.

Since metals have electrons and holes, free RM charge carrier densities n_{eff} and p_{eff} can be expressed as follows:

$$n_{\text{eff}} = g_{\text{el}}(E_F)kT, \tag{36}$$

$$p_{\text{eff}} = g_{\text{hole}}(E_F)kT, \tag{37}$$

then, taking into account Equations (34) – (37), Equations (32) and (33) can be rewritten as

$$\sigma = qn_{\text{eff}}\mu_{\text{drift}} + qp_{\text{eff}}\mu_{\text{drift}} = q\mu_{\text{drift}}g_{\text{total}}(E_F)kT, \tag{38}$$

$$R_{\text{H2}} = R_{\text{H1}}\left[1 - 2\cdot\frac{g_{\text{el}}(E_F)}{g_{\text{total}}(E_F)}\right] = \eta R_{\text{H1}} \tag{39}$$

where $g_{\text{total}}(E_F) = g_{\text{el}}(E_F) + g_{\text{hole}}(E_F)$, and $R_{\text{H1}} = 3/[2qg_{\text{total}}(E_F)E_F]$ is the Hall coefficient in the case of only a single type of randomly moving charge carriers in the sample, and

$$\eta = \frac{R_{\text{H2}}}{R_{\text{H1}}} = 1 - 2\cdot\frac{g_{\text{el}}(E_F)}{g_{\text{total}}(E_F)}, \tag{40}$$

is the quantity that plays the role of compensation in Hall voltages caused by electrons and holes in the Hall effect measurement. Thus, from the Hall coefficient R_{H2} data of metals (Figure 7a) and Equation (31) for R_{H1}, it is possible to determine the compensation quantity η and the relative parts of electron-like or hole-like densities of states at the Fermi surface. The relative electron-like and hole-like DOS parts are shown in Figure 9a,b, respectively.

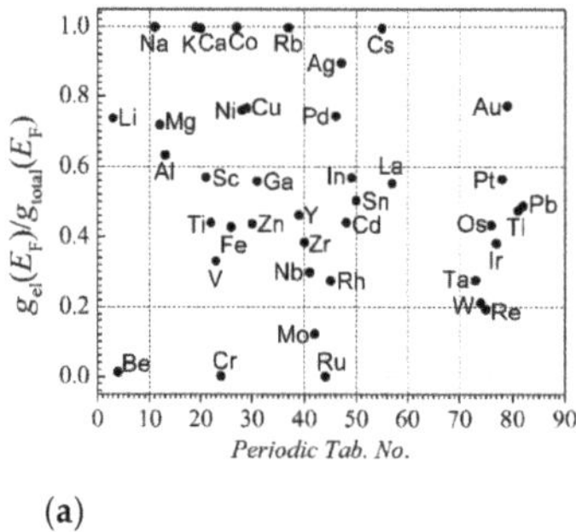
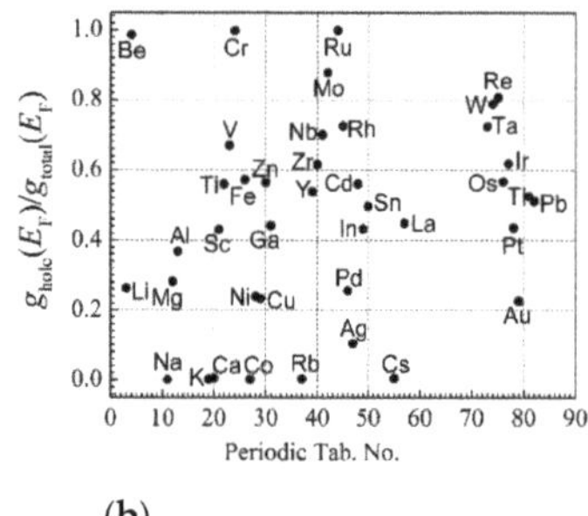

(a)   (b)

Figure 9. (a) Relative electron-like DOS part at the Fermi surface distribution in the periodic table of elemental metals; (b) relative hole-like DOS part at the Fermi surface distribution in the periodic table of elemental metals. The relative hole-like DOS part is determined as $g_{\mathrm{hole}}(E_F)/g_{\mathrm{total}}(E_F) = 1 - g_{\mathrm{el}}(E_F)/g_{\mathrm{total}}(E_F)$.

From Figure 9a,b, it can be seen that electronic-like DOS at the Fermi surface prevails for alkali metals, Ca, and Co. On the other hand, the hole-like DOS at the Fermi surface prevails for Be, Cr, and Ru.

The measurement results of the Hall coefficient show that its value depends on the magnetic field strength [38]. Figure 10a shows the Hall coefficient R_{H2} dependence of the magnetic field strength for aluminum, where the parameter of the magnetic field strength is expressed by the cyclotron frequency $\omega_c = (qB/m)$, where B is the magnetic flux density and τ is the charge carrier mean free flight time. It can be seen that at low magnetic field strengths ($\omega_c\tau \ll 0.1$) the Hall coefficient is negative and electrons are the dominant free charge carriers, but with a high magnetic field $\omega_c\tau \gg 10$ the Hall coefficient is positive and holes are the dominant free charge carrier.

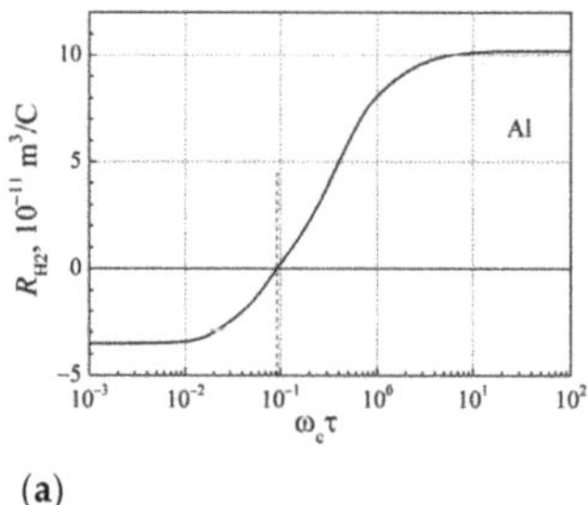
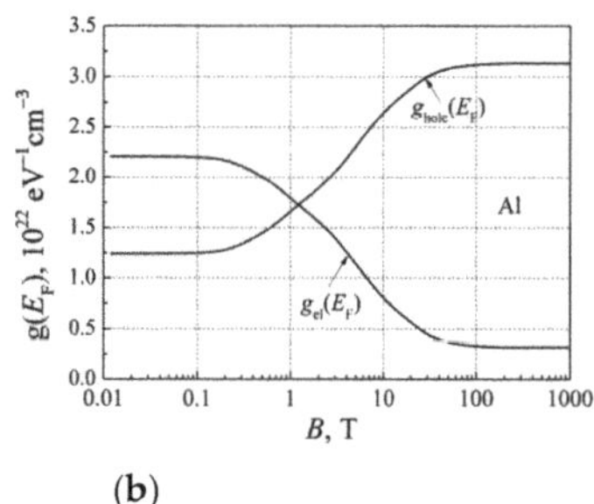

(a)   (b)

Figure 10. (a) The dependence of the Hall coefficient on $\omega_c\tau = (qB/m)\tau$ for aluminum (prepared with respect to data from [37]); (b) the dependences of the electron-like $g_{\mathrm{el}}(E_F)$ and hole-like $g_{\mathrm{hole}}(E_F)$ DOS on the magnetic field for aluminum.

This means that the Fermi surface nonuniformity is affected by the magnetic field due to changes in the energy derivative $\partial^2 E/\partial k^2$ at the Fermi surface. It causes the redistribution between electron-like and hole-like DOS at the Fermi surface.

Thus, the measurement of the Hall coefficient does not give the total density of the electrons in the conduction band, either at low or at high magnetic field strengths. Using Equations (31) and (39), and g(E_F) for Al, we have estimated the electron-like and hole-like DOS dependences on the magnetic field strength, which are shown in Figure 10b. It can be seen that these DOS densities are strongly dependent on the magnetic field strength, and that there is no relationship with the total electron density in the conduction band.

Equation (39) also allows us to explain the Hall mobility and the Hall coefficient measurement results in the normal state superconductor $Yba_2Cu_3\,O_{7-\delta}$ [39,40].

5. Magnetic Susceptibility of Free RM Electrons

The magnetic susceptibility χ is defined as

$$\chi = \mu_{\text{rel}} - 1, \tag{41}$$

where $\mu_{\text{rel}} = \mu/\mu_0$ is the relative permeability, and $\mu_0 = 4\pi \cdot 10^{-7}\,\text{H/m} = 1.257 \cdot 10^{-6}\,\text{V}\cdot\text{s}/\text{A}\cdot\text{m}$ is the magnetic constant. The magnetization of the material is given by

$$M = (B/H) - H = B\chi/\mu_0, \tag{42}$$

where B is the magnetic flux density and H is the magnetic field strength. Thus, the electron behaves as a two-state system with energies of $E_i = \pm\mu_B B$ (where $\mu_B = q\hbar/2m$ is the Bohr magneton), and the single electron partition function is given by

$$Z(1) = e^{\mu_B B/kT} + e^{-\mu_B B/kT} = 2\cosh(\mu_B B/kT). \tag{43}$$

The partition function for n free electrons per unit volume is

$$Z(n) = [Z(1)]^n = [2\cosh(\mu_B B/kT)]^n. \tag{44}$$

Then, the free energy F of the system is equal to [41]:

$$F = -kT\ln Z(n) = -nkT\ln[2\cosh(\mu_B B/kT)]. \tag{45}$$

The total magnetic moment M of the n free electron system can be obtained by calculating the derivative:

$$M = -(\partial F/\partial B)_T = n\mu_B\tanh(\mu_B B/kT). \tag{46}$$

For a small magnetic field $(\mu_B B/kT) << 1$, the paramagnetic susceptibility is

$$\chi_{\text{par}} = \frac{\mu_0 M}{B} = \frac{n\mu_B^2}{kT}, \tag{47}$$

The conduction band electrons can be expected to make a Curie-type paramagnetic contribution to the magnetization of the metal [9,42]:

$$M = \mu_0 n\mu_B^2 B/kT. \tag{48}$$

This formula is valid for free electrons. Considering the observed data of the M of the elemental non-ferromagnetic metals, the magnetization is almost independent of temperature, and it has been concluded that the Curie-type law is not valid for free electrons in metals. However, this erroneous conclusion has been reached by considering that all valence electrons are free, but only RM electrons are free, whose density is equal to $n_{\text{eff}} = g(E_F)kT$. Therefore, the Curie-type law is fulfilled if the real density of the free RM electrons is taken into account. Then, the paramagnetic susceptibility of the free RM electrons for non-ferromagnetic metals is given by

$$\chi_{\text{par}} = \frac{\mu_0 n_{\text{eff}}\mu_B^2}{kT} = \mu_0\mu_B^2 g(E_F) \tag{49}$$

It is known that when electrons are freely moving in a magnetic field, in addition to the paramagnetic effect of their spin, they exhibit a diamagnetic effect due to their motion. In accordance with Lenz's law this yields a magnetic field that is the opposite to the direction of an applied magnetic field. Considering that the rotational energy of the free randomly moving electron in a perpendicular magnetic field is $E_{\text{rot}} = \hbar\omega_c = \hbar qB/m$, where ω_c is the electron cyclotron frequency, one can write an analogous expression for the electron partition function, taking into account the electron spin (1/2):

$$Z(1) = e^{\hbar q B/2m} + e^{-\hbar q B/2m} = e^{\mu_B B/kT} + e^{-\mu_B B/kT} = 2\cosh(\mu_B B/kT) \tag{50}$$

Then, the diamagnetic susceptibility, taking into account the diamagnetic effect yielding a magnetic field opposed to the direction of the applied magnetic field, can be expressed as

$$\chi_{\text{dia}} = -\frac{\mu_0 n_{\text{eff}} \mu_B^2}{kT} = -\frac{1}{3}\mu_0 \mu_B^2 g(E_F), \tag{51}$$

because the effective density of free RM electrons in the plane perpendicular to the magnetic field is equal to $n_{\text{eff}} = (1/3)\, g(E_F)kT$. The absolute value of Landau diamagnetism is thus equal to one-third that of Pauli paramagnetism in the free-electron model [9,42]. Therefore, it produces diamagnetic susceptibility:

$$\chi_{\text{dia}} = -\frac{1}{3}\mu_0 \mu_B^2 g(E_F) = -\frac{1}{3}\chi_{\text{par}}. \tag{52}$$

Then, the resulting magnetic susceptibility of free RM electrons for non-ferromagnetic metals is equal to:

$$\chi_{\text{res}} = \frac{2}{3}\mu_0 \mu_B^2 g(E_F). \tag{53}$$

Though the magnetic susceptibility of free RM electrons is proportional to the DOS at the Fermi surface (Equation (53)), as for the electronic heat capacity, it cannot be used as a suitable quantity for evaluating the DOS at the Fermi surface because the magnetic susceptibility measurement results also depend on the susceptibilities of the neutral atoms and ions (Figure 11).

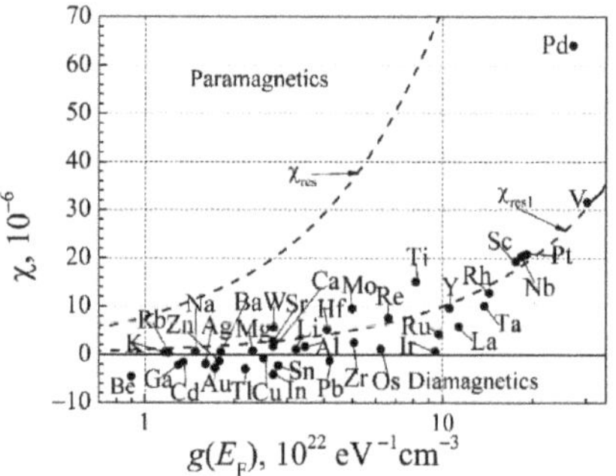

Figure 11. Magnetic susceptibility dependence on the DOS at the Fermi surface for non-ferromagnetic metals: the points are the experimental data from [35], the dashed curve χ_{res} is calculated with Equation (53), and the curve $\chi_{\text{res1}} = (1/7)\chi_{\text{res}}$ is plotted as an average value.

In addition, diamagnetism is also present in all atoms and molecules, and in gases of atoms and molecules that do not carry permanent magnetic moments. This leads to an additional diamagnetism of the ions, comparable to the Landau diamagnetism. There are therefore three temperature-independent contributions to the magnetic susceptibility of non-ferromagnetic metals: the free electron Pauli paramagnetism, the Landau diamagnetism, and the diamagnetism of the metal ions.

Therefore, the free RM electron model allows a very simple explanation of the diamagnetism and paramagnetism of non-ferromagnetic elemental metals.

6. Plasma Frequency of the Free RM Electrons in Elemental Metals

A.C. conductivity $\sigma(\omega)$ is usually presented in the following form [5]:

$$\sigma(\omega) = \frac{\sigma}{1 - j\omega <\tau>},\tag{54}$$

and the relative permittivity at $\omega<\tau> \gg 1$ as

$$\varepsilon_r = 1 - \frac{\omega_p^2}{\omega(\omega + j <\tau>)} \approx 1 - \frac{\omega_p^2}{\omega^2},\tag{55}$$

where

$$\omega_p^2 = \frac{\sigma}{\varepsilon_0 <\tau>},\tag{56}$$

where ω_p is the free electron oscillation plasma frequency. Taking into account Equation (28) for the d.c. conductivity, the plasma frequency for elemental metals can be described as:

$$\omega_p^2 = \frac{q^2}{\varepsilon_0 m^*} n_{\text{eff}} \alpha_\varepsilon = q^2 g(E_F) v_F^2 / (3\varepsilon_0)\tag{57}$$

It can be seen that the plasma frequency does not depend on the temperature. The dependence of the plasma frequency $fp = \omega p/2\pi$ on the DOS $g(E_F)$ at the Fermi surface for elemental metals is shown in Figure 12a. It can be seen that there is no correlation between plasma frequency and DOS at the Fermi surface for different metals.

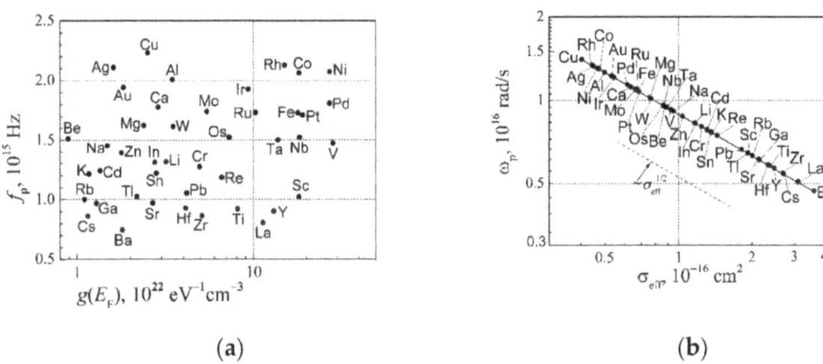

Figure 12. (a) The dependence of the plasma frequency f_p on the DOS $g(E_F)$ for elemental metals; (b) the dependence of the plasma frequency on the free electron effective scattering cross-section for elemental metals (Figure 5a).

The dependence of the plasma frequency on the effective free electron scattering cross-section of elemental metals is shown in Figure 12b. This figure shows that the dispersion of the results in Figure 12a is due to the different free electron scattering cross-sections of different metals.

7. Conclusions

A study of the main transport characteristics of the free electrons on the basis of the stochastic definition of the effective density of free RM electrons in elemental metals is presented. It is shown that thermal vibrations of the lattice atoms play a different role than has been explained in many publications: the thermal vibrations of the atom not only excite the free RM electrons, but also produce the same number of electronic defects (weakly shielded ions). The temperature dependence of the resistivity over a very wide temperature range is explained by the scattering of free RM electrons by the electronic defects accounting for the thermal energy exchange between the phonon and the free RM electron. The Hall coefficient of metals is explained by using electron-like and hole-like densities of states at the Fermi surface. The definition of the density of the free RM electrons allows a very simple explanation of the paramagnetism and diamagnetism of the free RM electron. It is shown that different values of the plasma frequency are caused by different values of the effective scattering cross-sections of different elemental metals.

Author Contributions: Conceptualization, V.P.; methodology and software, V.P. and V.J.; validation, V.P. and V.J.; investigation, V.P. and V.J.; writing-original draft preparation V.P.; review and editing V.P. and V.J. All authors have read and agreed to the published version of the manuscript.

Funding: This research received no external funding.

Data Availability Statement: Data are provided in the figures in the article. All data generated and analyzed during this study are included in the presented manuscript, and are available from the corresponding author on reasonable request.

Conflicts of Interest: The authors declare no conflict of interest.

References

1. Drude, P. Zur Elektronentheorie der Metale. *Ann. Phys.* **1900**, *306*, 441–624. [CrossRef]
2. Lorentz, H.A. *The Theory of Electrons*, 1st ed.; The Columbia University Press: New York, NY, USA, 1909.
3. Sommerfeld, A.; Bethe, H. *Elektronentheorie der Metals*; Springer: Berlin, Germany, 1967.
4. Abrikosov, A.A. *Fundamentals of the Theory of Metals*; Dover Publ.: Mineola, NY, USA, 2017.
5. Ashcroft, N.W.; Mermin, N.D. *Solid State Physics*; HRW Int. Ed.: New York, NY, USA, 1976.
6. Kaveh, M.; Wisser, N. Electron-electron scattering in conducting materials. *Adv. Phys.* **2001**, *33*, 257–372. [CrossRef]
7. Kittel, C. *Introduction to Solid State Physics*; John Wiley and Sons: New York, NY, USA, 1976.
8. Lundstrom, M. *Fundamentals of the Carrier Transport*; Cambridge Univ. Press Online: Cambridge, UK, 2014.
9. Mizutani, U. *Introduction to the Electron Theory of Metals*; Cambridge Univ. Press Online: Cambridge, UK, 2014.
10. Schulze, G.E.R. *Metallphysik*; Akademie: Berlin, Germany, 1967.
11. Sondheimer, E.H. The mean free path of electrons in metals. *Adv. Phys.* **2001**, *50*, 499–537. [CrossRef]
12. Ziman, J.M. *Principles of the Theory of Solids*; Cambridge Univ. Press: Cambridge, UK, 1972.
13. Sander, L.M. *Advanced Condensed Matter Physics*; Cambridge Univ. Press Online: Cambridge, UK, 2014.
14. Rossiter, P.L. *The Electrical Resistivity of Metals and Alloys*; Cambridge Univ. Press Online: Cambridge, UK, 2014.
15. Alloul, H. *Introduction of the Physics of Electrons in Solids*; Springer: Berlin/Heidelberg, Germany, 2011.
16. Sólyom, J. *Fundamentals of the Physics of Solids*; Springer: Berlin/Heidelberg, Germany, 2009.
17. Ibach, H.; Lüth, H. *Solid State Physics*, 4th ed.; Springer: Berlin/Heidelberg, Germany, 2009.
18. Belashchenko, D.K. The relationship between electrical conductivity and electromigration in liquid metals. *Dynamics* **2023**, *3*, 405–424. [CrossRef]
19. Palenskis, V. *A Novel View to Free Electron Theory*; LAP LAMBERT Acad. Publ.: Berlin, Germany, 2013.
20. Palenskis, V. Drift mobility, diffusion coefficient of randomly moving charge carriers in metals and other materials with degenerate electron gas. *World J. Cond. Matt. Phys.* **2013**, *3*, 73–81. [CrossRef]
21. Palenskis, V. The effective density of randomly moving electrons and related characteristics of materials with degenerate electron gas. *AIP Adv.* **2014**, *4*, 047119. [CrossRef]
22. Palenskis, V.; Žitkevičius, E. Study of the Transport of Charge Carriers in Materials with Degenerate Electron Gas, 123–186. In *Electron Gas: An Overview*; NOVA Science Publ., Inc.: New York, NY, USA, 2019.
23. Palenskis, V.; Žitkevičius, E. Summary of new insight into electron transport in metals. *Crystals* **2021**, *11*, 622. [CrossRef]
24. Stirzaker, D.R. *Elementary Probability*; Cambridge Univ. Press Online: Cambridge, UK, 2014.
25. Grimmet, G.R.; Stirzaker, D.R. *Probability and Random Processes*; Oxford Univ. Press: Oxford, UK, 2020.
26. Cracknell, A.P.; Wong, K.C. *The Fermi Surfaces: Its Concept, Determination, and Use in the Physics of Metals*; Clarendon Press: Oxford, UK, 1973.
27. Blokhintsev, D.L. *Quantum Mechanics*; Reidel Publ. Comp.: Dordrecht, The Netherlands, 1964.
28. Lide, D.R. (Ed.) *Handbook of Chemistry and Physics*, 84th ed.; CRC Press LLC: Boca Raton, FL, USA, 2004.
29. Kittel, C. *Thermal Physics*; John Wiley and Sons: New York, NY, USA, 1969.
30. Gall, D. Electron mean free path in elemental metals. *J. Appl. Phys.* **2016**, *119*, 085101. [CrossRef]
31. Palenskis, V. Free electron characteristic peculiarities caused by lattice vibrations in metals. *World J. Cond. Matt. Phys.* **2022**, *12*, 9–17. [CrossRef]
32. Palenskis, V.; Žitkevičius, E. Analysis of transport properties of the randomly moving electrons in metals. *Matt. Sci. (Medžiagotyra)* **2020**, *26*, 147–153. [CrossRef]
33. Smith, R.A. *Semiconductors*, 2nd ed.; Cambridge Univ. Press: Cambridge, UK, 1987.
34. Gray, D.E. *AIP Handbook*, 3rd ed.; McGraw-Hill: New York, NY, USA, 1972.
35. Grigoryev, I.S.; Meilihkov, E.Z. (Eds.) *Handbook of the Physical Quantities*; Energoatomizdat: Moscow, Russia, 1991.

36. Markiewicz, R.S. Simple model for the Hall effect in YBa$_2$Cu$_3$O$_{7-\delta}$. *Phys. Rev. B* **1988**, *38*, 5010–5011. [CrossRef] [PubMed]
37. Palenskis, V.; Pralgauskaitė, S.; Maknys, K.; Matulionis, A. Thermal Noise and Drift Mobility of Randomly Moving Electrons in Homogeneous Material with Highly Degenerate Electron Gas. In Proceedings of the IEEE of the 22nd Conference ICNF, Montpellier, France, 23–28 June 2013.
38. Lüc, K.R. Zur Temperatur- und Feldabhängigheit der galvanomagnetischen Eigenschaften von Aluminium und Indium. *Phy. Stat. Sol.* **1966**, *18*, 49–56. [CrossRef]
39. Palenskis, V. Transport characteristics of charge carriers in normal state superconductor YBa$_2$Cu$_3$O$_{7-\delta}$. *World J. Cond. Matt. Phys.* **2015**, *5*, 118–128. [CrossRef]
40. Palenskis, V. Description of the Transport Characteristic in Normal State Superconductor YBa$_2$Cu$_3$O$_{7-\delta}$, 52–64. In New Insights into Physical Science. Book Publ. Inter.: India. 2020. Available online: https://digiwire.co.in/description-of-the-transport-characteristics-of-charge-carriers-in-normal-state-superconductor-yba2cu3o7-%EF%81%A41/ (accessed on 1 July 2023).
41. Simon, S.H. *The Oxford Solid State Basics*; Clarendon Press: Oxford, UK, 2015.
42. Cottingham, W.N.; Greenwood, D.A. *Electricity and Magnetism*; Cambridge Univ. Press Online: Cambridge, UK, 2014.

Article

Effectiveness of Travelling Slice Modeling in Representing the Continuous Casting Process of Large Product Sections

Gianluca Bazzaro [1,*] and Francesco De Bona [2]

[1] Danieli & C. Officine Meccaniche spa, R&D Department, 33042 Buttrio, Italy
[2] DPIA, University of Udine, 33100 Udine, Italy; francesco.debona@uniud.it
* Correspondence: g.bazzaro@danieli.com; Tel.: +39-432-1958111

Abstract: It is critical in the metal continuous casting process to estimate the temperature evolution of the casted section along the machine from the meniscus (the point where liquid metal is poured) to the cutting machine, where the product is cut to commercial length. A convenient approximated model to achieve this goal with a feasible computational effort, particularly in the case of large sections, is the so-called travelling slice: the transversal section of casted product is subjected to different thermal boundary conditions (e.g., thermal flux, radiation, convection) that are found during the movement at constant speed from meniscus to the end of machine. In this work, the results obtained with the approximated travelling slice model are analyzed in the favorable case of an axisymmetric section. In this case, the reference model is 2D, whereas the travelling slice model degenerates in a simple 1D model. Three different casted shapes were investigated, rounds with diameters of 200 mm, 850 mm, and 1200 mm, spanning from traditional to only recently adopted product diameter sizes. To properly test the validity of the travelling slice model, other casting speeds were considered, even outside the industrial range. Results demonstrate the advantage of using the travelling slice, particularly the much lower computational cost without sacrificing precision, even at low casting speed and large dimensions.

Keywords: steel continuous casting process; travelling slice; FE transient thermal model; temperature evolution; metallurgical length

Citation: Bazzaro, G.; De Bona, F. Effectiveness of Travelling Slice Modeling in Representing the Continuous Casting Process of Large Product Sections. *Metals* **2023**, *13*, 1505. https://doi.org/10.3390/met13091505

Academic Editors: Wenchao Yang and Jiehua Li

Received: 19 July 2023
Revised: 18 August 2023
Accepted: 20 August 2023
Published: 22 August 2023

1. Introduction

Nowadays, the majority of metals with an engineering application, such as steel, are produced using a continuous casting process. The ingot casting process was commonly utilized at beginning of 20th century: an amount of molten metal was poured into a container which was then removed when the complete solidification was achieved. Following pioneering tests in the 1930s, the steel continuous casting process became a viable industrial solution from the 1950s and its share has not stopped growing in subsequent decades. In 2021, 97% of global steel was produced with the continuous casting technique. There are various reasons why the continuous casting process has taken over; the most important is the increased production when compared to other methods. Because most process events (e.g., solidification, segregation, defect production) are temperature driven, a tool to analyze the thermal field plays an important role in the development of this manufacturing approach. Theoretically, a model to describe such a process should be tridimensional; in this case, the computational cost, both in terms of resources as well as time needed to obtain the results, is significant (examples of 3D modeling applied to metal casting processes are found in refs. [1,2]). To overcome these issues, a faster bidimensional approach named travelling slice was developed and is now utilized frequently. The first attempts to study this problem date back to the 1970s, and FE thermomechanical models arose in the 1980s. In the years that followed, many people worked on this issue and several methods have been considered. The latest developments lead to the use of deep learning techniques in steel solidification [3]. One of the earliest studies where the travelling slice has been employed is [4]; later, other authors expanded the method to include phase transformations [5]; others developed hot tearing criteria in order to estimate crack formation [6–8]. Although such modeling is frequent in literature reviews, an assessment of its limitations remains lacking. Several decades ago, the continuous casting technology was limited to relatively small sections; nowadays, however, an increase of casted product size (up to 1200 mm diameter with a forecast to go beyond in forthcoming years) has been noted [9], which may call the validity of this modeling into doubt. The majority of authors rely on the accuracy of the assumption on the negligible heat conduction in casting direction. In this regard, [10,11] might be referenced. The purpose of this article is to verify the accuracy of this statement, which covers a wide variety of product sizes and casting speeds.

2. Numerical Modeling of the Continuous Casting Process

Due to the complexity of the phenomena that occur, the modeling of the steel continuous casting process must be multiphysics: thermal (solidification of molten metal, withdrawal of latent and sensible heat), mechanical (interaction between casted product and mould, friction, effect on the shape of pressure of the part which is still liquid), metallurgical (grain growth as a result of solidification, precipitation of secondary phases or carbides), and chemical (segregation of solutes, decomposition of lubricant). The fields stated above are interconnected and make the problem nonlinear; a common root for all of them is that they are temperature dependent. It must be noted that while the thermal problem can be tackled numerically with an almost feasible computational effort, this is not true when other fields, such as mechanical and metallurgical ones have to be taken into account; thus, a FE model reduced in dimensions should possibly be adopted. This article should be viewed as preparatory work of a certain modeling strategy that is intended to be used for offline calculations. The first step is to ensure its validity, beginning with the key aspect of the continuous casting process, the temperature distribution in the casted product, and then moving on with the other previously mentioned fields. Obviously, in the case of on-line calculation for process control purposes, a finite difference approach could be computationally more efficient at least for the thermal analysis. However, when casted profile becomes complex in shape (e.g., a beam blank in which both convex and concave zones are present), describing the boundary condition that a product is subjected to becomes challenging for finite differences models as well.

In a steel continuous casting plant (schematic representation of a continuous caster is given in Figure 1), the core equipment is represented by the mould, where in short time and small volume, compared to overall dimensions, a big amount of energy (order of magnitude: tens of MJ) is withdrawn from molten metal. This component is composed of copper due to its high thermal conductivity and is cooled by water. The thermal exchange between metal and mould is determined essentially by contact status; in early stages, when metal is too soft and not yet able to bear mechanical loads, the adherence and the withdrawal of energy are high. Moving downward stiffens the metal and creates a space between it and the mould, reducing thermal exchange and slowing the rate of growth of solidification of the shell. The solidified thickness at the mould exit must be sufficient to withstand the pressure imposed by the still-liquid inner metal.

As can be seen, heat exchange inside mould plays a significant role in the entire process; moreover, the majority of flaws on the casted material, such fractures and pinholes, are created here and are related to uneven heat extraction (see ref. [12]). It should be pointed out that uneven heat extraction results in irregular shell thickness growth for all casted sections: these phenomena are not predictable with a standard 3D FEM model, in which skin growth is kept as an average value. Other numerical models, i.e., CFD with RANS approaches, could address this issue, but are almost unfeasible for an industrial use due to long computational times.

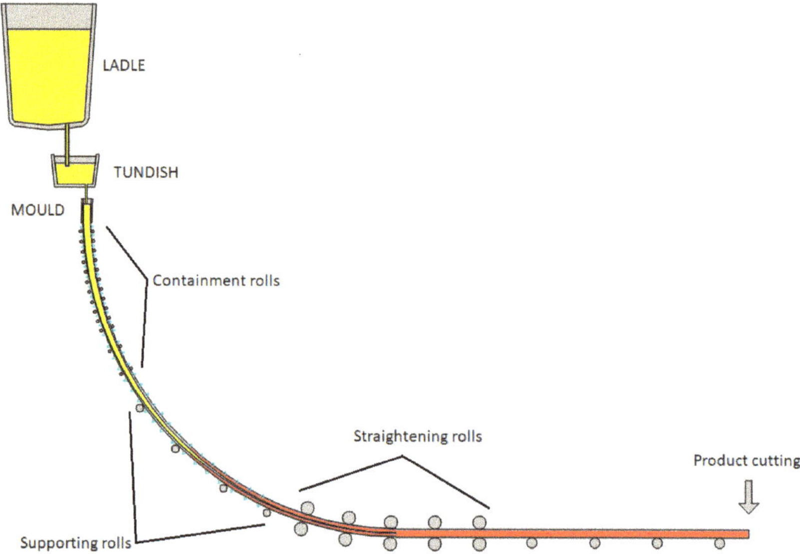

Figure 1. Sketch of a typical curved continuous casting machine. Molten steel in yellow, solidified steel in red. Cyan triangles: spray cooling.

What happens in the short space between the solidified shell and the mould surface is one of the most studied energy transfer mechanisms of the entire steel casting process; it can be represented by a thermal resistances model, in which each of them is the body (water, mould, lubricant, gap, shell, molten steel) crossed by heat. A comprehensive treatment can be found in ref. [13], while an interesting analysis on peak fluxes is given in ref. [14]. If each thermal resistance is known, starting from molten steel temperature all others in between and the thermal flux can be determined. This is the so-called fully coupled or direct approach, where both temperatures and flux are output; however, one of the difficulties of such modeling is to estimate the resistance value of the gap. Several studies on the gap creation in moulds have been carried out. Only in some specific circumstances, such as round shape, is the problem less complicated and a numerical approach can be used (see refs. [15,16]). For other common shapes such as squares (billet) and rectangles (bloom and slab), the computational cost for the gap estimation increases dramatically, necessitating a new approach: the problem decoupling. In the latter, once the thermal flux becomes an input, the complexity shifts onto the mathematical description of the heat withdrawn from the mould, which is dependent on several parameters such as casting speed, lubrication, material; e.g., for steel, two lubricants are typically used (powder or mineral oil) and specific grades (the peritectic ones) show differences in energy exchange (a quantification can be found in ref. [17]). To address this issue, mixed analytical-empirical models are used, see for example [18], so thermal flux is now established and no longer dependent from contact status; this is a strong assumption, but it allows the analysis to be performed in a reasonable amount of time. Typically, the mould flux trend decreases monotonically from meniscus to exit [13,18], although this statement in not widely accepted and a degree of uncertainty still remains.

3. Travelling Slice Model

3.1. Model Statement

As mentioned in the previous paragraph, the model has to take into account firstly the thermal aspect of continuous casting process. Analysis is transient and with temperature dependent thermal properties (taken from IDS software and then compared to [19] shows a very good agreement), and time/temperature dependent boundary conditions. In other words, Fourier's equation shall be solved over the entire calculation domain:

$$\frac{\partial H}{\partial t} = \nabla \cdot (\lambda \nabla T) \tag{1}$$

where T is temperature, $H(T)$ is enthalpy, t is time, and λ is the thermal conductivity of metal that is casted. It should be noted that enthalpy is temperature dependent,. Following the physics of the steel continuous casting process, the model should be tridimensional. In fact, even though geometry remains constant in the casting direction (z axis), boundary conditions changes along z.

The most natural choice Is to use a growing mesh strategy, where new elements are added at domain during simulation; however, this approach is less often adopted due to a higher complexity. As mentioned before, a good computational alternative is to use a plane model (in the plane perpendicular to the z axis containing the product section), which moves in z direction according to the following law:

$$z(t + \Delta t) = z(t) + v \cdot \Delta t \tag{2}$$

where t is time, Δt is time increment and v is casting speed. According to this approach, which is generally called travelling slice model, boundary conditions start below a certain level which can be chosen arbitrarily (e.g., 0), and casted products move in negative z direction; this means that all nodes above 0 shall have fixed temperatures in order to avoid thermal fluxes in the domain part, which has not yet been casted. As the nodes move below the aforementioned level, the fixed temperature condition disappears. For a generic non axisymmetric section, the alternative to a 3D model is a 2D travelling slice (also called r-θ model), see Figure 2. In the case of a casted round-shaped product, axisymmetry allows Equation (1) to be solved in a 2D model. In the latter case, the approximated travelling slice model degenerates into a 1D model (as visible in Figure 2).

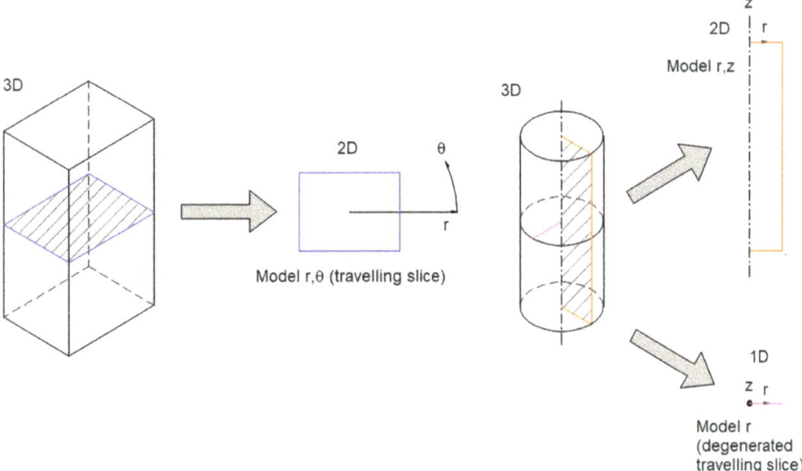

Figure 2. Different modeling strategies.

This favorable circumstance allows a comparison of the two models (r-z and degenerated travelling slice) to be easily performed with less computational effort, which permits a parametric analysis to be carried out. It is thus possible to verify in a wide range of product dimensions and casting speeds the correctness of the hypothesis that heat conduction, namely energy transfer, in casting direction is negligible.

3.2. Boundary and Initial Conditions

In a steel continuous casting machine there are three different kinds of thermal exchange, each predominant in a certain part of it (as seen in Figure 3).

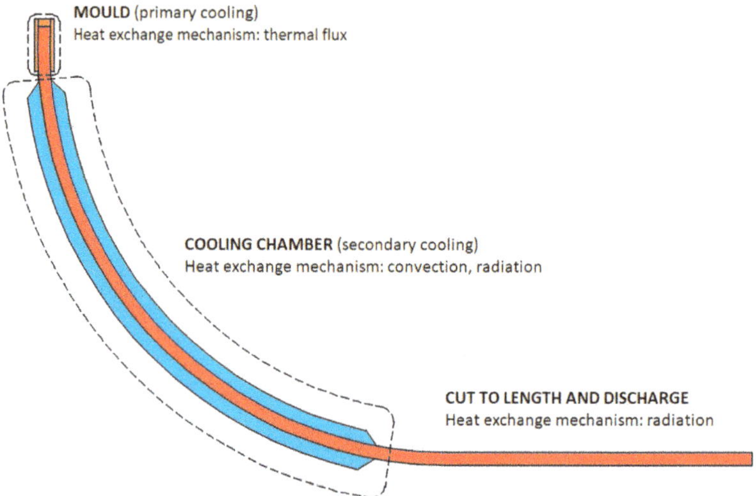

Figure 3. Prevailing heat exchange mechanisms in a continuous casting machine.

Thermal flux is withdrawn from the mould in order to start solidification; from a technical point of view, this is called primary cooling. When the mould is left, the casted product passes through a cooling chamber where energy is removed by arrays of water sprays (or air-mist for slabs). Here, two concurrent heat transfer mechanisms take place: convection with cooling media and radiation. The secondary cooling occurs in this zone. In the last part, machine cooling continues thanks to radiation only.

Since models are transient, an initial condition must be specified; this is a fixed temperature in all nodes of the calculation domain. For steel continuous casting, pouring temperature is the sum of liquidus temperature (which is dictated by chemical composition) and a specified superheat, namely the overheating relative to liquidus, typically in the range $20 \div 50\ °C$. This is done to avoid solidification in other devices (tundish and submerged entrance nozzle if present) prior to mould where steel shall remain in liquid phase, otherwise technological issues could halt or complicate the casting process.

3.2.1. Mould

As mentioned previously, thermal flux must be considered an input if a decoupled analysis is adopted, otherwise it becomes an output of the model; the first approach is employed in this research, as is done in the majority of published papers. From a global perspective, thermal flux acts on the domain borders; it is a Neumann's boundary condition, expressed as follows:

$$q = -\lambda \nabla T \cdot n \tag{3a}$$

where n is the unit vector perpendicular to surface and q is the thermal flux expressed in $[W/m^2]$.

From a mathematical point of view, the function above is one of time and coordinates (for shapes as billets, blooms and slabs):

$$q = f(time,\ coordinates) \tag{3b}$$

Thermal flux distributions are typically empiric: there are several references, however a first outlook can be found in [20,21].

3.2.2. Cooling Chamber

Cooling of the casted product that started in mould continues in this zone, where heat is withdrawn both by convection and radiation. A typical boundary condition could be written as

$$q = -\alpha(T, cooling\ flow\ rate)\left(T_{surf} - T_\infty\right) \tag{4}$$

where T_{surf} is the surface temperature and T_∞ is the temperature outside the boundary layer. Temperature dependence is due to include radiation effects avoiding the fourth power exponent, simplifying its numerical treatment; for this reason, α can be considered an equivalent heat transfer coefficient. As for mould heat fluxes, several measuring tests have been conducted for heat transfer coefficients also, resulting in a wide range of empirical correlations (see refs. [20–26] for details).

The cooling chamber is divided into sectors that are fed with different water-flow rates; the first one is just after the mould, and connected to it is Sector 1 (also called "foot rolls", in which cooling is intense), followed by Sector 2 ("mobile") and then by other sectors (called "fixed") until the end, depending on the machine layout and productivity. The three terms are enclosed in brackets because they refer to industrial jargon; for sake of clarity, the "mobile" sector does not move during casting its name comes from the fact that it is changed every time the cast section changes in order to get a more uniform cooling efficiency. The following sectors (the "fixed" ones) do not change because the cooling accuracy is less sensitive on the product quality when approaching the end of chamber. The longer the cooling chamber, the faster the casting pace (and thus productivity). The flow rate decreasing law is chosen to guarantee a smooth cooling path without excessive reheating on the product surface passing from a sector to the subsequent one. It shall be pointed out that the goal of secondary cooling is to achieve complete solidification prior to product cutting.

3.2.3. Cut to Length and Discharge

In this part, the casted product is not subjected to forced cooling anymore; it loses energy by radiation only. Hoods are used in some layouts to provide a passive control over how much temperature is left on the product prior to subsequent plastic deformation processes; in such cases, radiation losses are reduced in order to maintain a specified enthalpic level in it.

4. Case Study

Following the concepts shown until now, in the case of an axisymmetric product section, the plane model (or model r-z as in Figure 2) will be compared with the degenerated 1D travelling slice model. As mentioned previously, an increase of casted product size has been observed recently. It is therefore of great interest to analyze also these cases. For this purpose, three different casted shapes were investigated, including all rounds with diameters of 200 mm, 850 mm, and 1200 mm. The 200 mm round represents the traditional small-size product. Three casting speed values were created to vary in a range of industrial interest (from 2 m/min to 3 m/min). The 850 mm round represents a typical size of more a recent steelmaking plant used to feed the subsequent hot forging process. The adopted casting speed is 0.20 m/min. Nowadays, very big sections have been introduced as an alternative and more efficient process compared to ingot casting. The 1200 mm will be thus considered. Two values of casting speed are investigated: 0.08 m/min and 0.04 m/min; the latter one, although outside industrial range, has been considered to stress the travelling slice model to see if it fails.

All models have been implemented with commercial FE code MSC Marc and run on a 64-bit 8-processor (Intel(R) Core(TM) i7-8850H) with 64 GB installed RAM; Table 1 summarizes element and node counts. Degenerated travelling slice model adopts two node elements with linear interpolation along length (i.e., for Ø 200 mm, the radius is meshed with 100 linear elements, see Table 1); in these elements, heat can only flow along length. Model r-z, on the other hand, employs isoparametric four-node axisymmetric elements; in this case, the rectangular domain is discretized with a rather homogeneous mapping mesh, refined only towards the external radius to better simulate the high spatial temperature gradient associated with solidification. An automatic time stepping procedure has been selected for all simulation; equations are solved by using a multifrontal sparse solver in the framework of an implicit numerical scheme. More details can be found in [27].

Table 1. Characteristics of FE models.

Case	Degenerated 1D Travelling Slice		Model r-z	
	Elements	Nodes	Elements	Nodes
Ø 200 mm	100	101	67,756	70,389
Ø 850 mm	425	426	117,270	119,922
Ø 1200 mm	600	601	138,118	140,778

Steel composition is given in Table 2, as well as its characteristics in Table 3; it shall be highlighted that during solidification, the latent heat is released uniformly in the range between solidus and liquidus temperatures (also known as the mushy zone). Thermal conductivity of the steel should be divided into two phases: for the liquid one the Wiedemann–Franz-Lorenz rule is employed, while for solid one regression, formulas from experimental measurements are followed; explanation of both approaches can be found in [19]. In Figure 4, the values of enthalpy and volumic mass are reported for different temperatures. Enthalpy trend is very similar to ref. [28]. This is a common structural steel, widely used for round bar, rebar, and beam production.

Table 2. Steel chemical composition.

Element	Concentration [%]	Element	Concentration [%]
C	0.210	Mn	1.500
Cr	0.020	Ni	0.020
Mo	0.002	Si	0.200
Cu	0.035	Al	0.045
P	0.010	S	0.002
V	0.003	Fe	Balance

Table 3. Some physical parameters of selected steel.

Parameter	Value
Liquidus temperature	1541 °C
Solidus temperature	1457 °C
Solidification range	84 °C
Latent heat	297.66 kJ/kg

Boundary conditions are listed in Tables 4–6 for the three product diameters considered in this work. In particular, these indicate the thermal flux acting in mould, which is time dependent (the relation can be found in ref. [29]); since casting speed is constant during simulation, the dependence could be converted as position and thermal flux could be function of distance from meniscus. Moreover, the heat transfer coefficient (HTC) in the subsequent regions is reported; the latter one has been calculated according to ref. [20] and is speed dependent; for Ø 200 mm, examples in Table 4 reports the three HTCs values (respectively for 2 m/min, 2.5 m/min, and 3 m/min casting speed). For Ø 850 mm and Ø 1200 mm cases, a single HTC value is provided as only one casting speed has been considered. It must be observed that as a casted section increases, a less intense cooling is required due to lower casting speed and higher energy stored in the liquid steel. As pointed out previously, cooling is mainly governed by radiation in the discharge section.

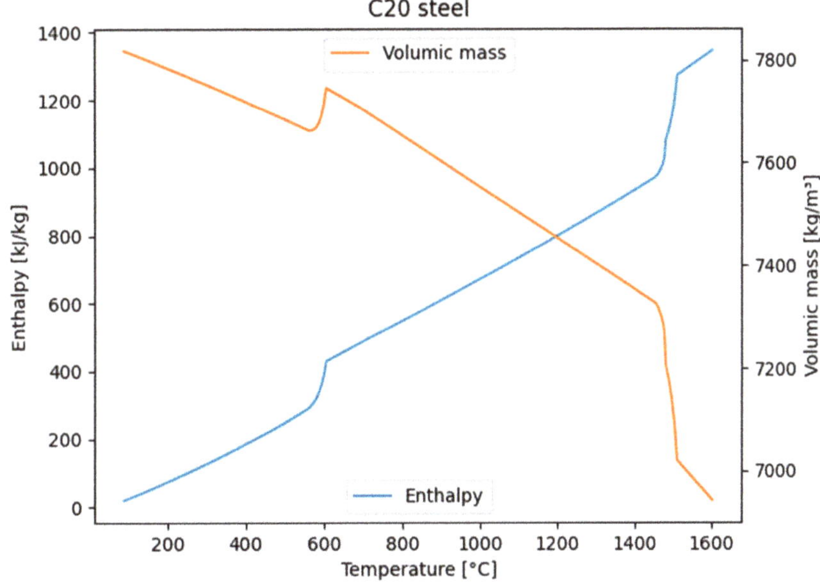

Figure 4. Enthalpy and volumic mass trend for the selected steel.

Table 4. Detail of applied boundary conditions for each zone for both models (cast section Ø 200).

Zone	Length [m]	Boundary Condition
Mould (steel level)	0.68	Thermal flux according to [29]
Sect. 1 (Foot rolls)	0.35	HTC = 1410, 1489, 1523 W/m^2·K (depending on speed)
Sect. 2 (Mobile)	1.90	HTC = 720, 759, 795 W/m^2·K (depending on speed)
Sect. 3 (Fixed 1)	2.30	HTC = 320, 351, 390 W/m^2·K (depending on speed)
Sect. 4 (Fixed 2)	1.15	HTC = 269, 280, 309 W/m^2·K (depending on speed)
Discharge	28.6	Radiation to environment

Table 5. Detail of applied boundary conditions for each zone for both models (cast section Ø 850).

Zone	Length [m]	Boundary Condition
Mould (steel level)	0.64	Average thermal flux = 548 kW/m^2
Sect. 1 (Foot rolls)	0.40	HTC = 331 W/m^2·K
Sect. 2 (Mobile 1)	0.75	HTC = 251 W/m^2·K
Sect. 3 (Mobile 2)	0.75	HTC = 242 W/m^2·K
Discharge	32.46	Radiation to environment

Table 6. Detail of applied boundary conditions for each zone for both models (cast section Ø 1200).

Zone	Length [m]	Boundary Condition
Mould (steel level)	0.64	Average thermal flux = 413 kW/m^2
Sect. 1 (Foot rolls)	0.40	HTC = 296 W/m^2·K
Discharge	33.96	Radiation to environment

5. Result and Discussion

Figures 5–7 use normalized distance as the abscissa; this is the ratio between local distance and the position of cutting devices, which is located 35 m after meniscus in all cases. Figure 5 represents the computed temperatures (surface and core) for the three cases of Ø 200 mm; both surface and core temperature are given. A similar pattern for all speeds is visible, although with different values; moreover, it can be observed that the complete solidification is reached before for lower speed, which is completely consistent with plant operations and reference [30].

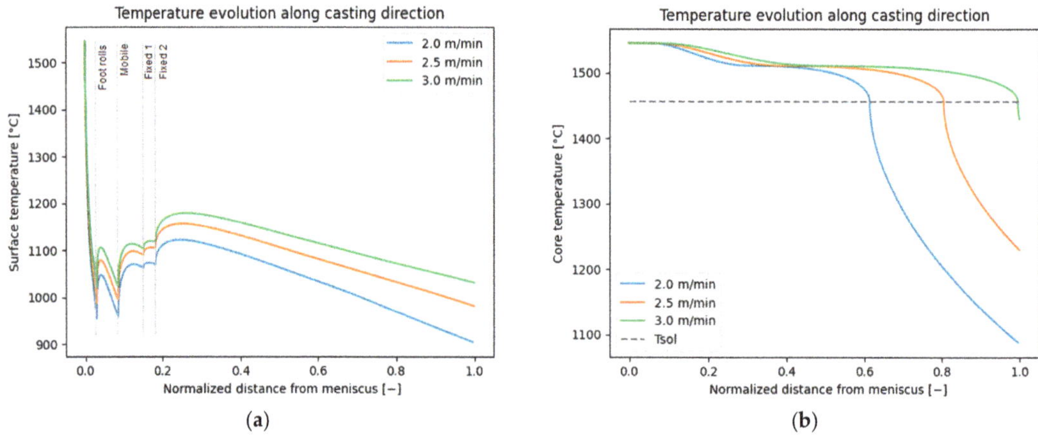

Figure 5. Surface (**a**) and core (**b**) temperatures for Ø 200 mm.

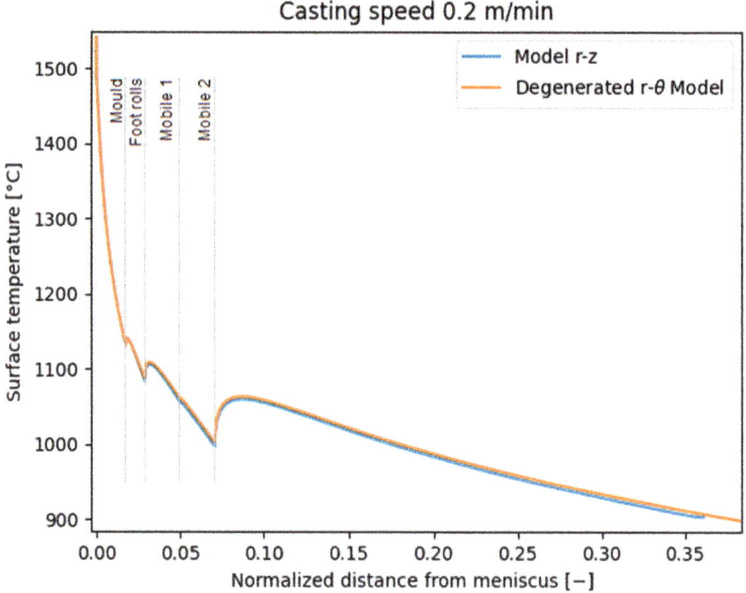

Figure 6. Comparison of both models on surface temperature (casted section Ø 850 mm).

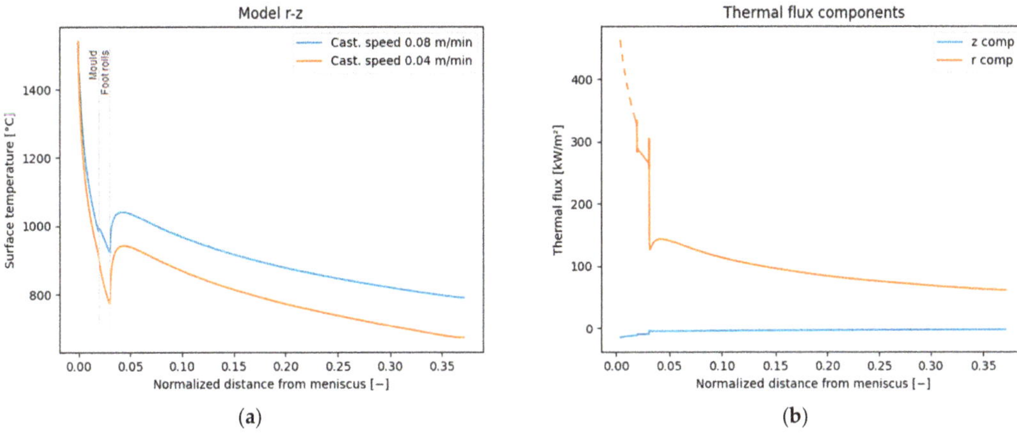

Figure 7. Surface temperature at different casting speeds (**a**) and thermal flux components at 0.04 m/min for Ø 1200 mm (**b**).

Figure 6 compares both models for Ø 850 mm casted at 0.20 m/min. The agreement between them is significant, since the two curves are virtually overlapped. In Table 7, the relative error for all sections and all casting speeds can be observed. The error is calculated as $(T_{reference} - T_{approx})/T_{reference}$, where subscript "reference" refers to model r-z while "approx" refers to degenerated travelling slice. The surface temperature is used as the reference temperature in all circumstances.

Table 7. Relative error on the computed surface temperature for all studied cases in two points.

Case	Mould Exit	Foot Rolls
Ø 200 mm–2.00 m/min	0.19%	−0.74%
Ø 200 mm–2.50 m/min	0.19%	0.51%
Ø 200 mm–3.00 m/min	0.28%	−0.59%
Ø 850 mm–0.20 m/min	0.09%	−0.28%
Ø 1200 mm–0.08 m/min	0.20%	0.43%
Ø 1200 mm–0.04 m/min	−0.87%	1.15%

Please keep in mind that the comparison is limited to the two points that appear in all cases: mould exit and foot rolls (both of which appear in Tables 4–6). The relative error is far under the acceptable threshold for industrial use purposes.

Figure 7 relates to the biggest section, the Ø 1200 mm. It must be pointed out that casting speed in such case, namely 0.08 m/min, is close to the lowest operational limit in the continuous casting process. As stated in the previous paragraph, a case at 0.04 m/min has been done only with the purpose of investigating the applicability of the travelling slice model at very low casting speed; boundary conditions were different in the mould (average heat flux 316 kW/m^2) but not in foot rolls, where same HTC (see Table 6) has been used. For this reason, the computed surface temperature results are low from an operational perspective. Moreover, it is visible that radial heat flux is always far bigger than axial one, confirming the correctness of neglecting this latter one (which is the fundamental hypothesis of travelling slice modeling). If this assumption is true for a such slow casted section, it will be even more so in the case of smaller and faster sections. Only in mould the radial thermal flux is imposed (with a descending law and its average value specified in Table 6), while it is the result of the computation in all other points; referring to Figure 7b, the dashed line is the imposed thermal flux, while the continuous one represents the calculated thermal flux (for both components).

Figure 8 shows the thermal field for Ø 1200 mm casted at 0.04 m/min, highlighting the "liquid pool end" that occurs, where the entire section has completely solidified (all domain is under solidus temperature).

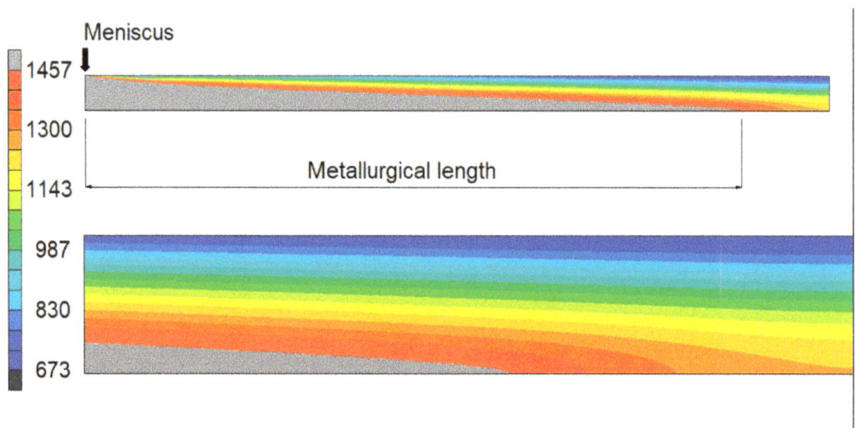

Figure 8. Thermal field in [°C] on whole domain (**top**) and particular of "liquid pool end" (**bottom**) for Ø 1200 mm.

It should be noted that travelling slice model yields the same metallurgical length of model bidimensional model, confirming its reliability in modeling the continuous casting process.

One of the last interesting aspects from a computational perspective, is the required time to run the models; it spans from 21 s to 123 s for a 1D degenerated travelling slice model (Ø 200 mm and Ø 1200 mm, respectively) to 22,285 s (Ø 200 mm) and 64,638 s (Ø 1200 mm) for model r-z.

6. Conclusions

In this work, the validity of the travelling slice model is investigated, notably in large dimension products and low casting speeds. Only in the case of a round section is a comparison with a reference model possible with a reasonable computational cost. Two different models (degenerated 1D travelling slice model and 2D axisymmetric reference model) have been developed and compared. The results show the validity of the travelling slice model in a wide range of product dimensions and casting speeds. Surface temperature comparison shows practically the overlap of profiles coming from both models with a relative error that is always less than 1.5%, also in case of the casting speed outside industrial range. Moreover, the component of thermal flux along casting direction is always a negligible portion of total flux, thus confirming the validity of the travelling slice model. It can be thus stated that also in case of big sections and low productivity, the travelling slice could be a reliable choice to model the continuous casting process. It should be highlighted that the computational cost of the reference 2D model is far higher the 1D approximated model; the ratio of calculation time for the same casted profile is proportional to 10^3, in favor of the travelling slice model (on the same hardware). In conclusion, it can be stated that the travelling slice approach could be considered a proven modeling technique in describing the continuous casting process. In future works, this will be extended to non-axisymmetric casted shapes, switching from 1D to 2D.

Author Contributions: Conceptualization, G.B.; methodology, G.B. and F.D.B.; validation, G.B.; formal analysis, G.B.; investigation, G.B. and F.D.B.; writing—original draft preparation, G.B.; writing—review and editing, F.D.B. All authors have read and agreed to the published version of the manuscript.

Funding: This research received no external funding.

Data Availability Statement: Restrictions are applied to the availability of the data used in this paper. Data were obtained from Danieli's subsidiaries and are available upon request to Danieli & C. Officine Meccaniche spa.

Conflicts of Interest: G.B. declares to be an employee of Danieli & C. Officine Meccaniche SpA headquartered in Buttrio (Italy). F.D.B. declares no conflict of interest.

References

1. Dong, Q.; Zhang, J.; Yin, Y.; Wang, B. Three-dimensional numerical modeling of macrosegregation in continuously cast billets. *Metals* **2017**, *7*, 209. [CrossRef]
2. Yang, J.; Xie, Z.; Meng, H.; Hu, Z.; Liu, W.; Ji, Z. 3D transient heat transfer simulation and optimization for initial stage of steel continuous casting process. *ISIJ Int.* **2023**, *63*, 862–869. [CrossRef]
3. Koric, S.; Abueidda, D. Deep learning sequence method in Multiphysics modeling of steel solidification. *Metals* **2021**, *11*, 494. [CrossRef]
4. Han, H.N.; Lee, J.; Yeo, T.; Won, Y.M.; Kim, K.; Oh, K.H.; Yoon, J. A finite element model for 2-dimensional slice of cast strand. *ISIJ Int.* **1999**, *39*, 445–454. [CrossRef]
5. Koric, S.; Thomas, B.G. Efficient thermo-mechanical model for solidification processes. *Int. J. Numer. Meth. Engng.* **2006**, *66*, 1955–1989. [CrossRef]
6. Heger, J. Finite element modelling of mechanical phenomena connected to the technological process of continuous casting of steel. *Acta Polytech.* **2004**, *44*, 15–20. [CrossRef]
7. Kong, Y.; Chen, D.; Liu, Q.; Long, M. A prediction model for internal cracks during slab continuous casting. *Metals* **2019**, *9*, 587. [CrossRef]
8. Li, C.; Thomas, B.G. Thermo-mechanical finite element model of shell behavior in continuous casting of steel. In Proceedings of the Modeling of Casting, Welding and Advanced Solidification Processes X, San Destin, FL, USA, 25–30 May 2003; pp. 385–392.
9. Nian, Y.; Zhang, L.; Zhang, C.; Ali, N.; Chu, J.; Li, J.; Liu, X. Application status and development trend of continuous casting reduction technology: A review. *Processes* **2022**, *10*, 2669. [CrossRef]
10. Wang, E.; He, J. FE numerical simulation for influence of mold taper on thermomechanical behavior of steel billet in continuous casting process. *J. Mater. Sci. Technol.* **2001**, *17* (Suppl. 1), s8–s12.
11. Fang, Q.; Ni, H.; Zhang, H.; Wang, B.; Liu, C. Numerical study on solidification behavior and structure of continuously cast U71Mn steel. *Metals* **2017**, *7*, 483. [CrossRef]
12. Kwon, S.H.; Won, Y.M.; Back, G.S.; Kim, H.; Lee, J.S.; Kim, D.G.; Heo, Y.U.; Yim, C.H. Prediction model for degree of solid-shell unevenness during initial solidification in the mold. *ISIJ Int.* **2021**, *61*, 2534–2539. [CrossRef]
13. Saraswat, R.; Maijer, D.M.; Lee, P.D.; Mills, K.C. The effect of mould flux properties on thermo-mechanical behavior during billet continuous casting. *ISIJ Int.* **2007**, *47*, 95–104. [CrossRef]
14. Mills, K.C.; Fox, A.B. The role of mould fluxes in continuous casting—So simple yet so complex. *ISIJ Int.* **2003**, *43*, 1479–1486. [CrossRef]
15. Vynnycky, M. Air gaps in vertical continuous casting in round moulds. *J. Eng. Math.* **2010**, *68*, 129–152. [CrossRef]
16. Vynnycky, M. Applied mathematical modelling of continuous casting process: A review. *Metals* **2018**, *8*, 928. [CrossRef]
17. Jolivet, J.M.; Le Papillon, Y.; Bellavia, L. Development of high productivity casting of conventional and thin slabs. EUR 23887 Final Report. *RFCS Publ.* **2009**. [CrossRef]
18. Alizadeh, M.; Jahromi, A.J.; Abouali, O. New analytical model for local heat flux density in the mold in continuous casting of steel. *Comput. Mater. Sci.* **2008**, *44*, 807–812. [CrossRef]
19. Mills, K.C.; Karagadde, S.; Lee, P.D.; Yuan, L.; Shahbzian, F. Calculation of physical properties for use in models of continuous casting processes—Part 2: Steels. *ISIJ Int.* **2016**, *56*, 274–281. [CrossRef]
20. Cramb, A.W. *The Making, Shaping and Treating of Steel (MSTS)*; The AISE Steel Foundation: Pittsburg, PA, USA, 2003.
21. Assuncao, C.; Tavares, R.; Oliveira, G. Improvement in secondary cooling of continuos casting of round billets through analysis of heat flux distribution. *Ironmak. Steelmak.* **2015**, *42*, 1–8. [CrossRef]
22. Wang, W.; Zhang, H.; Nakajima, K.; Lei, H.; Tang, G.; Wang, X.; Mu, W.; Jiang, M. Prediction of final solidification position in continuous casting bearing steel billets by slice moving method combined with Kobayashi approximation and considering MnS and Fe_3P precipitation. *ISIJ Int.* **2021**, *61*, 2703–2714. [CrossRef]
23. Long, M.; Chen, H.; Chen, D.; Yu, S.; Liang, B.; Duan, H. A combined hybrid 3-D/2-D model for flow solidification prediction during slab continuous casting. *Metals* **2018**, *8*, 182. [CrossRef]

24. Milkowska-Piszczek, K.; Falkus, J. Control and design of the steel continuous casting process based on advanced numerical models. *Metals* **2018**, *8*, 581. [CrossRef]
25. Li, L.; Zhang, Z.; Luo, M.; Li, B.; Lan, P.; Zhang, J. Control of shrinkage porosity and spot segregation in Ø195 mm continuously casted round bloom of oil pipe steel by soft reduction. *Metals* **2021**, *11*, 9. [CrossRef]
26. Yang, J.; Ji, Z.; Liu, W.; Xie, Z. Digital-twin-based coordinated optimal control for steel continuous casting process. *Metals* **2023**, *13*, 816. [CrossRef]
27. MSC Marc. *Volume A: Theory and User Information*; MSC Software Corporation: Irving, CA, USA, 2022.
28. Ramirez-Lopez, A.; Davila-Maldonado, O.; Najera-Bastida, A.; Morales, R.D.; Rodriguez-Avila, J.; Muniz-Valdez, C.R. Analysis of non-symmetrical heat transfer during the casting of steel billets and slabs. *Metals* **2021**, *11*, 1380. [CrossRef]
29. Cai, S.W.; Wang, T.M.; Xu, J.J.; Li, J.; Cao, Z.Q.; Li, T.J. Continuous casting novel mould for round steel billet optimized by solidification shrinkage simulation. *Mater. Res. Innov.* **2011**, *15*, 29–35. [CrossRef]
30. Chen, Y.; Peng, Z.; Wu, L.; Zhao, L.; Wang, M.; Bao, Y. High-precision numerical simulation for effect of casting speed on solidification of 40Cr during continuous billet casting. *Metall. Ital.* **2015**, *1*, 47–51.

Article

Experimental and Numerical Investigation of Hot Extruded Inconel 718

Stefano Bacchetti [1], Michele A. Coppola [1], Francesco De Bona [2,*], Alex Lanzutti [2], Pierpaolo Miotti [1], Enrico Salvati [2] and Francesco Sordetti [2]

[1] Pietro Rosa TBM, 33085 Maniago, Italy; sbacchetti@pietrorosatbm.it (S.B.); mcoppola@pietrorosatbm.it (M.A.C.); pmiotti@pietrorosatbm.it (P.M.)

[2] Department Polytechnic of Engineering and Architecture, University of Udine, Via delle Scienze 208, 33100 Udine, Italy; alex.lanzutti@uniud.it (A.L.); enrico.salvati@uniud.it (E.S.); francesco.sordetti@uniud.it (F.S.)

* Correspondence: debona@uniud.it; Tel.: +39-0432558269

Abstract: Inconel 718 is a widely used superalloy, due to its unique corrosion resistance and mechanical strength properties at very high temperatures. Hot metal extrusion is the most widely used forming technique, if the manufacturing of slender components is required. As the current scientific literature does not comprehensively cover the fundamental aspects related to the process–structure relationships, in the present work, a combined numerical and experimental approach is employed. A finite element (FE) model was established to answer three key questions: (1) predicting the required extrusion force at different extrusion speeds; (2) evaluating the influence of the main processing parameters on the formation of surface cracks using the normalized Cockcroft Latham's (nCL) damage criterion; and (3) quantitatively assessing the amount of recrystallized microstructure through Avrami's equation. For the sake of modeling validation, several experimental investigations were carried out under different processing conditions. Particularly, it was found that the higher the initial temperature of the billet, the lower the extrusion force, although a trade-off must be sought to avoid the formation of surface cracks occurring at excessive temperatures, while limiting the required extrusion payload. The extrusion speed also plays a relevant role. Similarly to the role of the temperature, an optimal extrusion speed value must be identified to minimize the possibility of surface crack formation (high speeds) and to minimize the melting of intergranular niobium carbides (low speeds).

Keywords: extrusion; Inconel 718; metal-forming simulation

Citation: Bacchetti, S.; Coppola, M.A.; De Bona, F.; Lanzutti, A.; Miotti, P.; Salvati, E.; Sordetti, F. Experimental and Numerical Investigation of Hot Extruded Inconel 718. *Metals* **2023**, *13*, 1129. https://doi.org/10.3390/met13061129

Academic Editor: José Valdemar Fernandes

Received: 19 May 2023
Revised: 12 June 2023
Accepted: 13 June 2023
Published: 16 June 2023

1. Introduction

Inconel 718 is probably one of the most important among the nickel-based superalloys, due to its exceptional properties such as high oxidation resistance, corrosion resistance and high mechanical strength, even at high temperatures. For this reason, Inconel 718 is currently widely employed in the aircraft engine industry [1,2]. Specifically, it is used in many critical engine components (i.e., diffusers, combustion chambers, shells of gas generators, rocket engines and gas turbines), accounting for over 30% of the total finished component mass of a modern aircraft engine [3].

Despite its outstanding properties, the workability of Inconel 718 shows several unsolved critical issues. Its peculiar physics characteristics, such as its lower thermal conductivity, work hardening, presence of abrasive carbide particles, hardness, affinity for reacting with the tool material, etc., make Inconel 718 difficult to machine [4–6]. Recent advances in various material processing techniques, e.g., turning, milling and drilling, for machining Inconel 718 are investigated and discussed in [7]. Although Inconel 718 is supposed to possess good weldability, cracking problems may still arise [8,9]. As highlighted in a welding process review [10], strain cracking during the post-weld heat treatment, solidification cracking and liquation cracking may still arise. In the last decade, the possibility to obtain Inconel 718 components by using additive manufacturing techniques was deeply explored [11]. In [12], Inconel 718 semi-finished parts were produced by selective laser melting (SLM) and tensile tests were performed, showing the capability of the SLM process to produce parts with mechanical properties better than forged and cast materials at room temperature and equal to properties to forged material at high temperatures. Recently, additively manufactured metamaterials were also obtained for the thermal stress accommodation of metal heat pipes [13]. Inconel 718 can be also coated to improve its wear resistance, as suggested in [14].

From the industrial end-user point of view, hot plastic deformation manufacturing processes remain the most common way to obtain Inconel 718 semi-finished parts. It is well-known that, in these cases, a detailed processing setup is mandatory to control recrystallization and grain growth phenomena. The main challenge during the hot working process of Inconel 718 is related to its high deformation resistance during forging and its tendency to form a duplex grain structure. These issues can be mitigated by precisely controlling both temperature and deformation during the manufacturing process [15]. As far as hot forging is concerned, a better understanding of the competition between several mechanisms (dynamic recovery, dynamic recrystallization and plasticity hardening) is presented in [16], where several thermo-mechanical parameters are taken into account. Additionally, an interesting study on the spatial microstructural evolution of the as-forged and heat-treated Inconel 718 disks can be found in the literature [17]. In this case, due to the different thermomechanical conditions across the thicknesses, a variation in the grain size and characteristics of the reinforcing precipitates were observed.

When the final product is characterized by a high aspect ratio, i.e., turbine blades, hot metal extrusion techniques are generally preferred. This process induces higher plastic deformation in comparison to forging, thus a finer process parameter tuning is required to avoid unwanted microstructural effects [18]. A typical failure occurring during hot metal extrusion is due to the formation of surface cracks. Such a phenomenon was widely investigated only in the case of aluminum alloy extrusion, where T-shape extrusion experiments were performed and compared with finite element (FE) simulation, in order to evaluate the possible predictive criteria to assess material recrystallization and surface crack formation [19].

Nevertheless, to the best of the authors' knowledge, no works seem to be available in the literature dealing with the prediction of the above-mentioned microstructural aspects in the hot extrusion process of Inconel 718 superalloy, along with other technological characteristics, such as determining the extrusion force, speed and temperature given the geometry of the mold. To tackle this outstanding problem, a FE-based numerical model of the extrusion process of an Inconel 718 component was devised. In parallel, a series of experimental tests were designed and carried out to examine the extrusion of pre-heated Inconel 718 billets, thus allowing for the comprehensive validation of the numerical model and respective results. The validated model was then used to perform parametric analyses in order to pinpoint ranges of processing conditions that avoid the formation of unwanted features.

2. Materials and Methods

To accomplish all the envisaged tasks, the study focuses on the hot extrusion of an axisymmetric-shaped component presenting a high aspect ratio ($\cong$1:8). Therefore, both experimental and modeling activities were based upon this assumption. In this section, all the experimental and numerical details regarding the analyses of the present study are reported.

2.1. Extrusion Process Set-Up

The studied extrusion setup consists of three main parts, i.e., a mould, a punch and a processed material (billet), as illustrated in Figure 1. According to this figure, the billet shows a cylindrical shape at the beginning of the extrusion process (a), to turn into a nail-like shape at the end (b). It is important to point out that the billet initial diameter is slightly smaller than that of the accommodating mold. This implies that the punch compression against the billet produces a radial expansion of the billet in the initial stage of the process. At the stage when the material can no longer radially expand, then the material flows through the narrower section of the mold. The punch arrests prior to reaching its mechanical travel stop in order to obtain the head of the extruded piece.

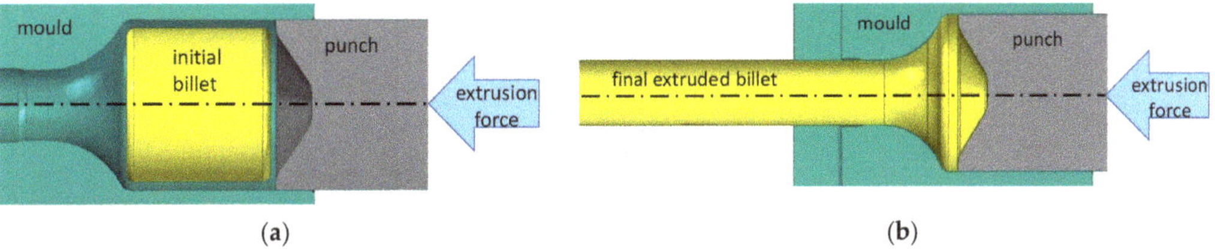

Figure 1. Billet, mold and punch, respectively, (**a**) at the beginning and (**b**) at the end of the extrusion process.

2.2. Process Simulation and Material Modelling

At present, several commercial products are available for metal-forming simulation [20], based on FE approaches. The possibility of using a digital model, for the preliminary set-up of the technological process, would permit considerable time and cost reductions. Nevertheless, the reliability of these tools is still an object of debate, particularly concerning the material properties to be introduced in the model [21,22]. In the present work, Q-form (10.1.6 version) was used. A metal-forming process simulation of a structural component made of Inconel 718 can be found in the literature, using the same FE code [23]. Such a reference study presented a partial comparison with experiments, suggesting that the material model proposed herein is sufficiently accurate.

The simulation of hot plastic deformation processes generally requires huge computational time. In this case, an axisymmetric geometry was chosen, so that the dimension of the problem could be reduced by employing a plane model. The billet was modelled by 6-node triangular isoparametric elements (initial mesh: 9000 nodes, 17,000 elements). Generally, a plastic deformation process involves very high plastic strain, therefore, to avoid excessive element distortion, re-meshing was performed when strain increments higher than 0.1 among subsequent steps occurred. Moreover, the mesh was automatically updated where high strain and temperature gradients were detected. As the deformation of the mould is negligible, a fine mesh of this component was only required close to the inner surface where the thermal gradients could be relevant.

As the study is mainly related to the extrusion process simulation, the mold and punch were modelled as rigid bodies; their initial temperature was 200 °C and the initial temperature of the air was 20 °C. The dimension of the mesh was defined according to a sensitivity analysis performed referring to the damage parameter D.

The numerical model for the FE simulation is based on the usual flow formulation [24], where the material is considered as an incompressible rigid-viscoelastic continuum. The contact between the punch, mould and billet is dealt with according to a friction model proposed by [25]. It can be considered as a combination of constant friction and Coulomb friction models; the first one is dominant in the case of a high value of contact pressure, whereas, for low values, friction forces are approximately linearly dependent on the normal contact pressure.

The billet was made of Inconel 718, showing Ni (51.5%), Cr (18.4%), Fe (20%), Mo (3%) and Nb (5%) as major alloying elements (by weight %) with balanced other elements. The mold was made of a chromium-vanadium-molybdenum alloyed steel.

The physical and mechanical properties of the studied Inconel in the range 20–1100 °C were obtained from [26].

The flow stress of Inconel 718 at different strain rates and temperatures are reported for the case at $\varepsilon = 0.7\%$ in [27].

In a more general case, flow stresses can be obtained according to the following relations [28]:

$$\frac{\dot{\varepsilon}}{\bar{D}(T)} = A\left(sinh\alpha\frac{\sigma}{E(T)}\right)^{0.5} \tag{1}$$

where $\dot{\varepsilon}$ and σ are the strain rate and the flow stress, respectively, while $\bar{D}(T)$ is the self-diffusion coefficient, given by:

$$\bar{D}(T) = \bar{D}_0 exp\left(\frac{-Q_{st}}{RT}\right) \tag{2}$$

where R is the universal gas constant, T the absolute temperature, Q_{st} is the activation energy and $\bar{D}_0$ is the maximal diffusion coefficient. In the present study, the latter two constants are set as: $\bar{D}_0 = 1.6 \times 10^{-4}$ m^2/s and $Q_{st} = 285$ kJ/mol.

The dependence of the Young's modulus E to temperature T can be obtained through the following relation [29]:

$$E = 2(1+v)G \tag{3}$$

$$G = G_0\left(1 - 0.5\frac{v}{T_M}\right) \tag{4}$$

where G is the shear modulus, G_0 is the shear modulus at 300 K (8.31 $\times$ 10^4 MPa), T_M is the meeting temperature (1673 K) and v is the Poisson coefficient (0.33).

To predict the formation of surface cracks, a damage criterion can be adopted. Several fracture models are proposed in the literature. In [30], a thorough review of the most common ductile damage criteria is presented. Probably the most widely used phenomenological fracture model is the original Cockcroft Latham's criteria [31]. It assumes that the material fracture is controlled by maximum principal stress. This approach was improved in the normalized Cockcroft Latham's (nCL), also called Oh et al., a criterion where an equivalent stress is introduced to normalize the maximum principal stress. In the modified Cockcroft Latham's (called also Clift et al.) criterion, it is assumed that the ductile fracture starts or initiates when a critical value of plastic work per unit volume is achieved. Other fracture models (see [30] for more details), such as Mc Clintock's, Brozzo's and RT (also called Rice and Tracey) models, have been proposed and reported in the literature.

In [19,32], the nCL criterion was indicated as the best amongst the various existing criteria for the prediction of fracture initiation during extrusion. Recently, in [30], a comparative assessment of failure strain predictions using ductile damage criteria in the metal-forming of Inconel 718 was performed, concluding that the nCL criterion seems the most suitable option to predict fracture locus with high accuracy both at room and at high temperatures. The nCL was also adopted in [33], where a FE simulation of an Inconel 718 machining process was performed with the aim of predicting chip segmentation during orthogonal cutting. In particular, another study pointed out that the critical value of D in Equation (5) should be dependent on the strain path and, therefore, it cannot be considered as a material constant [34]. Thus, the nCL does not allow for a quantitative prediction of the damage for an arbitrary technological process of metal forming. The criterion can be used only for a qualitative comparison of similar technological processes. In this case, the nCL gives substantial support in evaluating the most critical zones of the final extruded component.

According to the previously mentioned literature assessments, in this work, the nCL damage criteria was used:

$$D = \int_0^{\bar{\varepsilon}_f} \frac{\sigma_I}{\sigma_{VM}} d\bar{\varepsilon} = \int_0^{t_f} \frac{\sigma_I}{\sigma_{VM}} \dot{\varepsilon} dt \tag{5}$$

where:

σ_I = maximum principal tensile stress;

$\bar{\varepsilon}_f$ = total equivalent strain;

σ_{VM} = von Mises equivalent stress;

$\dot{\varepsilon}$ = equivalent strain rate;

D = damage index.

The microstructure evolution can be simulated by means of a semi-empirical model [35]. The model takes into account both the recrystallization (dynamic and static) and grain growth, providing the average grain size and the fraction of recrystallized grains [36]. In the considered alloy, dynamic recrystallization usually occurs starting on deformed grain boundaries after the critical strain is reached for a certain applied strain at a certain processing temperature [37–39]. The fraction of recrystallized crystals X_{dyn} can be calculated by means of Equation (6). The formula is obtained from Avrami's equation:

$$X_{dyn} = 1 - exp\left[-ln2\left(\frac{\varepsilon}{\varepsilon_{0.5}}\right)^{1.9}\right] \tag{6}$$

The strain condition $\varepsilon_{0.5}$ at which 50% of grains are recrystallized (Equation (7)), that is a parameter necessary to determine the fraction of dynamically recrystallized grains, depends on the initial grain size d_0 and on the Zener–Hollomon parameter Z, which is calculated on the basis of processing conditions (strain rate and temperature).

$$\varepsilon_{0.5} = 0.029 \, d_0^{0.2} Z^{0.06} \tag{7}$$

The grain size during dynamic recrystallization $\bar{\bar{D}}_{dyn}$ can be determined by Equation (8), which considers the Zener–Hollomon parameter as the most important to describe the grain size evolution:

$$\bar{\bar{D}}_{dyn} = 1.35 \cdot 10^3 d_0^{-0.01} \cdot Z^{-0.124} \tag{8}$$

Further recent work concerning the simulation of microstructure evolution and dynamic crystallization in hot plastic deformation of metals can be found in [40,41].

The main process parameters analyzed in the simulations are the punch speed and the initial temperature of the billet T_b in the range 1020–1090 °C, within the classic deformation temperature limits for Inconel 718 hot working [26]. In fact, these alloys are limited in terms of processing temperatures due to the intergranular precipitations of Nb compounds at temperatures above 1150 °C that reduce the ductility of the material, while, at temperatures below 950 °C, the material is prone to generate adiabatic shear bands [27].

The mold geometry is characterized by an extrusion ratio of ~1:6. The punch and mold undergo relevant loads during the extrusion process and, thus, they must be designed thoroughly, in order to avoid possible failures. Nevertheless, this aspect is strongly related to the specific product shape and thus it is not the objective of this work.

2.3. Extrusion Process Test

The parent billets were obtained by cutting a circular bar and turned to remove sharp edges, thus facilitating its self-centering at the first phase of the process. To lubricate and avoid oxidation at the surface of the billets, a glass-based coating was subsequently applied (see Figure 2a).

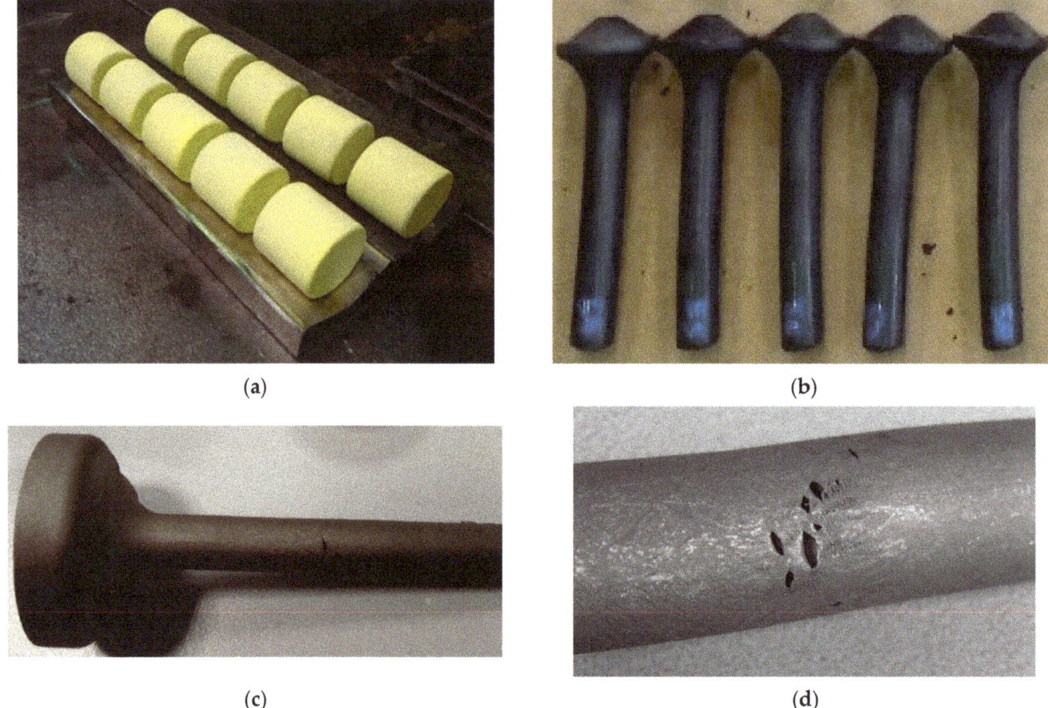

Figure 2. Set of glass-coated billets (**a**) ready for the extrusion test and (**b**) after the test. Surface crack formation after preliminary tests (**c**) and close up view (**d**) of the damaged zone.

To perform the extrusion tests, the following procedure was carried out:

- The billet was heated to a temperature T_b (holding time 1500 s).
- The punch and mold were lubricated with a graphite and water mixture.
- The billet was transferred from the oven to the mold (duration time: 3 s).
- The billet was maintained in position for 3 s (in this phase, heat was transferred from the billet to the mold)
- The punch was moved forward to perform the extrusion of the billet at constant speed v.
- Finally, the punch was retracted and the extruded billet was extracted.

A total of 63 tests were performed at different extrusion speeds, spanning from $0.27v_{max}$ to $1.0v_{max}$ and at an initial temperature of the billet of T_b = 1090 °C. v_{max} is a normalizing velocity. The selected speed range and temperature was set, such that the formation of surface cracks was inhibited. In particular, in the present study, higher temperatures (>1095 °C and lower extrusion speeds ($<0.270v_{max}$) were discarded as preliminary experimental tests showed the formation of surface cracks on final extruded components at $0.19v_{max}$ and at T = 1110 °C. An example is shown in Figure 2c,d.

The value of the force was obtained by measuring the hydraulic fluid pressure at a sampling speed of 100 Hz.

2.4. Macro and Micro-Structure Analysis

The extruded specimen underwent a macro- and micro-structural characterization. In particular, one extruded sample was cut into two halves along the longitudinal direction to obtain flat surfaces, above which the microstructural analyses were performed. The target surface was then manually ground and polished to obtain a mirror-like surface. Afterwards, the sample was macro-etched (submerged) by using a solution of HCl and iron chloride for

30 min. The sample was finally rinsed in deionized water with 1 M of sodium bicarbonate and visually inspected to analyze the macroscopic structural features.

Following this, metallographic samples were extracted from the previously analyzed sample at the areas circled in Figure 3. These samples were ground and polished again to reach a mirror-like surface and then etched using Marble's etchant for 1–2 min to reveal the microstructure. The microstructure was then analyzed by light microscopy and the grain size was measured through the line intercept method. The exact locations of analyses are indicated in Figure 3. In detail, the microstructural characterization was performed in the centerline of the test specimen and in the proximity of the extruded surface. The filleted region, indicated by the large circle in Figure 3, was analyzed in detail from a microstructural point of view in order to detect the critical deformation that triggers the grain crystallization. In this area, an evaluation of the recrystallized fraction was carried out by using the image analysis approach (phase partition analysis).

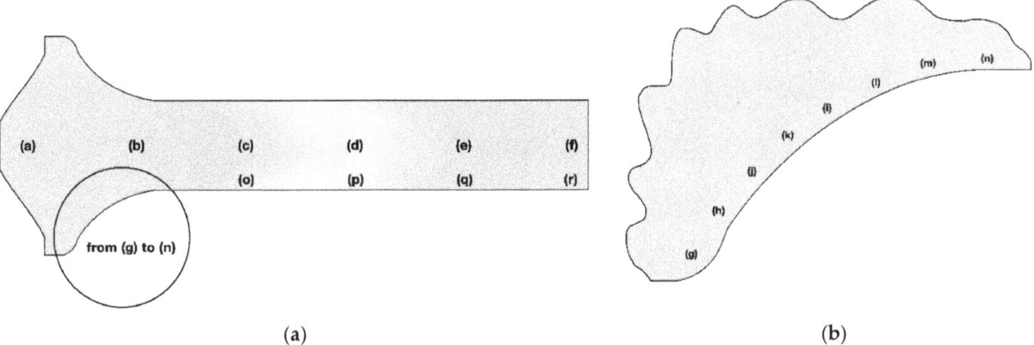

Figure 3. Microstructural characterization map of analysis: (**a**) overall view of the component and (**b**) close up view of the fillet region.

3. Results and Discussion

3.1. Extrusion Process and Force Evolution: Numerical Simulation and Experiments

Thanks to the numerical simulation of the hot extrusion process, it is possible to graphically distinguish three phases of the process (see Figure 4), referring to an extrusion speed $v = 0.67v_{max}$. At $t = 0$, the punch touches the billet, producing its radial expansion up to $t = 0.09$ s. After that time, the material starts flowing inside the matrix up to $t = 0.18$ s. During this phase, the high friction between the billet and the mold produces heat and, consequently, a significant increase in the temperature of the billet. From $t = 0.18$ s up to the end of the process ($t = 0.35$ s), the billet shows its final forms and starts flowing more freely.

For the sake of the validation of the FE model, the extrusion force for two different extrusion speeds is compared (i.e., simulation vs. experiment). Figure 5 shows the evolution of the simulated extrusion force versus time for two different values of the extrusion speed. An interpretation of the force evolution can be proposed; during the first phase of the process (1), (see Figure 5), the slope of the curve is constant and relatively low, up to the beginning of the second phase (2), where a rapid increase is observed. This is mainly related to the large plastic deformation occurring in the billet. Moreover, the extrusion force has to overcome the high frictional forces due to the increase in the contact pressure occurring between the billet and the mold. In this phase, once the frictional force cannot increase any further, the peak force is reached. In the last phase (3), a slight decrease in the force is observed. This is due to the heat produced by the sliding of the mold with respect to the billet that increases the temperature of the material, thus reducing the flow stress. Moreover, as the extrusion process evolves, the contact between the extruded part and the mold shows a progressive area reduction at the head of the billet, with a consequent decrease in friction, thus a decrease in the extrusion force.

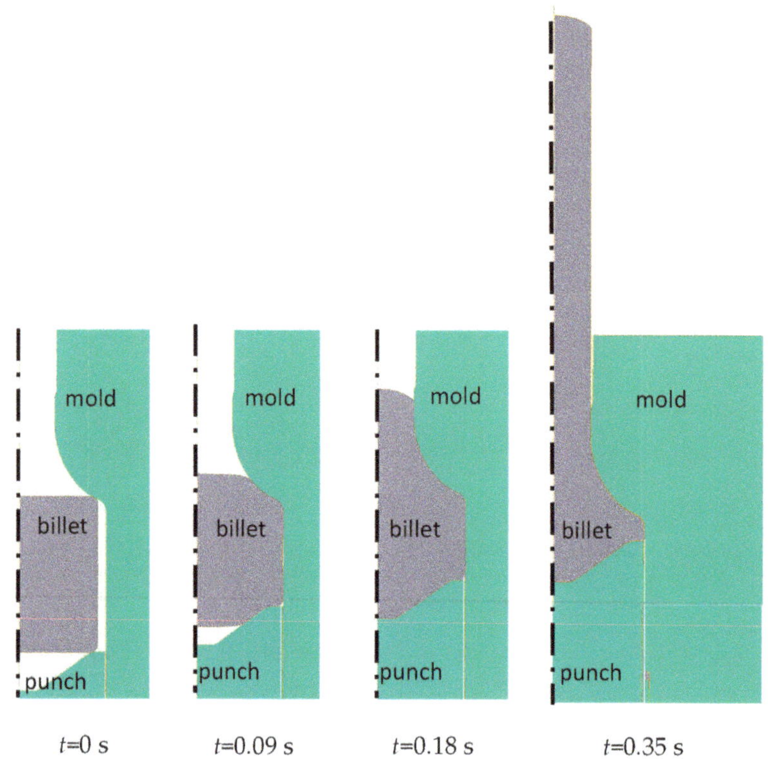

Figure 4. The three different phases of the process: simulated results.

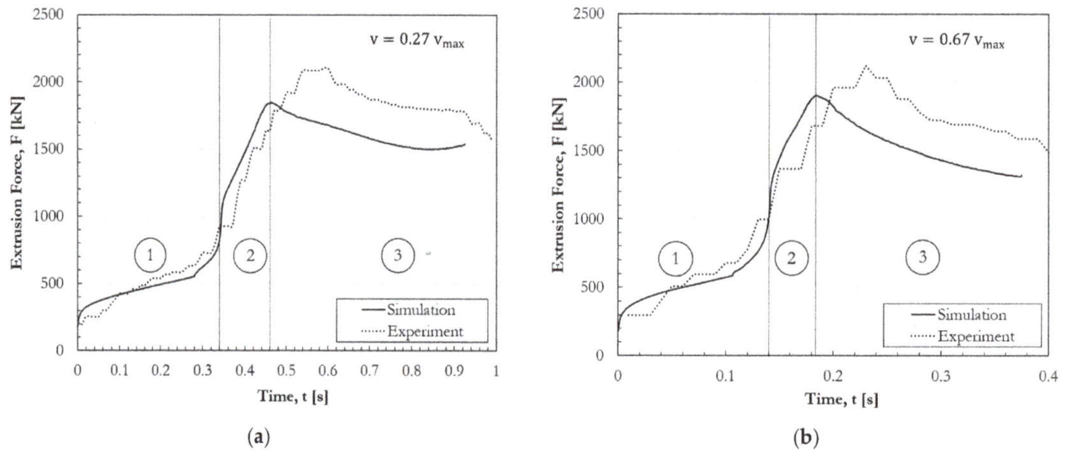

Figure 5. Extrusion force versus time for (**a**) an extrusion speed of $0.27v_{max}$ and (**b**) of $0.67v_{max}$.

The measured values of forces are also reported. The agreement is satisfactory in the first two extrusion phases (1 and 2, in Figure 5); in the third phase a higher difference is clearly observed. This is probably related to the high extrusion force reached in this phase. A large compression force thus acts on the punch, producing a radial expansion due to Poisson's effect. It follows that a slight shrink fit in the punch–mold coupling is likely to arise. Therefore, an increase in the friction—and consequent force—takes place. In the simulation, the punch and mold were modeled as rigid bodies and, therefore, this effect could not be captured.

To even further support the validation concerning the extrusion force, a set of experiments and simulations were carried out at different extrusion speeds. Figure 6 shows the values of maximum extrusion forces (F_{max}) at different extrusion speeds; furthermore, the simulated results are compared with the experiments (each point is the average of 6–10 different tests). It can be observed that the extrusion force remains almost constant with respect to the extrusion speed. This is justified by considering that the increase in the flow stress (and also the force), with the strain rate (and the speed), is counterbalanced by the heat exchange from the billet to the mold that is reduced for high speeds, promoting high temperatures and subsequent low stress flow values. Simulated extrusion forces are always slightly lower with respect to the measurements; this is probably due to the uncertainty of the value of the friction parameters to be used in the simulation and also due to the shrink fit phenomena mentioned above.

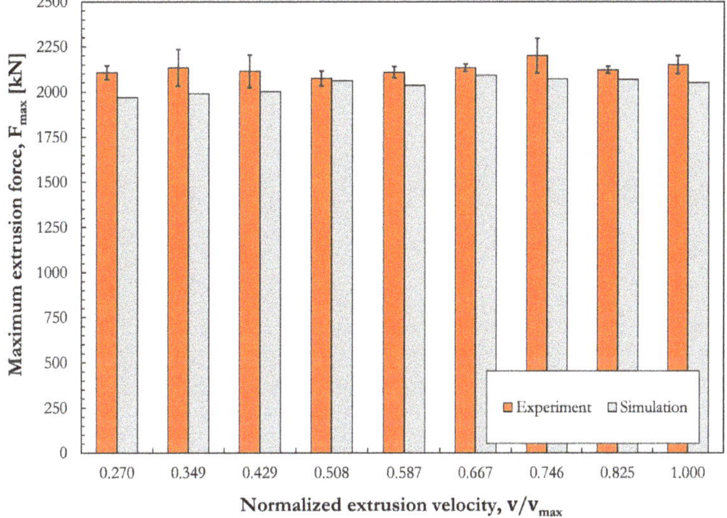

Figure 6. Maximum extrusion force for different extrusion speed: simulated and experimentally measured results.

3.2. Parametrical Analyses and Surface Crack Formation

In this section, the results of the parametric simulations to assess the influence of the billet temperature, and of the extrusion velocity, are shown.

Concerning the maximum extrusion force, Figure 7a shows its inverse proportionality to the initial billet temperature, T_b, at a constant extrusion speed of $0.67v_{max}$. This is due to the flow stress of the material that is higher at lower temperatures, making the extrusion force significantly increase. On the other hand, an increase in the D parameter also takes place when T_b increases, which nullifies the beneficial effect of the reduction in the extrusion force at higher T_b. Consequently, T_b must be chosen as the minimum temperature suitable to guarantee the extruded forces are compatible with the technological limits due to extruder payloads (T_b = 1090 °C in this case). Moreover, the influence of the extrusion speed on the maximum surface temperature T_b experienced at the surface of the billet and the damage parameter D are displayed in Figure 7b. Regarding the surface temperature, it is possible to observe its increase at high magnitudes of extrusion velocities. This can be imputed to the thermal exchange with the mold that is drastically reduced at high velocities and, thus, the process becomes almost adiabatic with a significant increase in the billet temperature. Conversely, the greater the extrusion speed, the greater the damage parameter D. Nevertheless, as $T_{s,max}$ above 1150 °C has to be avoided, due to the possible intergranular melting of Niobium precipitates, extrusion speeds higher than 0.67 should be avoided.

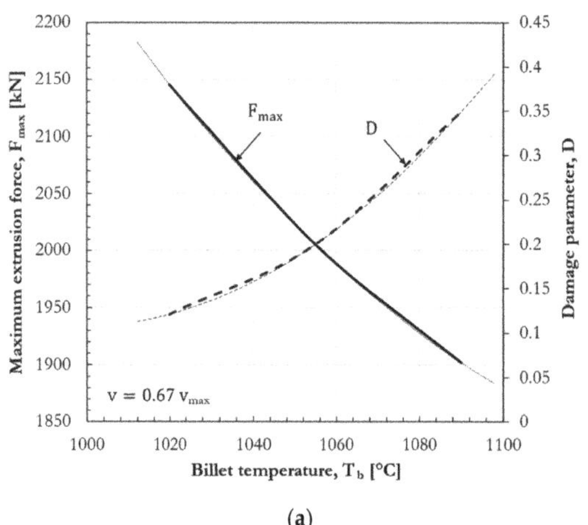
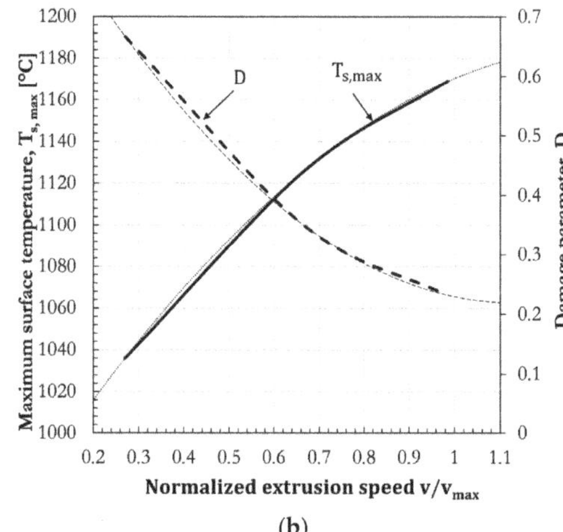

(a)

(b)

Figure 7. Influence (**a**) of the initial billet temperature T_b on the maximum extrusion force and on the damage parameter D. Influence (**b**) of the normalized extrusion speed on the maximum surface temperature of the billet $T_{s,max}$ and on the damage parameter D.

For the sake of illustrating a significant full-field distribution of the damage parameter D over the analyzed domain, we decided to choose the following process parameters: $T_b = 1090$ °C and $v = 0.67 v_{max}$. Following this, Figure 8 shows the result of this analysis. It can be observed that the most critical values occur over the surface of the upper part of the extruded workpiece (maximum value of D at the point labelled as "A"). In this zone, the risk of surface crack formation is higher than that of the remaining regions of the model. Although the damage parameter does not provide a quantitative evaluation of the crack formation condition, the proposed model is certainly capable of identifying the most critical regions, as verified by the preliminary experimental tests conducted in this work (see Figure 2c,d). Therefore, D should be minimized as much as possible to avoid material cracking within the limits of the other technological constraints.

In order to better understand the obtained results, point "A" was monitored throughout the whole extrusion process to observe the evolution of the damage parameter D (see Figure 9). It can be observed that ~50% of damage D accumulates during the first extrusion step that involves the billet expanding radially (1). At $t = 0.08$ s, the radial expansion of the billet in the mold is concluded and point "A" is located by the entrance of the mold diameter reduction. From $t = 0.08$ s to $t = 0.18$ s, mostly the inner part of the billet material starts to flow within the smaller section of the mold (2). In fact, in this time interval, point "A" does not undergo major displacement and deformation, therefore the damage parameter D remains almost constant. At $t = 0.18$ s, the severe plastic deformation of the flowing billet starts also involving the external regions of the billet, which slides with respect to the mold wall, including point "A". At $t = 0.19$ s, point "A" exits the mold and detaches from the mold wall (3). In this phase, a further increase in D occurs, mainly due to the positive stress in the axial direction induced by the friction forces and counterbalanced by the reaction of the mold. The evolution of the ratio $\frac{\sigma_L}{\sigma_{VM}}$ is also reported to be made clear, according to the expression reported in Equation (5), when the increase in D is related to an increase in the strain rate (1).

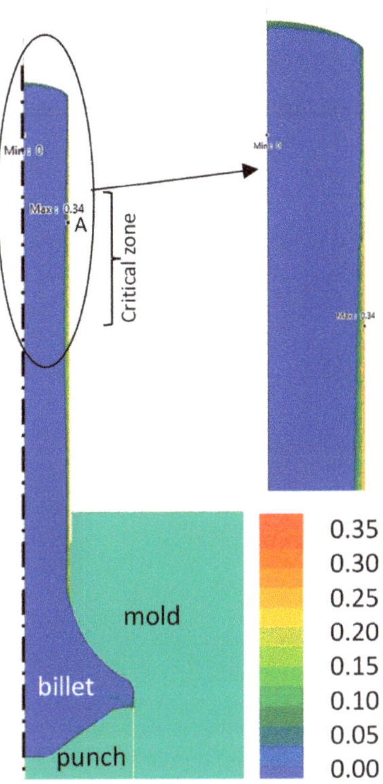

Figure 8. Contour plot of the damage parameter D over the billet at the end of the extrusion process for T_b = 1090 °C, v = 0.67v_{max}.

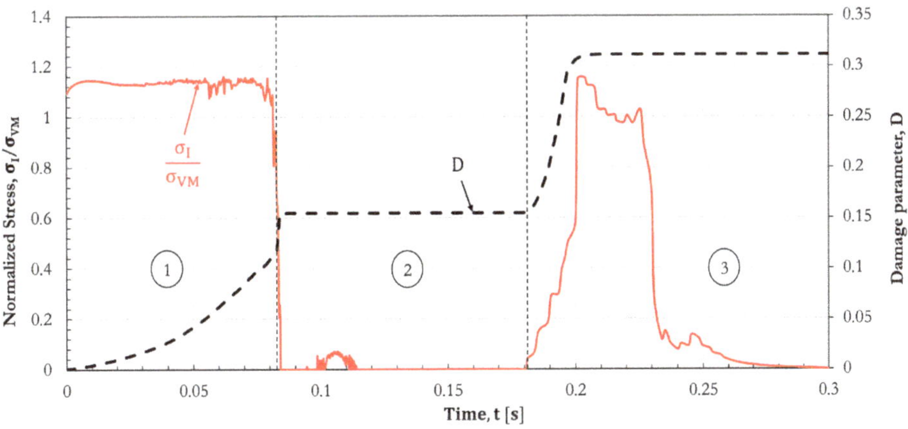

Figure 9. Evolution of the damage parameter D at point "A" during the extrusion process (T_b = 1090 °C) and of the ratio $\frac{\sigma_I}{\sigma_{VM}}$.

3.3. Experimental and Numerical Macro and Micro-Structure Analysis

The macro-etched sample (process parameters: T_b = 1090 °C; v = 0.67v_{max}) shows the presence of a uniform microstructure along the main body of the billet (Figure 10). Some differences are well highlighted in the proximity of the ends of the extruded sample. In detail, the two ends present a coarser grain distribution in comparison to the core of the billet. This grain distribution is linked to both low material deformation at the larger end, and at the material first undergoing extrusion at the thinner end. The extruded sample does not show any fibering due to both a possible homogenous chemical composition (absence of segregated areas) and to a probable homogenous monophasic recrystallized microstructure that is not prone to highlight the presence of microstructural bands.

Figure 10. Macro structure of the analyzed sample obtained after macro-etching.

The material, in any case, is composed of a monophasic gamma microstructure with no apparent precipitates at the micron-scale. Going into the details of the metallographic characterization, it is possible to observe a colony of grains along the centerline and close to the extruded free surface. In the centerline of the billet, the grain size is equiaxed and completely re-crystallized, although, in some regions, re-crystallization did not take place. The presence of these microstructural features probably indicates a meta-dynamic recrystallization process. No appreciable variations of the grain size were observed in the non-re-crystallized grains, although significant plastic deformation induced grain rotation and shearing effects. At the billet free-surface, the grains seem to be finer in comparison to the centerline due to higher cooling rates. Such experimental evidence allowed for the validation of the numerical model regarding the microstructure evolution, as shown in Figure 11. These results are in good agreement with the simulations in terms of the average grain size at the end of the process. In fact, finer grains can be easily observed at the extruded external surface with respect to the centerline, whereas, at the two ends, where the material is nearly undeformed, their dimension is unchanged ($\cong$170 μm) in comparison to the pre-extrusion conditions.

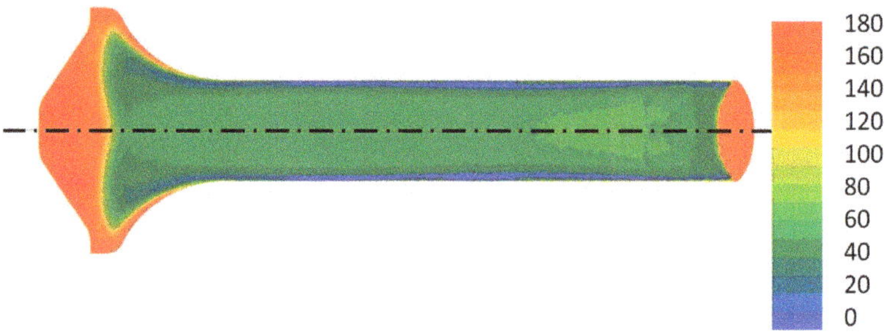

Figure 11. Average grain size (simulated) at the end of the process (in μm).

The analysis of the microstructure in the proximity of the billet large fillet in Figure 12 shows the presence of a gradual deformation of the grains leading to a re-crystallization of the material that is fully accomplished before the material flows outside the smaller section of the mold.

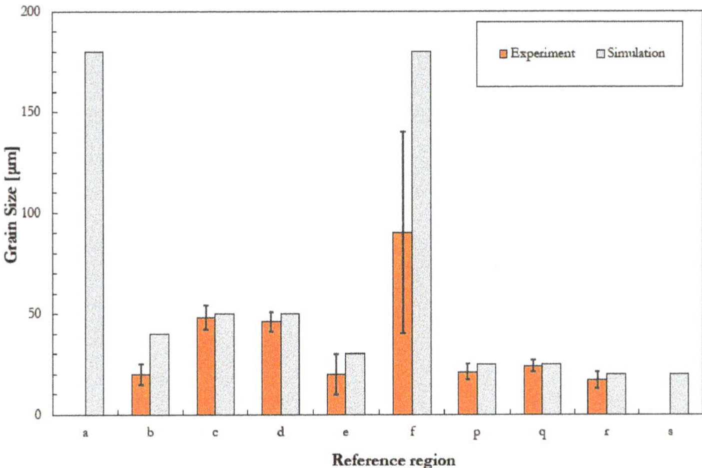

Figure 12. Grain size evaluation: experimentally measured vs. calculated. Labels refer to the regions listed in Figure 3. Note that regions (a) and (s) do not present a measured grain size.

A graph that compares the calculated grain dimension and the measured grain size is shown in Figure 12. The graph confirms that the regions with a fully recrystallized grain show a good agreement as far as the grain size is concerned, between the values evaluated by using the numerical methods previously presented and the experimental observations. On the other hand, the same match could not be fully verified at the regions where partial recrystallization occurred. In this case, the differences are more related to the difficulty in measuring the grain size by image analysis (intercept method).

Specifically, new grain nuclei, at approximately half of the fillet, are formed, where a mixture of re-crystallized and un-recrystallized grains are gradually observed. The position where around 50% of grains are recrystallized can be easily identified, and it corresponds, by comparing the position from microstructural characterization (Figure 13), to the plastic strain map obtained from the simulation (Figure 14), to a total strain between 1.2 and 1.5.

In this region, the plastic strains can also be experimentally evaluated by analyzing the length of undeformed grains and comparing this measurement to the initial dimension of the austenitic grain to assess the total strain in this position. A very good match was found with the total strain obtained from the numerical model.

Finally, the length of the extruded samples was measured and compared with the analogous dimensions obtained via simulation, observing a difference that was always lower that 5%. This value fulfills the technological requirements for the considered component.

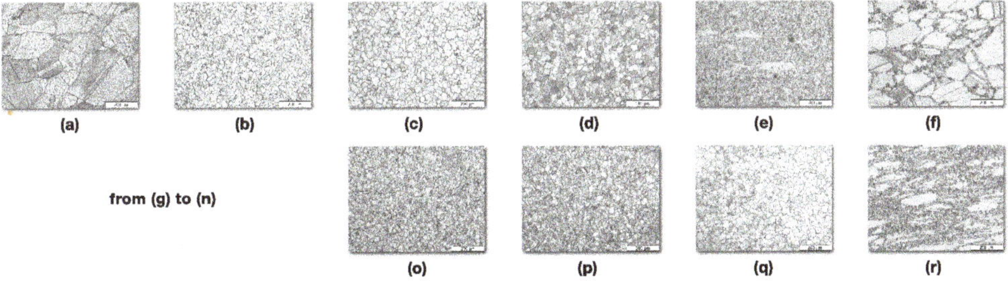

Figure 13. *Cont.*

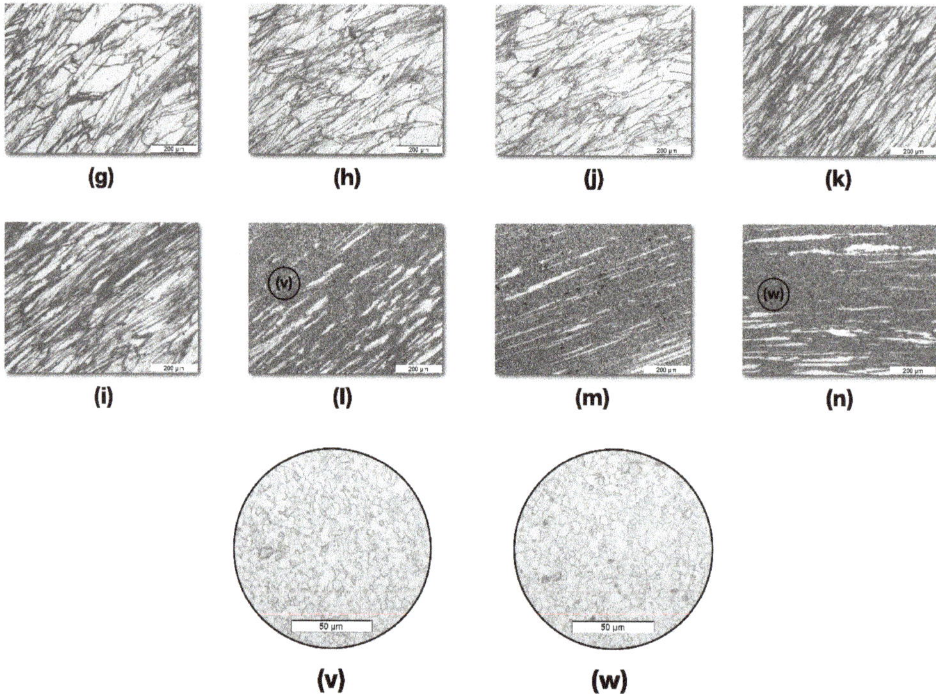

Figure 13. Microstructural characterizaion of the final billet (refer to Figure 3).

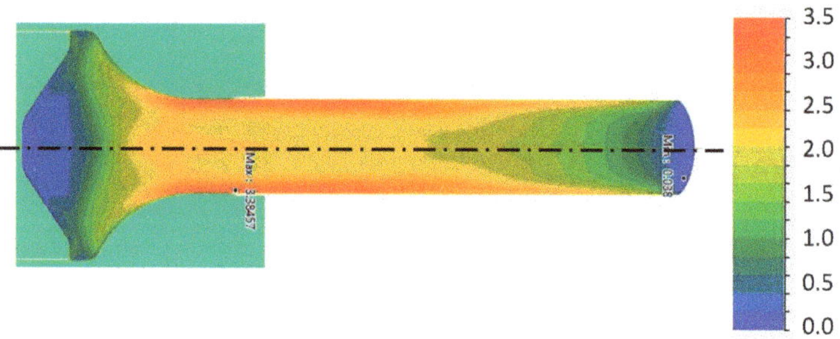

Figure 14. Strain distribution at the end of the extrusion process.

4. Conclusions

In this work, an extensive study was carried out to tackle the problem of Inconel 718 hot extrusion by employing both numerical and experimental methods, a subject that is scarcely addressed by the current literature. The development of a numerical approach, supported by experimental evidence, allowed for the evaluation of how key technological parameters (i.e., initial temperature of the billet, extrusion speed) affect both the required extrusion force and the final quality of the product (i.e., surface cracks, microstructure). The results show a very good agreement between the measurements and simulation. In particular, the proposed model was able to accurately predict the evolution of the extrusion force and the location of the zone of possible surface crack formation. Additionally, the model was capable of accurately capturing the microstructural grain deformation and re-crystallization.

Thanks to the fully validated developed model, the following recommendations can be drawn to attain the structural integrity and the sought mechanical performance of the final product. A higher initial temperature of the billet permits the extrusion force to be reduced, although the surface quality of the final product may be impaired if such a temperature is excessively increased. For the case presented in this work, a temperature around 1090 °C could be a reasonable trade-off. An increase in the punch speed is also advisable, as it lowers nCL's damage parameter, reducing the probability of surface cracks formation. Nevertheless, excessive values of speed could produce an increase in the temperature of the billet, which may be detrimental as it may induce the local melting of niobium carbide. An extrusion speed of $0.6 \div 0.7 \, v/v_{max}$ could fulfill the abovementioned requirements.

The proposed procedure aims at providing a robust tool for the setting up of hot extrusion processes, regardless of the complexity of the geometry. This aspect is pivotal to reducing the design-to-market time and massively reduces the costly trial and error approaches.

Author Contributions: Conceptualization, P.M. and F.D.B.; methodology, S.B. and A.L.; software, M.A.C. and S.B.; validation, S.B. and A.L.; formal analysis, S.B. and F.S.; investigation, S.B. and A.L.; resources, F.S.; data curation, S.B.; writing—original draft preparation, F.D.B.; writing—review and editing, F.D.B. and E.S.; visualization, E.S.; supervision, F.D.B. and P.M.; All authors have read and agreed to the published version of the manuscript.

Funding: This research received no external funding.

Data Availability Statement: Restrictions apply to the availability of these data. Data were obtained from Pietro Rosa TBM and are available from the authors with the permission of Pietro Rosa TBM.

Conflicts of Interest: The authors declare no conflict of interest.

References

1. Salvati, E.; Lunt, A.J.G.; Ying, S.; Sui, T.; Zhang, H.J.; Heason, C.; Baxter, G.; Korsunsky, A.M. Eigenstrain reconstruction of residual strains in an additively manufactured and shot peened nickel superalloy compressor blade. *Comput. Methods Appl. Mech. Eng.* **2017**, *320*, 335–351. [CrossRef]
2. Tang, Y.T.; Panwisawas, C.; Jenkins, B.M.; Liu, J.; Shen, Z.; Salvati, E.; Reed, R.C. Multi-length-scale study on the heat treatment response to supersaturated nickel-based superalloys: Precipitation reactions and incipient recrystallisation. *Addit. Manuf.* **2023**, *62*, 103389. [CrossRef]
3. Qi, H. Review of inconel 718 alloy: Its history, properties, processing and developing substitutes. *J. Mater. Eng.* **2012**, *2*, 92–100.
4. Rahman, M.; Seah, W.K.H.; Teo, T.T. The machinability of Inconel 718. *J. Mater. Process. Technol.* **1997**, *63*, 199–204. [CrossRef]
5. Thakur, D.G.; Ramamoorthy, B.; Vijayaraghavan, L. Study on the machinability characteristics of superalloy Inconel 718 during high speed turning. *Mater. Des.* **2009**, *30*, 1718–1725. [CrossRef]
6. Felusiak-Czyryca, A.; Madajewski, M.; Twardowski, P.; Wiciak-Pikuła, M. Cutting forces during Inconel 718 orthogonal turn-milling. *Materials* **2021**, *14*, 6152. [CrossRef]
7. De Bartolomeis, A.; Newman, S.T.; Jawahir, I.S.; Biermann, D.; Shokrani, A. Future research directions in the machining of Inconel 718. *J. Mater. Process. Technol.* **2021**, *297*, 117260. [CrossRef]
8. Gordine, J. Some problems in welding Inconel 718. *WELD J.* **1971**, *50*, 480–484.
9. Chen, J.; Salvati, E.; Uzun, F.; Papadaki, C.; Wang, Z.; Everaerts, J.; Korsunsky, A.M. An experimental and numerical analysis of residual stresses in a TIG weldment of a single crystal nickel-base superalloy. *J. Manuf. Process.* **2020**, *53*, 190–200. [CrossRef]
10. Tharappel, J.T.; Babu, J. Welding processes for Inconel 718-A brief review. *IOP Conf. Ser. Mater. Sci. Eng.* **2018**, *330*, 012082. [CrossRef]
11. Salvati, E.; Lunt, A.J.; Heason, C.P.; Baxter, G.J.; Korsunsky, A.M. An analysis of fatigue failure mechanisms in an additively manufactured and shot peened IN 718 nickel superalloy. *Mater. Des.* **2020**, *191*, 108605. [CrossRef]
12. Trosch, T.; Strößner, J.; Völkl, R.; Glatzel, U. Microstructure and mechanical properties of selective laser melted Inconel 718 compared to forging and casting. *Mater. Lett.* **2016**, *164*, 428–431. [CrossRef]
13. Hu, L.; Shi, K.; Luo, X.; Yu, J.; Ai, B.; Liu, C. Application of additively manufactured pentamode metamaterials in sodium/Inconel 718 heat pipes. *Materials* **2021**, *14*, 3016. [CrossRef] [PubMed]
14. Wu, K.; Chee, S.W.; Sun, W.; Tan, A.W.-Y.; Tan, S.C.; Liu, E.; Zhou, W. Inconel 713C Coating by cold spray for surface enhancement of Inconel 718. *Metals* **2021**, *11*, 2048. [CrossRef]
15. Bauccio, M. *ASM Metals Handbook Vol. 14: Forming and Forging*, 9th ed.; ASM International: Tokyo, Japan, 1996.
16. De Jaeger, J.; Solas, D.; Baudin, T.; Fandeur, O.; Schmitt, J.H.; Rey, C. Inconel 718 single and multipass modelling of hot forging. In *Superalloys 2012: 12th International Symposium on Superalloys*; John Wiley & Sons Inc.: Chichester, UK, 2012; pp. 663–672.

17. Chamanfar, A.; Sarrat, L.; Jahazi, M.; Asadi, M.; Weck, A.; Koul, A.K. Microstructural characteristics of forged and heat treated Inconel-718 disks. *Mater. Des.* **2013**, *52*, 791–800. [CrossRef]
18. Jackman, L.A. Forming and Fabrication of Superalloys. In Proceedings of the Symposium on Properties of High Temperature Alloys, Las Vegas, NV, USA, 18–21 October 1976; pp. 42–58.
19. Duan, X.; Velay, X.; Sheppard, T. Application of finite element method in the hot extrusion of aluminium alloys. *Mater. Sci. Eng. A* **2004**, *369*, 66–75. [CrossRef]
20. Nielsen, C.V.; Martin, P.A.F. *Metal Forming: Formability, Simulation, and Tool Design*, 1st ed.; Academic Press: London, UK, 2021.
21. Moro, L.; Srnec Novak, J.; Benasciutti, D.; De Bona, F. Thermal distortion in copper moulds for continuous casting of steel: Numerical study on creep and plasticity effect. *Ironmak. Steelmak.* **2019**, *46*, 97–103. [CrossRef]
22. De Bona, F.; Novak, J.S.; Lanzutti, A.; Lucacci, G. Distortion after solubilization treatment of X12CrNiMoV12-3 beam-like samples: A novel FE modelling technique supported by experiments. *Eng. Fail. Anal.* **2022**, *135*, 106141. [CrossRef]
23. Biba, N.; Borowikow, A.; Wehage, D. Simulation of recrystallisation and grain size evolution in hot metal forming. *AIP Conf. Proc.* **2011**, *1353*, 127–132.
24. Zinkiewitcz, O.C. Flow formulation for numerical solution of metal forming processes. In *Numerical Analysis of Forming Processes*; Pittman, J.F.T., Zienkiewicz, O.C., Woof, R.D., Alexander, J.M., Eds.; Wiley: Chichester, UK, 1984; pp. 1–44.
25. Levanov, A.N.; Kolmogorov, V.L.; Burkin, S.P.; Kartak, B.R.; Ashpur Yu, V.; Spassky Yu, I. *Kontaktnoe Trenie v Protsessakh Obrabotki Metallov Davleniem (Contact Friction in Metal Forming Processes)*; Metallurgiya Publish: Moscow, Russia, 1976.
26. *ASM Handbook Vol. 2: Properties and Selection: Nonferrous Alloys and Special-Purpose Materials*; ASM International: Tokyo, Japan, 1990.
27. Sui, F.L.; Xu, L.X.; Chen, L.Q.; Liu, X.H. Processing map for hot working of Inconel 718 alloy. *J. Mater. Process. Technol.* **2011**, *211*, 433–440. [CrossRef]
28. Thomas, A.; El-Wahabi, M.; Cabrera, J.M.; Prado, J.M. High temperature deformation of Inconel 718. *J. Mater. Process. Technol.* **2006**, *177*, 469–472. [CrossRef]
29. Frost, H.J.; Ashby, M.F. *Deformation—Mechanism Maps. The Plasticity and Creep of Metals and Ceramics*; Pergamon Press: Oxford, UK, 1982; pp. 54–55.
30. Mahalle, G.; Kotkunde, N.; Gupta, A.K.; Singh, S.K. Comparative assessment of failure strain predictions using ductile damage criteria for warm stretch forming of IN718 alloy. *J. Mater. Form.* **2021**, *14*, 799–812. [CrossRef]
31. Cockcroft, M.G.; Latham, D.J. Ductility and the workability of metals. *J. Inst. Met.* **1968**, *96*, 33–39.
32. Clift, S.E.; Hartley, P.; Sturgess, C.E.N.; Rowe, G.W. Fracture prediction in plastic deformation processes. *Int. J. Mech. Sci.* **1990**, *32*, 1–17. [CrossRef]
33. Jafarian, F.; Ciaran, M.I.; Umbrello, D.; Arrazola, P.J.; Filice, L.; Amirabadi, H. Finite element simulation of machining Inconel 718 alloy including microstructure changes. *Int. J. Mech. Sci.* **2014**, *88*, 110–121. [CrossRef]
34. Stebunov, S.; Vlasov, A.; Biba, N. Prediction of fracture in cold forging with modified Cockcroft-Latham criterion. *Procedia Manuf.* **2018**, *15*, 519–526. [CrossRef]
35. Sellars, C.M.; Whiteman, J.A. Recrystallization and grain growth in hot rolling. *Met. Sci.* **1979**, *13*, 187–194. [CrossRef]
36. Biba, N.; Stebunov, S.; Vlasov, A. Material forming simulation environment based on QForm3D software system. *Energy* **2017**, *2*, 4.
37. Na, Y.S.; Yeom, J.T.; Park, N.K.; Lee, J.Y. Prediction of microstructure evolution during high temperature blade forging of a Ni−Fe based superalloy 718. *Met. Mater. Int.* **2003**, *9*, 15–19. [CrossRef]
38. Gruber, C.; Raninger, P.; Stanojevic, A.; Godor, F.; Rath, M.; Kozeschnik, E.; Stockinger, M. Simulation of dynamic and meta-dynamic recrystallization behavior of forged alloy 718 Parts using a multi-class grain size model. *Materials* **2021**, *14*, 111. [CrossRef]
39. Soufian, E.; Darabi, R.; Abouridouane, M.; Reis, A.; Bergs, T. Numerical predictions of orthogonal cutting–induced residual stress of super alloy Inconel 718 considering dynamic recrystallization. *Int. J. Adv. Manuf. Technol.* **2022**, *122*, 601–617. [CrossRef]
40. Churyumov, A.Y.; Pozdniakov, A.V. Simulation of microstructure evolution in metal materials under hot plastic deformation and heat treatment. *Phys. Metals Metallogr.* **2020**, *121*, 1064–1086. [CrossRef]
41. Churyumov, A.Y.; Pozdnyakov, A.V.; Churyumova, T.A.; Cheverikin, V.V. Hot plastic deformation of heat-resistant austenitic aisi 310s steel. Part 1. simulation of flow stress and dynamic recrystallization. *Chernye Met.* **2020**, *9*, 48–55.

Article

Numerical Simulation of Low-Pressure Carburizing and Gas Quenching for Pyrowear 53 Steel

Bartosz Iżowski [1,2,*], Artur Wojtyczka [1,2] and Maciej Motyka [2,3]

[1] Research and Development Center, Pratt & Whitney Rzeszów S.A., 35-078 Rzeszów, Poland
[2] Faculty of Mechanical Engineering and Aeronautics, Rzeszow University of Technology, 12 Powstańców Warszawy Ave., 35-959 Rzeszów, Poland
[3] Research and Development Laboratory for Aerospace Materials, Rzeszow University of Technology, 4 Żwirki i Wigury Str., 35-959 Rzeszów, Poland
[*] Correspondence: bartosz.izowski@prattwhitney.com

Abstract: The hardness and phase composition are, among other things, the critical material properties considered in the quality control of aerospace gears made from Pyrowear 53 steel after high-pressure gas quenching. The low availability of data on and applications of such demandingstructures justify investigating the choice of the material and the need to improve its manufacturability. In this study, computational finite-element analyses of low-pressure carburizing followed by oil and gas quenching of Pyrowear 53 steel were undertaken, the objective of which was to examine the influence of the process parameters on the materials' final phase composition and hardness. The material input was prepared using JMatPro. The properties computed by the CALPHAD method were calibrated by the values obtained from physical experiments. The heat transfer coefficient was regarded as an objective variable to be optimized. A 3D model of the Standard Navy C-ring specimen was utilized to predict the phase composition after the high-pressure gas quenching of the steel and the hardness at the final stage. These two parameters are considered good indicators of the actual process parameters and are used in the industry. The results of the simulation, e.g., optimized heat transfer coefficients, cooling curves, and hardness and phase composition, are presented and compared with experimental values. The accuracy of the simulation was validated, and a good correlation of the data was found, which demonstrates the quality of the input data and setup of the numerical procedure. A computational approach to heat treatment processes' design could contribute to accelerating new procedures' implementation and lowering the development costs.

Keywords: steel; modeling; simulation; Pyrowear 53; inverse heat transfer; DEFORM3D; low pressure carburizing; high-pressure gas quenching; simufact; phase transformations; latent heat; Koistinen–Marburger

Citation: Iżowski, B.; Wojtyczka, A.; Motyka, M. Numerical Simulation of Low-Pressure Carburizing and Gas Quenching for Pyrowear 53 Steel. *Metals* **2023**, *13*, 371. https://doi.org/10.3390/met13020371

Academic Editors: Jelena Srnec Novak, Francesco De Bona and Francesco Mocera

Received: 31 December 2022
Revised: 6 February 2023
Accepted: 6 February 2023
Published: 12 February 2023

1. Introduction

Quenching is a critical heat treatment operation in the manufacturing process, which determines the enhancement of the fatigue durability and wear resistance of carburized steel components, such as power-transmitting gears. Aerospace gearbox development requires the application of modern materials with more resistance to tempering at elevated temperatures. In the 1970s, the Pyrowear 53 low-alloy steel was developed when a new material for power transmission was needed due to the limitations of the commonly used 9310 steel and its tendency to score and scuff under high-temperature conditions. This steel was designed to operate mainly in power transmission components such as gears due to its surface durability and fatigue resistance, enhanced by the compressive residual stress generated during carburizing and related to the delayed formation of martensite near the surface. Gears made of Pyrowear 53 are able to maintain their properties even when no lubrication is present due to the material's wear resistance. The outstanding properties of Pyrowear 53 are achieved thanks to, among other things, the high content of silicon, providing resistance to softening during tempering, nickel, responsible for added toughness and fabricability, molybdenum, to ensure hot hardness, and vanadium, providing secondary hardening, whichalso increases the material's susceptibility to grain refinement [1]. In contrast to typical steel grades for carburizing, Pyrowear 53 contains molybdenum and copper. The addition of these two elements distinguishes it among the variety of steel grades. Molybdenum is responsible for increasing the heat and wear abrasion resistance, and copper improves the lubrication properties and shock load [2]. The carburization process of Pyrowear 53 is recommended at temperatures between 870 and 927 °C, while hardening from 904 to 921 °C. To obtain the maximum hardness and dimensional stability, subzero treatment is recommended at −73 °C. Double-tempering is recommended for applications requiring elevated dimensional stability [3,4]. For parts with tight dimensional tolerances, such as power transmission components, distortions, being an effect of heat treating, have a substantial economic impact on the following machining processes, increasing the overall production costs by 20–40% to eliminate dimensional deviations [5].

The analysis of the metallurgical effects during quenching is of great importance. Due to the complexity of the interrelation between the phenomena and their transient characteristics, it is challenging to capture the sequence and magnitude of these effects experimentally. Numerical simulations of heat treatment processes have been developed and used in the industry to analyze transient processes such as quenching thanks to constantly growing computational capabilities and commercial software availability. There are multiple systems based on the finite-element method, specifically dedicated to heat treatment processes' design and optimization; among them, DANTE [6], DEFORM2D/3D [7,8], and Simufact [9] are either fully dedicated or provide specific modules to solve the thermo-mechano-metallurgical problems often coupled with diffusion models for carburization or nitriding analyses. For the computational approach to be efficient, this requires detailed input data, which, in the case of steels' heat treatment analysis, may be difficult to find in the literature or to determine experimentally, as the key to numerical simulation is predicting that the final property of the material is an effect of the stress evolution caused by the phase transformation, as well as the difference between the properties of the coexisting phases. Therefore, the precise definition of the input data for specific phases present in the system is crucial. To define the temperature-dependent thermophysical and mechanical properties of each phase or the TTT/CCT diagrams, software such as JMatPro [10–12] or Thermo-Calc [13] based on the CALPHAD method can be used.

Witness specimens are commonly used to eliminate the need to cut up production parts, leading to increased costs for development efforts. Standard Navy C-ring-type specimens are known to reflect the dimensional change of the actual components under investigation. Its asymmetric geometry prevents uniform cooling and delays the beginning of phase transformation in thicker sections, causing distortion. The relatively simple geometry of the specimens makes it easy to use as, usually, three characteristics are measured: the outer diameter, inner diameter, and gap. Numerous research works utilizing the computational approach and using the C-ring-type specimens as the baseline have been conducted to predict the distortions that are an effect of the heat treatment, confirming the accuracy of simulations' capabilities and the geometry of the specimen [5,14–16].

This paper describes the simulation procedure to predict the influence of the carbon content and quenching process parameters on the phase composition and hardness distribution after heat treatment. Because of the relatively new process of high-pressure gas quenching of Pyrowear 53 for applications in demanding structures, such as aerospace gears, and the limited amount of literature references related to its properties, the objective of this work was to research them. To achieve this, numerical simulations were used to predict the material's behavior under specific process conditions. Quenching by high-pressure gas quenching (HPGQ) and OH70 quenching oil was employed to validate the computational results.

2. Materials, Methods, and Modeling

2.1. Test Specimen

Twelve Navy C-ring-type specimens, as presented in Figure 1, were analyzed in this study. The choice of the specimen geometry was dictated by the fact that it is widely known and accepted in the heat treatment industry as having numerous advantages, among which the most important is the ability to analyze rapid temperature changes during quenching and their effects on various cross-sections of the treated parts [17–19]. The Navy C-ring is a short cylinder with an eccentric hole open at one extreme. Each cross-section has a different size and, therefore, cooling rate, which leads to variations in the metallurgical effects and, finally, the properties of the material being analyzed. Due to the transient size change, the C-ring specimen is a good indicator of actual parts' behavior during heat treatment. Six specimens were carburized, while another six were left without carburization. All specimens underwent cryogenic treatment and tempering. Quenching was performed in OH70 quenching oil. Low-pressure carburizing (LPC) and high-pressure gas quenching (HPGQ) were performed in an ALD MonoTherm HK.446.VC.10.gr vacuum furnace. The details of the heat treatment procedure for all samples are presented in Table 1.

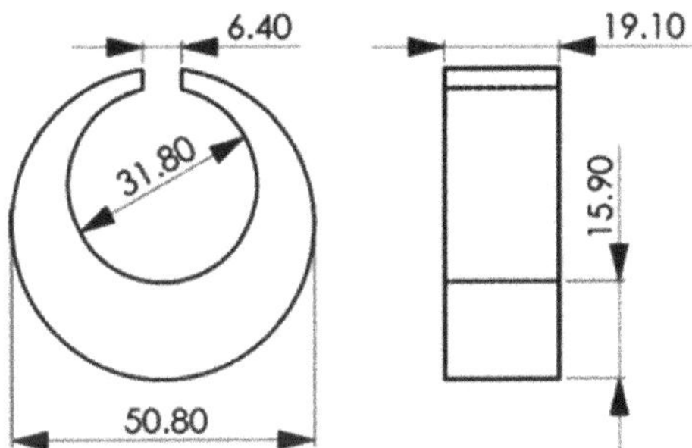

Figure 1. The geometry of the C-ring specimen used in the research. Dimensions are in mm.

Table 1. Test plan description for all specimens.

Specimen No.	LPC	HPGQ	Oil Q.
1	+	+	
2	+	+	
3	+	+	
4	+		+
5	+		+
6	+		+
7		+	
8		+	
9		+	
10			+
11			+
12			+

Specimen Nos. 1 to 6 were processed by LPC at 921 °C, which consisted of 16 boost–diffusion cycles. The carbon carrier boosts' total time was 17 min, while the entire process time was approximately 9.5 h. After the LPC was finished, specimens were cooled down slowly to room temperature. The load, containing three carburized specimens, Nos. 1, 2, and 3, and three that were kept uncarburized, Nos. 7, 8, and 9, was then heated up to 913 °C and held at this temperature for a sufficient period of time to ensure obtaining a homogeneous temperature and, therefore, the austenitic microstructure in the whole volume, then quenched. A standard industrial procedure for parts requiring dimensional stability cryogenic treatment at −75 °C was performed to finish the martensitic transformation.

Tempering was performed at 230 °C for 4 h. As two different techniques of quenching were employed in this research, a variety of heat-transfer-related parameters had to be incorporated into the simulation. A protective copper cladding was employed for specimens hardened in oil to prevent surface oxidation and decarburization. The copper coating and stripping were performed electrochemically. Oil quenching was performed in the RIVA automatic quenching system, which includes a rotary furnace RDLS with a heating power of 10 kW, a quenching bath OQ-1000 equipped with a 1.1 kW agitation pump, and vertical collectors to ensure vertical stirring of the OH70 oil. The oil temperature was 45 °C. The heat transfer coefficient of the oil is shown in Figure 2. The heat transfer in the gas quenching was considered convective and almost constant [20], which is not realistic and constitutes the need for optimization to find realistic values. For the sake of this research, the value of the heat transfer coefficient (HTC) to the environment during gas quenching was optimized utilizing parabola optimization methods. The cooling curves captured during an experiment were the objectives. As shown in Figure 3, three heat transfer zones were defined, and specific upper and lower bounds and initial guesses were initialized for each. The HTC_{HPGQ} optimization results were further used in the simulations.

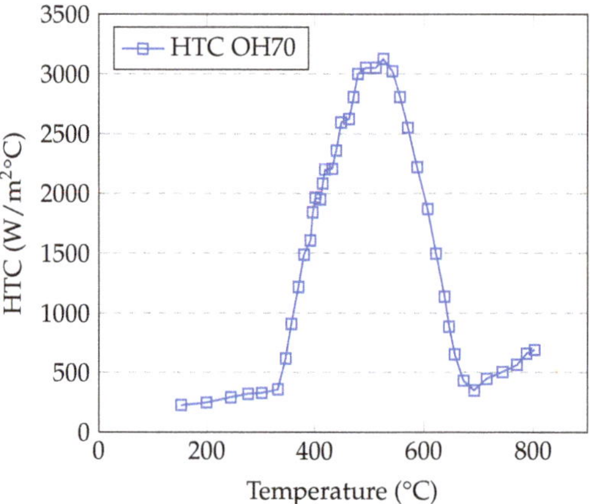

Figure 2. Heat transfer Heat transfer coefficients of OH70 oil's dependence on temperature (adapted with permission from Ref. [21], Copyright 2019, Stal, Metale & Nowe Technologie).

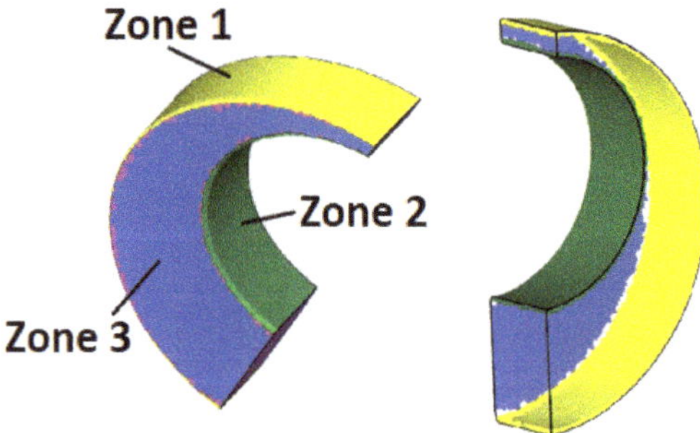

Figure 3. Heat transfer zones' definition.

Hardness was measured using the NEXUS 4303 tester and Vickers method under a 500 g load in the transverse and longitudinal cross-sections of the specimens, as shown in Figure 4, performed according to aerospace standards to confirm the effective case depth.

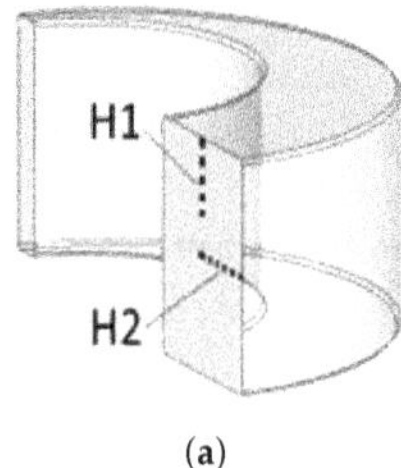
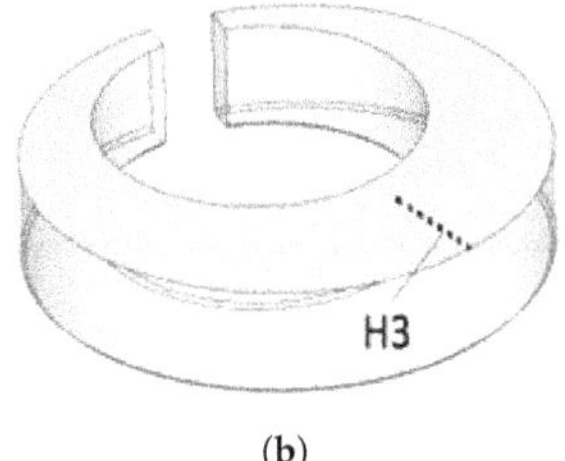

Figure 4. Hardness check locations in C-ring cross-sections: transverse (**a**) and longitudinal (**b**).

Carbon concentration measurement using glow discharge optical emission spectroscopy (GD-OES) was performed to provide validation data to correlate the simulation results with the empirical ones.

2.2. The Material

The material analyzed in this study was the Pyrowear 53 low-alloy steel. The steel's chemical composition (wt.%) was measured using a Thermo ARL 3460 optical emission spectrometer. Table 2 presents the main alloying elements' content.

Table 2. Chemical composition of Pyrowear 53 steel.

Item	C	Si	Ni	Cu	Mn	Cr	Mo	V	S	P	Al
Result	0.138	0.945	1.873	1.776	0.319	1.060	3.320	0.092	0.005	0.007	0.002
SD *	0.004	0.008	0.011	0.011	0.001	0.041	0.019	-	-	-	-

* Standard deviation.

In order to obtain an accurate description of the materials' properties for each phase present in the material during the whole thermal cycle, which may be further used, the JMat-Pro simulation was utilized. The CAPLHAD results were then validated and adjusted by the experimental results. The main adjustments of the inputs made by the authors consisted of defining the carbon concentration's influence on the martensite properties, such as hardness, as shown in Figure 5.

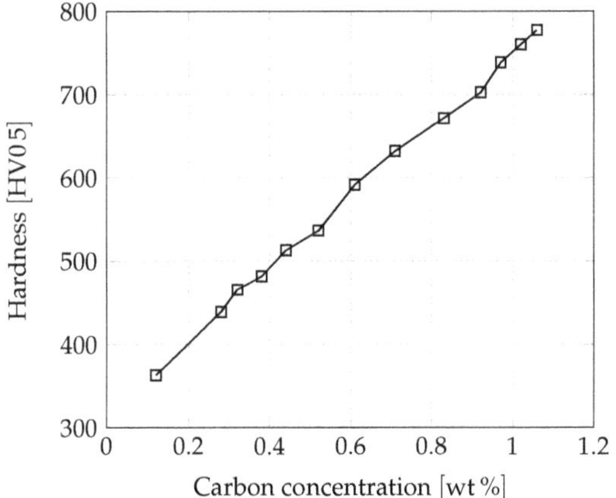

Figure 5. Dependency of the martensite hardness on carbon concentration for Pyrowear 53 steel.

The model of carbon diffusivity Equation (1) described in [22] and supplemented with an alloy correction factor Equation (2) [23] that incorporates the influence of alloying elements on the diffusivity of carbon was used in this research. Thermal data such as thermal conductivity and heat capacity also required adjustments; therefore, experimental data were used [24]. A simulation that utilized the diffusion coefficient as a function of carbon concentration, temperature, and alloying elements' concentration was performed. The results are presented in this paper.

$$D = 0.47 \cdot exp(-1.6\%C) \cdot exp\left(\frac{-(37,000 - 6600\%C)}{R \cdot T}\right) \cdot q \tag{1}$$

where R and T are the Boltzmann constant and the carburizing temperature, respectively. Equation (2) describes the alloy correction factor.

$$\begin{aligned}
q = 1 + [\%Si] &\cdot (0.15 + 0.033[\%Si]) - [\%Mn] \cdot 0.0365 + \\
&- [\%Cr] \cdot (0.13 - 0.0055 \cdot [\%Cr]) + [\%Ni] \cdot (0.03 - 0.03365 \cdot [\%Ni]) + \\
&- [\%Mo] \cdot (0.025 - 0.01[\%Mo]) - [\%Al] \cdot (0.03 - 0.02[\%Al]) + \\
&- [\%Cu] \cdot (0.016 + 0.0014[\%Cu]) - [\%V] \cdot (0.22 - 0.01[\%V])
\end{aligned} \tag{2}$$

Heat Treatment

The continuous cooling transformation (CCT) diagram shown in Figure 6 was determined experimentally [24]. According to it, martensitic transformation starts at 438 °C. Other data may be found in the literature where values of 437 °C [25], 510 °C [4], or 460 °C [26] are mentioned. Because of the untypical chemical composition of Pyrowear 53, which may exceed the JMatPro limits, it was found that, compared to the empirical results, the data lack precision. Therefore, attempts were made to adjust and validate a numerical model of the material.

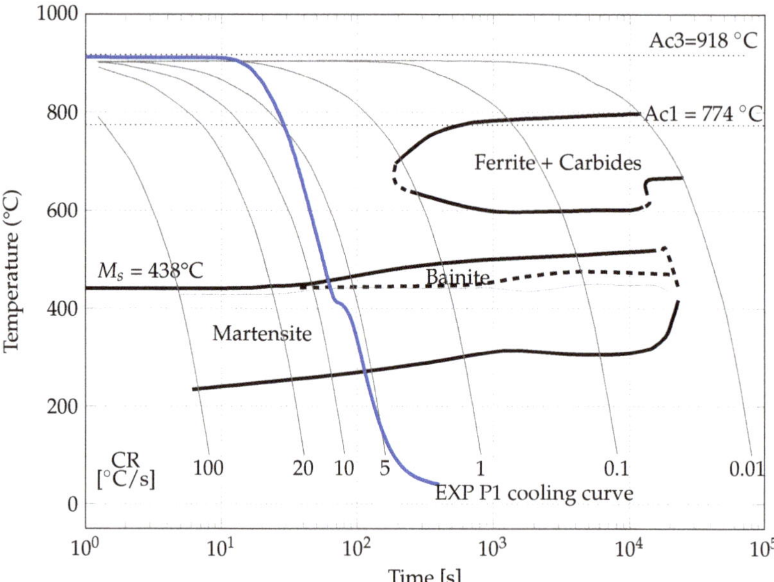

Figure 6. CCT diagram for Pyrowear 53.

The adjustment of the martensite start temperature (M_s) consisted of a modification of the model presented in [27] by the author. The results of the modification were compared to the experimental results and showed a good fit, as illustrated in Figure 7. It must be noted that the carbides present in Pyrowear 53 have a remarkable influence on the material's hardness; however, in this simulation, martensite is the only phase formed from austenite's decomposition; its hardness depends on the carbon content. Both the carbon concentration and hardness shown in Figure 5 are empirical data, and their origin, whether from structural martensite or carbides, are negligible from the numerical point of view in this case.

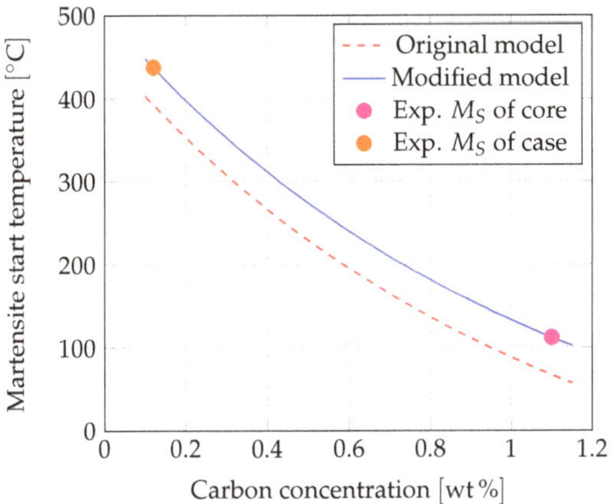

Figure 7. Comparison of martensite start temperature models. Original vs. modified by the author.

2.3. Modeling and Simulation

The simulation presented in this paper was based on the finite-element method. Two computational systems, DEFORM3D and Simufact Forming, were used in this study. An analysis of the thermo-mechano-metallurgical problem coupled with the carbon diffusion phenomenon is presented, and due to the model's symmetry, a quarter of the specimen was used, as shown in Figure 8. The models were meshed using 895,845 tetrahedral elements. Surface refinement was implemented to ensure results' accuracy close to the computational domain boundary. Boundary conditions, both thermal and diffusional, were assigned to the outer surface of the model. The elasto-plastic model of the material was used in the simulation. Hardness computations were based on a mixture of the volume fractions of specific phases.

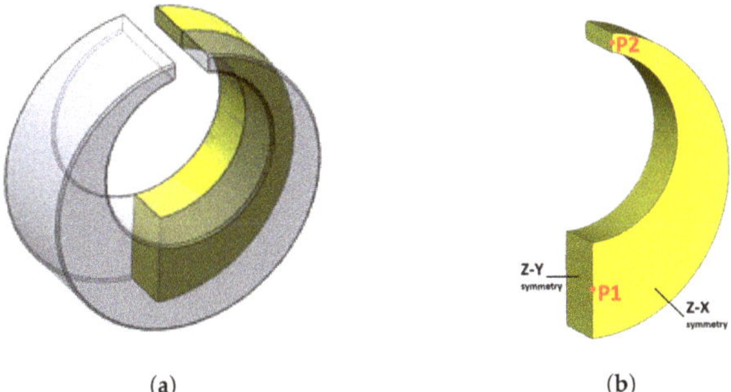

Figure 8. Computational domain (**a**) and FE model with control points P1 and P2 and symmetry planes (**b**).

2.3.1. Inverse Heat Transfer Analysis

Heat transfer capabilities of gaseous coolants depend on factors such as gas type, pressure, flow rates, turbulent behavior, or the complexity of parts being treated. The present research deals with two different quenching procedures. The first is well-researched and uses liquid quenchant OH70 quenching oil in this case, while the second is high-pressure gas quenching (HPGQ)—a technique in which convection is the primary mode of heat transfer during quenching. Because no precise data are available, the heat transfer coefficients for the high-pressure gas quenching were optimized, using the DEFORM3D Inverse Heat Transfer module, to obtain a realistic input for the quenching simulation. The cooling curves captured from the experiment were used as the optimization objectives, Broyden–Fletcher–Goldfarb–Shannon as the optimization algorithm, and B-spline as the interpolation method [8].

2.3.2. Thermal Field and Latent Heat

The thermal field is described by Fourier's heat conduction equation for transient heat transfer. Considering the fact that the phase transformation taking place during rapid cooling of overcooled austenite, i.e., martensite creation, is exothermic, the latent heat of the phase transformation alters the thermal field, and therefore, Fourier's equation can be written as

$$\rho C_p \frac{\partial T}{\partial t} = \nabla(\nabla(\lambda T)) + L \tag{3}$$

where ρ, C_p, and λ are the density, specific heat, and thermal conductivity, respectively [28]. L is the latent heat of transformation considered as an additional heat source that influences the thermal field during quenching.

$$L = \Delta E \frac{\zeta_{n+1} - \zeta_n}{t_{n+1} - t_n} = \Delta E \cdot \frac{\Delta \zeta}{\Delta t} \tag{4}$$

where ΔE is the enthalpy of austenite transformation to a child phase (see Figure 9) and ζ_{n+1} and ζ_n are the volume fractions of the specific phase that transformed at t_n and t_{n+1} simulation time steps, respectively [29]. The initial condition at time step $t = 0$ is defined as

$$T \mid_{t=0} = T_0(x, y, z) \tag{5}$$

and the boundary condition for temperature field calculation during quenching can be expressed as

$$-\lambda \frac{\partial T}{\partial n} = h_c(T_s - T_e) + R \cdot \epsilon \left(T_s^4 - T_e^4 \right) \tag{6}$$

where T_e and T_s are the temperatures of the environment and the steel surface. Because the product of the Boltzmann constant R and the emissivity ϵ is considered as the radiation heat transfer coefficient,

$$R \cdot \epsilon = h_r \tag{7}$$

Equation (6) can be written as

$$-\lambda \frac{\partial T}{\partial n} = h_c(T_s - T_e) + h_r \left(T_s^4 - T_e^4 \right) = h(T_s - T_e) \tag{8}$$

where h_c and h_r are the convection and radiation coefficients, respectively, and where h is the total heat transfer coefficient [30,31], which combines the effect of convective and radiative heat transfer, as both of these phenomena take place simultaneously with a varied intensity along the quenching time. The determination of the total heat transfer coefficient was the main goal of the inverse heat transfer analysis.

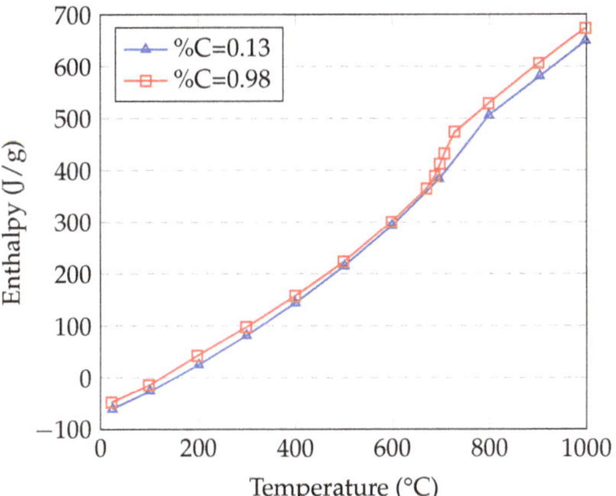

Figure 9. Enthalpies of Pyrowear 53 calculated in JMatPro [24].

2.3.3. Carbon Diffusion Modeling

A low-pressure carburizing process was used to create the gradient profile of the carbon concentration. The Neumann-type boundary conditions were used to calculate the carbon content on the surface [2]. The necessary data were the diffusion coefficient, carbon content at the surface in the equilibrium condition, and the mass transfer coefficient.

$$\frac{\partial C}{\partial t} = \nabla(D(\nabla C)) \tag{9}$$

where D is the diffusion coefficient, t is time, and C is the carbon concentration at a specific position. This type of equation requires initial and boundary conditions' definition. The initial condition for $t = 0$ is given as

$$C(x, y, z) = C_0 \tag{10}$$

where C_0 is the initial carbon content in the steel before carburizing. In this work, carbides' precipitation was neglected, and the assumption of no soot formation on the steel surface was made; therefore, the equation for the Neumann-type boundary condition expressing the mass balance at the surface may be simplified [32,33] as

$$-D(\nabla C) = \beta(C_e - C_s) \tag{11}$$

where C_e is the carbon concentration in equilibrium with the atmosphere and C_s is the carbon content in the gas–solid interface that reacts with the hot surface of the steel [33]. The surface carbon concentration C_e is required to calculate the mass transfer coefficient β. The value of this coefficient was determined from the experiment, where the carbon saturation was constant in time and equal to 60 min. Then, for processes realized at a constant temperature, T = 921 °C in this research, the value of the mass transfer coefficient β can be calculated from Equation (12).

$$\beta = \frac{\left(\frac{\Delta m}{A}\right)}{t(C_e - C_s)} \tag{12}$$

where Δm is the mass of carbon that the specimen could absorb with no soot formation on the surface and A is the specimen's surface that underwent carburizing. For cyclic processes such as low-pressure carburizing, the determinative parameters are the number of cycles and the total time of the saturation kinetic model given as a function of temperature and time factors may be used [33–35]. The approach described above allows determining the carbon concentration in equilibrium with the atmosphere C_e and the mass transfer coefficient β as 1.47 (wt.%) and $9.87 \times 10^{-3} \left(\frac{\text{mm}}{\text{s}} \right)$, respectively.

2.3.4. Phase Transformation Kinetics

Depending on the cooling rate, phase transformations may be classified as diffusion-controlled and diffusionless. For transformation during which diffusion is the governing phenomenon, the Johnson–Mehl–Avrami–Kolmogorov (JMAK) model is widely accepted to describe the kinetics of isothermal phase transformation through the nucleation and growth of the new phase. Cooling the austenite leads to phase transformation into ferrite (F), pearlite (P), or bainite (B), and the volume fraction of the newly transformed phase can be predicted according to the JMAK model [36,37].

$$\zeta_{(F,P,B)} = 1 - e^{-b \cdot t^n} \tag{13}$$

where $\zeta_{(F,P,B)}$ stands for the volume fraction of ferrite, pearlite, and bainite, respectively, and b and n are material kinetic parameters defined as:

$$n = \frac{ln \frac{ln(1-\zeta_1)}{ln(1-\zeta_2)}}{ln \left(\frac{t_1}{t_2} \right)} \tag{14}$$

$$b = -\frac{ln(1 - \zeta_1)}{t_1^n} \tag{15}$$

The isothermal times of a certain temperature, t_1 and t_2, and the corresponding volume phase fractions ζ_1 and ζ_2 are defined by means of TTT diagram analysis [29]. Volume fractions of 1% and 99% of the specific phase are usually used as the beginning and the end of certain phase transformations.

The martensitic transformation is a diffusionless transformation that occurs upon quick quenching of the austenite phase, and it is characterized by the martensite start and martensite finish temperatures. As the martensitic transformation is athermal, that is not being controlled by the thermal history of the material, the volume fraction of the transformed phase is calculated based on an equation incorporating the degree of under-cooling of the material. For transformation's kinetics description, a Koistinen–Marburger model [38] was used.

$$\zeta_M = \zeta_A \left(1 - e^{-K_M(M_S - T(t))} \right) \tag{16}$$

where

$$K_M = \frac{M_S - M_f}{\ln 0.1} \tag{17}$$

While both the martensite start and martensite finish temperatures depend on the alloying elements' concentration, the K_M parameter is considered [39] as

$$\sum_i S_i \cdot x_i = 0.14 x_{Mn} + 0.21 x_{Si} + 0.11 x_{Cr} + 0.08 x_{Ni} + 0.05 x_{Mo}. \tag{18}$$

where $T(t)$ is the temperature at a specific time step and ζ_M and ζ_A represent the volume phase fractions of martensite and austenite, respectively.

2.3.5. Hardness Prediction

Several technological parameters are considered when the heat treatment process is designed. Hardness is the most-important among them because it is a good indicator of the overall strengthening of the material while being simple to measure simultaneously.

The overall hardness of the material is calculated based on the rule of phases mixture, which defines the total material's hardness as a weighted sum of the volume fraction of

$$HV = \zeta_M HV_M + \zeta_B HV_B + \zeta_P HV_P + \zeta_F HV_F \tag{19}$$

where ζ_i is the volume fraction of the i-th phase and HV_i is its hardness for $i = M, B, P, F$, representing martensite, bainite, pearlite, and ferrite, respectively.

Multiple models are available in the literature describing models for hardness calculations, among which the Maynier model Equation (20) is considered reliable, although limited to carbon concentration up to 0.5%wt [40]. An additional Equation (21) for martensite hardness containing over 0.5% of carbon was proposed by Leslie [41]:

- For %C < 0.5,

$$\begin{aligned} HV = &\zeta_M \cdot (949\%C + 156.73 + 21 \cdot log(T85time[s])) + \\ &+ \zeta_B \cdot (185\%C + 29.14 + (53 \cdot \%C + 36.39) \cdot log(T85time[s])) + \\ &+ (\zeta_A + \zeta_F + \zeta_P) \cdot (223 \cdot \%C + 84.368 + 16.47 \cdot log(T85time[s])) \end{aligned} \tag{20}$$

- For %C > 0.5,

$$\begin{aligned} HV = &\zeta_M \cdot (1667\%C - 926\%C^2 + 150) + \\ &+ \zeta_B \cdot (185\%C + 29.14 + (53 \cdot \%C + 36.39) \cdot log(T85time[s])) + \\ &+ (\zeta_A + \zeta_F + \zeta_P) \cdot (223 \cdot \%C + 84.368 + 16.47 \cdot log(T85time[s])) \end{aligned} \tag{21}$$

Both cases were used for the computations of the carburized case in steel, as the appropriate carbon concentration point has a trespassed point of 0.5% of carbon concentration.

3. Results and Discussion

3.1. Heat Transfer Coefficient Optimization

The inverse thermal analysis and optimization result were that heat removal from the top, side, and bottom surfaces proceeds at different rates. Such behavior is related to the transient characteristic of cooling and the turbulent flow of the gas, which heats up at different rates during the cooling of the sample, changing its ability to accumulate energy. As a result, different heat transfer coefficient values are locally active on the surface. The DEFORM3D Inverse Heat Transfer module was utilized for this part of the simulation due to its capability to determine heat transfer coefficients based on an objective function, which in the case of quenching are the cooling curves at specific points and an environmental temperature drop as a boundary condition [42]. Optimization revealed that the heat transfer coefficient values vary depending on the temperature and the specimen's location. The highest values were found for the outer diameter surface (Zone 1) reaching its maximum, i.e., 1.2 $\frac{kW}{m^2 \cdot \degree C}$ for a temperature of approximately 800 $\degree$C. Intermediate values of around 1.1 $\frac{kW}{m^2 \cdot \degree C}$ were determined for the specimen's flat faces (Zone 3). In contrast, the lowest heat transfer of a maximum of 0.87 $\frac{kW}{m^2 \cdot \degree C}$ was found for the inner diameter (Zone 2).

The results showed the heat transfer coefficients as a function of temperature at the computational boundary, so influenced by the thermal field only with no accounting of time. The dashed lines in Figure 10 represent the results of the simulation, which did not take the latent heat of martensitic transformation into account; therefore, a dimple in the curves, reaching the minimum at 0.5 $\frac{kW}{m^2 \cdot \degree C}$, is visible. This is a numerical artifact caused by the apparent heat capacity of the material. Further analyses of the heat transfer allowed us to optimize the apparent heat capacity and compute the value of the latent heat of the martensitic transformation by integrating the function described by the heat capacity peak's data points. Integration with the trapezoid method and 100 steps were utilized, and eventually, the latent heat of 244 $\frac{J}{g}$ was determined and used in further simulations.

Figure 11 shows the optimization results for the heat capacity and the cooling, where the determined value of martensitic transformation latent heat was used.

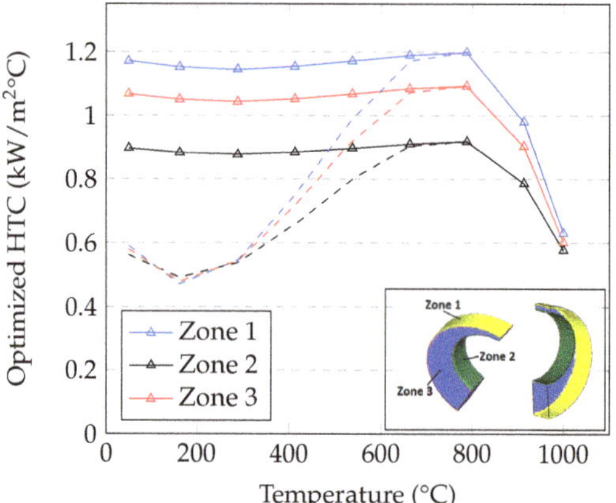

Figure 10. Optimized values of the heat transfer coefficients for specific zones.

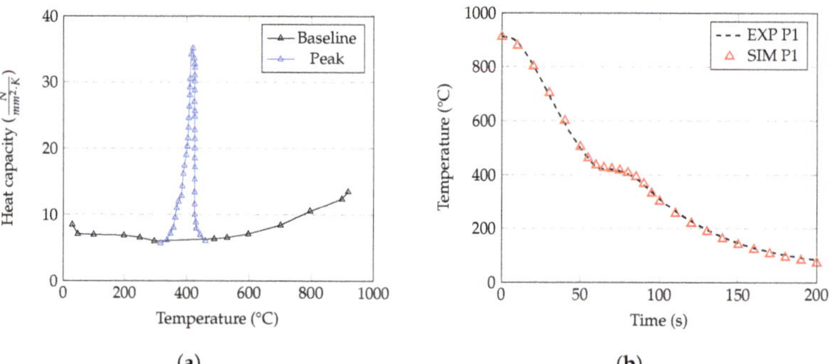

(**a**)   (**b**)

Figure 11. The latent heat of martensitic transformation's optimization results. Optimized apparent heat capacity (**a**) and simulated cooling curve with optimized latent heat (**b**).

A more detailed analysis, which may include a time factor on the change of the quenching gas temperature and, therefore, impacting the heat transfer between the gas and the specimen's surface, would require a CFD approach and computations of the transient thermal field in the whole volume of the vacuum chamber, which was not the goal of the current research.

3.2. Low-Pressure Carburizing

Carbon diffusion was simulated and compared with the GD-OES experimental results. Figure 12 compares the carbon profiles. Sixteen cycles of carbon boosts followed by a period of diffusion were modeled. The slight discrepancy of the carbon concentration curves in Figure 12b is an effect of modeling the diffusion coefficient in the steel and may be improved in further efforts.

The results of the carbon concentration simulation did not vary between the oil and high-pressure gas quenching processes. Diffusion coefficient model Equation (1) supplemented with the q factor implementing the chemical composition's influence on material's diffusivity Equation (2) and mass transfer coefficient Equation (12) allowed computing the diffusion of carbon with a good fit to the experimental values for Pyrowear 53 steel.

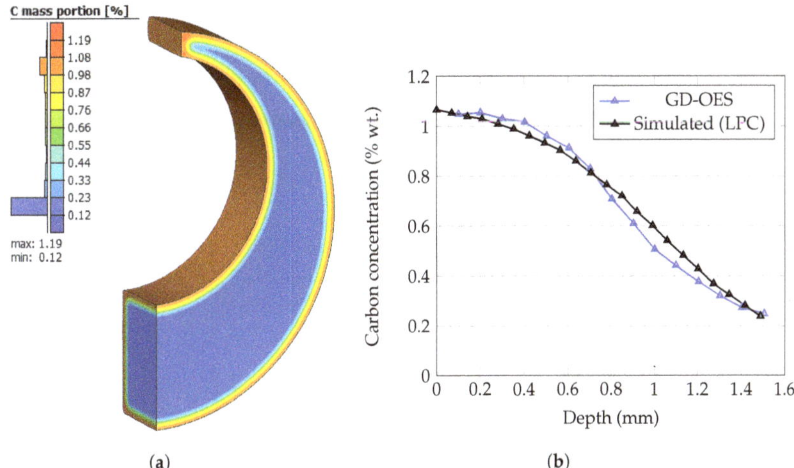

Figure 12. Simulated carbon concentration. Visualization of the carburized case (**a**) and comparison of the experimental and predicted concentration profiles (**b**) .

3.3. Cooling Behavior

Cooling was analyzed at two control points of the specimen marked in Figure 8. Point P1 was located in the middle of the thickest cross-section and was the indicator of the core material behavior. At the same time, the position of point P2 was aimed at carburized case analysis and, therefore, was located in the thinnest zone 0.05 mm circumferentially under the surface. Quenching started at time point $t = 0$. As presented in Figure 13a, in the first 10 s, the temperature did not drop, which is related to the location of the control point P1 in the center of the thickest cross-section, 7.95 mm from the surface. Because of the material's heat conductivity, a temperature dropin the experimental curve may be observed. Martensitic transformation starts at approximately 421 °C, and because it is exothermic, the latent heat released during transformation contracts the quenching extraction, causing a plateau in the cooling curve. The amount of latent heat for Pyrowear 53 was calculated as presented in Section 3.1, and it was found to be equal to 244 $\frac{J}{g}$. Figure 13b presents a simulated cooling curve for control point P2 located 0.05 mm from the surface. Here, the temperature decreases much faster with no initial period of constant temperature. Because of the carbon concentration in the surface layer is equal to approximately 1.1% wt. and the martensitic transformation starts slightly above 100 °C (see Figure 7), the plateau in the cooling curve may not be observed.

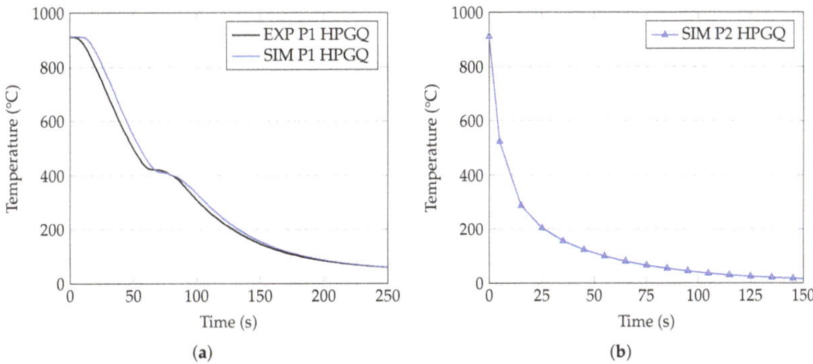

Figure 13. Cooling curves for the core material at P1 (**a**) and the carburized case at P2 (**b**).

The computed cooling rates presented similarity to those obtained from the experiment. The choice of a cooling rate of 11 °C/s was found for a temperature T = 715 °C, which corresponds to a time point of t = 26 s of the quenching process. The decrease of the cooling rate value corresponding to a temperature of 422 °C was related to the temperature dropduring the exothermic martensitic transformation, as may be observed in Figure 14.

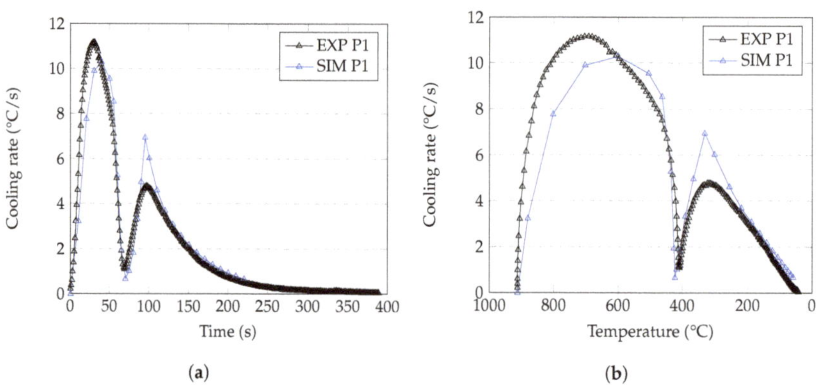

Figure 14. Cooling rates as a function of the process time (**a**) and the temperature (**b**).

3.4. Phase Composition and Hardness

All specimens were tested against hardness. Both longitudinal and transverse cross-sections, as shown in Figure 4, were measured and showed similar results. The carburized material presented a similar carbon concentration profile, which did not depend on the quenching technique. The average hardness at a distance of 0.1 mm from the surface for carburized specimens was 760 HV. According to the AGMA standards, the effective case depth (ECD) is the distance from the surface of the specimen where the hardness is 50 HRC, corresponding to approximately 513 HV. For the analyzed process parameters, an effective case depth of 1.1 mm can be obtained for Pyrowear 53. Figure 15 compares the hardness distribution curves obtained from the simulation with the experimental curves, being the arithmetic average of all carburized specimens. The presented hardness profiles showed that the differences were barely noticeable, which means the carbon concentration profile was not affected by the quenching method. Figure 16 presents the transformation and final distribution of the phases after the quenching. Only two phases were present in Pyrowear 53 after heat treatment by the analyzed process. Although the cooling curve measured at P1 shown in the CCT diagram in Figure 6 crosses the point of the beginning of bainite transformation, the bainite was not observed, probably because of its content in the range of a few percent. Based on that, it was found that no products of austenite decomposition other than martensite formed with the cooling rates as analyzed, which is in agreement with other research [26].

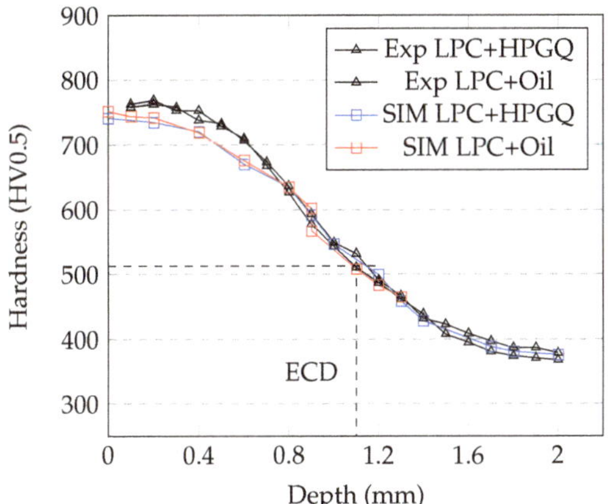

Figure 15. Results of hardness profile simulation comparison with the experimentally measured after the HPGQ and oil quenching processes.

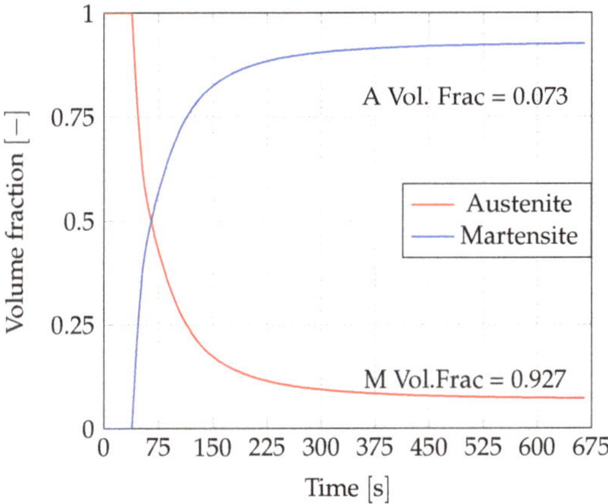

Figure 16. Computed phase volume fraction changes during quenching for a control point P1.

4. Conclusions

This paper presented the results of heat-treated specimens made from Pyrowear 53 low-alloy steel. The simulation of the quenching of previously carburized and non-carburized specimens was analyzed. The analysis of the C-ring, which went through LPC, followed by different quenching techniques, also concluded that high-pressure gas quenching provides results comparable to standard oil quenching. The numerical simulation of the heat treatment processes could accelerate new procedures' implementation and lower development costs. The simulation of the heat-treatment process using Simufact Forming supplied with the input of JMatPro material's properties, although requiring validations and some adjustments, provided a good correlation with the experimental measurements for all analyzed variables. A computational approach as presented in this paper's procedure for phase composition and hardness analysis may be successfully applied to the analysis of production parts made from different alloys to optimize the process by providing metallurgical analysis with a limited testing campaign.

Author Contributions: Conceptualization, B.I. and M.M.; methodology, B.I. and A.W.; software, B.I.; validation, B.I. and M.M.; formal analysis, M.M.; investigation, B.I. and A.W.; resources, B.I. and

A.W.; data curation, B.I.; writing—original draft preparation, B.I.; writing—review and editing, B.I.; visualization, B.I.; supervision, M.M.; project administration, B.I.; funding acquisition, M.M. All authors have read and agreed to the published version of the manuscript.

Funding: This research received no external funding.

Data Availability Statement: Not applicable.

Acknowledgments: For calculations and simulations in this work, some of the results gathered within the framework of the development project in the TECHMATSTRATEG2/406725/1/NCBR/2020 program of the National Centre for Research and Development were used. In particular, the CCT diagram (Figure 6) was prepared by W. Zalecki from the Łukasiewicz Research Network - Upper Silesian Institute of Technology. The authors would also like to express their gratitude to FIN Sp. z o.o. for the preparation of the C-ring specimens that were used in the experiments described in this article.

Conflicts of Interest: The authors declare no conflict of interest.

References

1. National Academy of Sciences. *Materials for Helicopter Gears, Final Report No. NMAB-351*; Technical Report; Committee on Helicopter Transmission Gear Materials: Washington, DC, USA, 1979.
2. Korecki, M.; Wołowiec-Korecka, E.; Bazel, M.; Sut, M.; Kreuzaler, T. Thomas Detlef. Outstanding Hardening of Pyrowear® Alloy 53 with Low Pressure Carburizing. In Proceedings of the 70° Congresso Anual da ABM, Rio de Janeiro, Brazil, 17–21 August 2005. [CrossRef]
3. AMS6308 rev. *F-Steel, Bars and Forgings 0.90Si-1.0Cr-2.0Ni-3.2Mo-2.0Cu-0.10V (0.07–0.13C) Vacuum Arc or Electroslag Remelted*; SAE International: Warrendale, PA, USA, 2018
4. Technology, C. CarTech® Pyrowear® Alloy 53—Technical Datasheet.
5. Easton, D.; Perez, M.; Huang, J.; Rahimi, S. Effects of Forming Route and Heat Treatment on the Distortion Behaviour of Case-Hardened Martensitic Steel type S156. *Heat Treat* **2017**, *2017*,
6. Dowling, W.E. Development of a Carburizing and Quenching Simulation Tool: Program Overview. In Proceedings of the 2nd International Conference on Quenching and Control of Distortion, Materials Park, OH, USA, 9–13 September 1996; pp. 349–355.
7. Arimoto, K. Development of heat treatment simulation system Deform-HT. In Proceedings of the 18th Heat Treating Conference, ASM International, Materials Park, OH, USA, 12–15 October 1998.
8. Scientific Forming Technologies Corporation. *DEFORM V13.0.1 Manual*; Scientific Forming Technologies Corporation: Columbus, OH, USA, 2022.
9. Farivar, H.; Rothenbucher, G.; Prahl, U.; Bernhardt, R. *ICME-Based Process and Alloy Design for Vacuum Carburized Steel Components with High Potential of Reduced Distortion, Proceedings of the 4th World Congress on Integrated Computational Materials Engineering (ICME 2017)*; Springer: Berlin/Heidelberg, Germany, 2017; pp. 133–144.
10. Schillé, J.P.; Guo, Z.; Saunders, N.; Miodownik, A.P. Modeling phase transformations and material properties critical to processing simulation of steels. *Mater. Manuf. Process.* **2011**, *26*, 137–143. [CrossRef]
11. Guo, Z.; Saunders, N.; Schillé, J.; Miodownik, A. Material properties for process simulation. *Mater. Sci. Eng. A* **2009**, *499*, 7–13. [CrossRef]
12. Guo, Z.L.; Turner, R.; Da Silva, A.D.; Sauders, N.; Schroeder, F.; Cetlin, P.R.; Schillé, J.P. Introduction of materials modeling into processing simulation. *Trans Tech. Publ.* **2013**, *762*, 266–276.
13. Jo, A.; Thomas, H.; Lars, H.; Pingfang, S.; Bo, S. Thermo-Calc & DICTRA, computational tools for materials science. *Calphad* **2002**, *26*, 273–312.
14. da Silva, A.; Pedrosa, T.; Gonzalez-Mendez, J.; Jiang, X.; Cetlin, P.; Altan, T. Distortion in quenching an AISI 4140 C-ring – Predictions and experiments. *Mater. Des.* **2012**, *42*, 55–61. [CrossRef]
15. Nunes, M.M.; da Silva, E.M.; Renzetti, R.A.; Brito, T.G. Analysis of Quenching Parameters in AISI 4340 Steel by Using Design of Experiments. *Mater. Res.* **2019**, *22*, 1–7. [CrossRef]
16. Sims, J.; Li, Z.; Ferguson, B.L. Causes of Distortion during High Pressure Gas Quenching Process of Steel Parts. In Proceedings of the Heat Treat 2019: 30th ASM Heat Treating Society Conference, Detroit, MI, USA, 15–17 October 2019; pp. 228–236. [CrossRef]
17. Manivannan, M.; Northwood, D.; Stoilov, V. Use of Navy C-rings to study and predict distortion in heat treated components: Experimental measurements and computer modeling. *Int. Heat Treat. Surf. Eng.* **2014**, *8*, 168–175. [CrossRef]
18. Boyle, E.; Bowers, R.; Northwood, D.O. The use of navy C-ring specimens to investigate the effects of initial microstructure and heat treatment on the residual stress, retained austenite, and distortion of carburized automotive steels. *SAE Trans.* **2007**, *116*, 253–261.
19. Yu, H.; Yang, M.; Sisson, R.D. Application of C-Ring Specimen for Controlling the Distortion of Parts During Quenching. *Met. Sci. Heat Treat.* **2021**, *63*, 220–228. [CrossRef]
20. Heuer, V. Gas Quenching. In *Steel Heat Treating Fundamentals and Processes*; ASM Handbook; ASM International: Almere, The Netherlands, 2013; Volume 4A.
21. Adrian, H.; Karbowniczek, M.; Dobosz vel Sypulski, A.; Kowalski, J.; Kozdroń, S. Badania wpływu temperatury na zdolność chłodzącą wybranych olejów hartowniczych. *Stal. Met. Nowe Technol.* **2019**, *9*, 75–80.

22. Tibbets, G.G. Diffusivity of carbon in iron and steels at high temperatures. *J. Appl. Phys.* **1980**, *51*, 4813–4816. [CrossRef]
23. Neumann, F.; Person, B. A contribution to the metallurgy of gas carburization. The effects of alloying elements on the relationship between the carbon potential of the gas phase and that of the workpiece. *Haerterei-Tech. Mitteilungen* **1968**, *23*, 296–310.
24. Research project "Development of high pressure gas quenching technology for the satellite gears of the FDGS engine's epicyclic gearbox, made of Pyrowear 53 steel and operating under long-term and cyclically fluctuating loads" funded by National Center for Research and Development (Narodowe Centrum Badan i Rozwoju) under the TECHMATSTRATEG2/406725/1/NCBR/2020 program. Unpublished Research Data.
25. Freborg, A.M. Investigating and Understanding the Role of Transformation Induced Residual Stress to Increase Fatigue Life of High Strength Steel Used in Transmission Gears. Ph.D. Thesis, University of Akron, Akron, OH, USA, 2013.
26. Horstemeyer, M.F. *Integrated Computational Materials Engineering (ICME) for Metals: Concepts and Case Studies*; John Wiley & Sons: Hoboken, NJ, USA, 2018.
27. van Bohemen, S.; Sietsma, J. Effect of composition on kinetics of athermal martensite formation in plain carbon steels. *Mater. Sci. Technol.* **2009**, *25*, 1009–1012. [CrossRef]
28. Şimşir, C.; Gür, C.H. Simulation of quenching. In *Handbook of Thermal Process Modeling of Steels*; Routledge: Abingdon, UK, 2009; pp. 341–425.
29. Li, J.; Feng, Y.; Zhang, H.; Min, N.; Wu, X. Thermomechanical Analysis of Deep Cryogenic Treatment of Navy C-Ring Specimen. *J. Mater. Eng. Perform.* **2014**, *23*, 4237–4250. [CrossRef]
30. Li, H.; Zhao, G.; Huang, C.; Niu, S. Technological parameters evaluation of gas quenching based on the finite element method. *Comput. Mater. Sci.* **2007**, *40*, 282–291. [CrossRef]
31. Sugianto, A.; Narazaki, M.; Kogawara, M.; Shirayori, A.; Kim, S.Y.; Kubota, S. Numerical simulation and experimental verification of carburizing-quenching process of SCr420H steel helical gear. *J. Mater. Process. Technol.* **2009**, *209*, 3597–3609. .: 10.1016/j.jmatprotec.2008.08.017. [CrossRef]
32. Khan, D.; Gautham, B. Integrated modeling of carburizing-quenching-tempering of steel gears for an ICME framework. *Integr. Mater. Manuf. Innov.* **2018**, *7*, 28–41. [CrossRef]
33. Zajusz, M.; Tkacz-Śmiech, K.; Dychtoń, K.; Danielewski, M. Pulse carburization of steel–model of the process. *Trans. Tech. Publ.* **2014**, *354*, 145–152. [CrossRef]
34. Wierzba, B.; Romanowska, J.; Kubiak, K.; Sieniawski, J. The cyclic carburization process by bi-velocity method. *High Temp. Mater. Process.* **2015**, *34*, 373–379. [CrossRef]
35. Smirnov, A.; Ryzhova, M.Y.; Semenov, M.Y. Choice of boundary condition for solving the diffusion problem in simulation of the process of vacuum carburizing. *Met. Sci. Heat Treat.* **2017**, *59*, 237–242. [CrossRef]
36. Avrami, M. Kinetics of phase change. I General theory. *J. Chem. Phys.* **1939**, *7*, 1103–1112. [CrossRef]
37. William, J.; Mehl, R. Reaction kinetics in processes of nucleation and growth. *Trans. Metall. Soc. AIME* **1939**, *135*, 416–442.
38. Dp, K.; Re, M. A general equation prescribing the extent of the austenite-martensite transformation in pure iron-carbon alloys and plain carbon steels. *Acta Metall.* **1959**, *7*, 59–60.
39. Simufact Engineering GmbH. *Simufact Forming Users Manual*; Simufact Engineering: Hamburg, Germany, 2022.
40. Maynier, P.; Dollet, J.; Bastien, P. Hardenability concepts with applications to steels. In Proceedings of the International Conference on Artificial Intelligence in Medicine, New York, NY, USA, 1978; Volume 518.
41. Leslie, W. Interstitial atoms in alpha iron. *Leslie WC—The Physical Metallurgy of Steels*; McGraw-Hill: New York, NY, USA, 1982; pp. 85–91.
42. Wu, C.; Xu, W.; Wan, S.; Luo, C.; Lin, Z.; Jiang, X. Determination of Heat Transfer Coefficient by Inverse Analyzing for Selective Laser Melting (SLM) of AlSi10Mg. *Crystals* **2022**, *12*, 1309. [CrossRef]

Article

Understanding Uncertainty in Microstructure Evolution and Constitutive Properties in Additive Process Modeling

Matthew Rolchigo [1,*], Robert Carson [2] and James Belak [2]

[1] Oak Ridge National Laboratory, Oak Ridge, TN 37830, USA
[2] Lawrence Livermore National Laboratory, Livermore, CA 94550, USA; carson16@llnl.gov (R.C.); belak1@llnl.gov (J.B.)
* Correspondence: rolchigomr@ornl.gov

Abstract: Coupled process–microstructure–property modeling, and understanding the sources of uncertainty and their propagation toward error in part property prediction, are key steps toward full utilization of additive manufacturing (AM) for predictable quality part development. The Open-FOAM model for process conditions, the ExaCA model for as-solidified grain structure, and the ExaConstit model for constitutive mechanical properties are used as part of the ExaAM modeling framework to examine a few of the various sources of uncertainty in the modeling workflow. In addition to "random" uncertainty (due to random number generation in the orientations and locations of grains present), the heterogeneous nucleation density N_0 and the mean substrate grain spacing S_0 are varied to examine their impact of grain area development as a function of build height in the simulated microstructure. While mean grain area after 1 mm of build is found to be sensitive to N_0 and S_0, particularly at small N_0 and large S_0 (despite some convergence toward similar values), the resulting grain shapes and overall textures develop in a reasonably similar manner. As a result of these similar textures, ExaConstit simulation using ExaCA representative volume elements (RVEs) from various permutations of N_0, S_0, and location within the build resulted in similar yield stress, stress–strain curve shape, and stress triaxiality distributions. It is concluded that for this particular material and scan pattern, 15 layers is sufficient for ExaCA texture and ExaConstit predicted properties to become relatively independent of additional layer simulation, provided that reasonable estimates for N_0 and S_0 are used. However, additional layers of ExaCA will need to be run to obtain mean grain areas independent of build height and baseplate structure.

Keywords: additive manufacturing; microstructure; properties

Citation: Rolchigo, M.; Carson, R.; Belak, J. Understanding Uncertainty in Microstructure Evolution and Constitutive Properties in Additive Process Modeling. *Metals* **2022**, *12*, 324. https://doi.org/10.3390/met12020324

Academic Editors: Francesco De Bona, Jelena Srnec Novak and Francesco Mocera

Received: 1 January 2022
Accepted: 8 February 2022
Published: 12 February 2022

Publisher's Note: MDPI stays neutral with regard to jurisdictional claims in published maps and institutional affiliations.

1. Introduction

Additive manufacturing (AM) processes for alloys are an exciting frontier in metallurgical and materials science and engineering, in part due to the ability to manufacture unique geometries and compositions not possible via traditional metallurgical processing [1,2], as well as the potential for local control over microstructure and properties. However, realizing this potential for local microstructure and property control is difficult in practice. As discussed extensively in recent review articles on the subject, microstructure and properties within AM builds result from complex multiscale relationships between material parameters and processing conditions [3–7]. By taking advantage of these relationships, a large number of studies have attempted to control features of AM microstructure (and resulting properties) for various alloys and specific AM processes. One method of controlling microstructure is through altering aspects of the energy source itself, such as changing scan rotation between layers [8], modifying the beam shape [9], using nonlinear spot or island-based scan strategies [10–12], varying volumetric energy density [13], and using a pulsed beam with varied frequency and energy [14]. Other aspects of the process can also be altered to affect microstructure and properties; for example, the application of an external magnetic field [15], the use of ultrasonic vibration [16,17], or the variation in the supports and machining conditions in the case of a deposition that requires support rods [18]. The feedstock material composition will also affect microstructure and properties. Examples include the variation of Al content in a steel alloy [19], adding trace quantities of grain refining solutes and inoculants to titanium alloy melt pools [20], and various combinations of alloying elements for Ni-base alloys [21]. Whether varying aspects of the process or the material, all of the aforementioned studies demonstrated some degree of control over various aspects of microstructure (particularly grain size distribution, texture, and phase fractions) and/or properties (particularly elastic modulus, tensile strength, and crack susceptibility).

As manufacturing and characterizing microstructure and properties for all sets of process conditions and materials is not practical, many of the aforementioned studies included materials modeling components. For example, Ref. [9] used a computational fluid dynamics model along with analytical modeling of grain structure transitions to rationalize observed grain shapes, Ref. [18] used mechanical response modeling to aid in part design, and [13,21] used thermodynamic calculations to understand and control the phases present in manufactured parts. However, modeling AM can be challenging due to the large degree of uncertainty surrounding various input parameters and the details of multiscale physical phenomena. Coupled modeling of processing, microstructure development, and constitutive properties can provide insights into AM, but the sensitivity of the models to the various sources of uncertainty may lead to inaccurate predictions and obscure aspects of process–microstructure–property relationships. The present work takes advantage of various codes capable of running on pre-Exascale hardware that were developed in part through the Exascale Additive Manufacturing project (ExaAM) [22] to address this challenge. Specifically, we use an ensemble of simulations to examine the sensitivity of microstructure prediction to uncertain input parameters and physics associated with solidification and the sensitivity of constitutive properties to the predicted microstructure. Figure 1 shows the relevant portion of the ExaAM workflow: the model of temperature field and melt pool evolution (OpenFOAM) [23,24], the model of as-solidified microstructure development (ExaCA) [25], and the model of constitutive properties (ExaConstit) [26]. It is noted that input uncertainties, model error, and error associated with model coupling will propagate downstream to property prediction; the focus of the present work is on ExaCA and how variation in ExaCA predictions due to uncertainty in initial and boundary conditions impact ExaConstit results.

The following section gives more background into coupled melt pool–microstructure–property models for AM processing and further defines specific goals of the present modeling effort that address gaps in the literature. An overview of the cellular automata (CA) algorithm, parameters, conditions, and microstructure output of interest from ExaCA is then given along with a brief overview of ExaConstit and the procedure for model coupling. This is followed by predicted microstructure results and analysis of predicted properties to put the modeled microstructure variations in context. Finally, the main takeaways of the work are given, followed by future work planned using ExaCA. While this study focuses on a specific scan pattern (90 degree rotations between layers) and material (Inconel 625), it is not designed to reproduce a specific result from experiment but rather to demonstrate model capabilities and their usefulness in investigating the sensitivity of microstructure and properties to uncertain inputs.

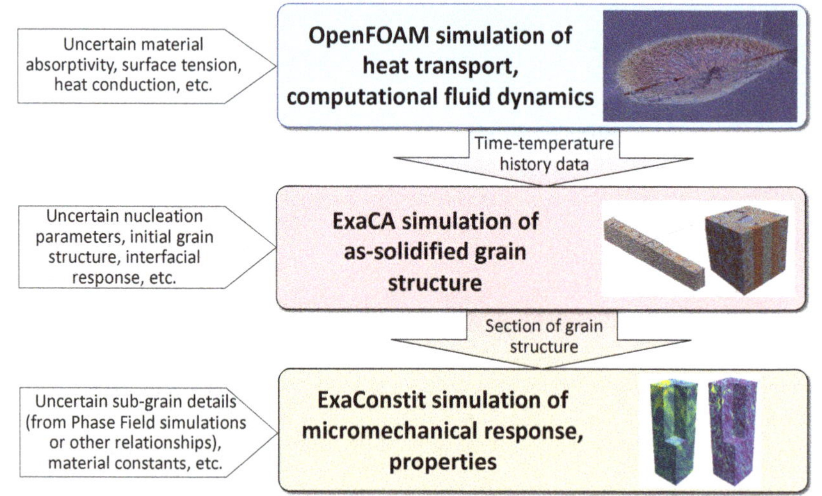

Figure 1. Schematic of the workflow associated with OpenFOAM modeling of AM processing, ExaCA modeling of microstructure, and ExaConstit modeling of properties as part of the ExaAM project, noting sources of uncertainty in model inputs.

2. Background

The evolution of grain structure in AM depends on the complex fluid flow and heat transport processes common to AM melt pools, as the temperature field in the material and grain structure evolution at the solid–liquid interface are inherently linked. As-solidified microstructure is controlled by the conditions in the vicinity of the melt pool as well as the alloy composition; the morphology of the solidification front is often cells or dendrites growing perpendicular to the melt pool boundary, though it may also include primary or secondary phase nucleation and growth [27]. This solidification behavior is dependent not only on the magnitude of thermal gradient G and cooling rate $\dot{T}$, but on underlying substrate conditions, variation in the direction of G as a function of location in the melt pool and time, and solute and interfacial effects at the solid–liquid interface [4,5,28]. For cubic crystal systems, including the as-solidified FCC Inconel 625 modeled in the present work, dendritic solidification proceeds along the <001> crystallographic directions. Dendritic grains with <001> directions closely aligned with the direction of the thermal gradient at the solidification front will therefore be geometrically favored for advance [29]. In turn, this leads to highly textured grain structures that are highly anisotropic in the build direction [30,31], with correspondingly anisotropic properties [32–34].

Modeling of as-solidified microstructure during alloy solidification can be accomplished through various methods, each with their own benefits and drawbacks. Recent review articles have discussed many of these methods, including those at the scale of individual cells and dendrites such as phase field and dendritic needle network models, and those at the scale of the grains themselves, such as the cellular automata and kinetic Monte Carlo models [35,36]. While subgrain features such as microsegregation, cell or dendrite morphology, and dendrite arm spacing are useful insights from the higher fidelity models such as phase field, these models are often limited to small fractions of large problems (such as AM melt pools) due to their high computational cost. ExaCA is based on the grain-scale cellular-automata (CA)-based method for modeling alloy solidification originally proposed by Gandin and Rappaz, which makes subscale assumptions on dendrite growth velocity as a function of local undercooling and models the grain envelopes, rather than the cellular or dendritic structure explicitly [37–39]. While at a lower fidelity than phase field (i.e., subgrain level details such as microsegregation are not explicitly modeled), the grain-scale

CA method can account for the key problem physics, namely nucleation, and undercooling- and orientation-dependent dendrite growth kinetics, while being sufficiently computationally simple to simulate large problem sizes and ensembles of simulations. Several CA models for AM grain structure have been developed via coupling to semianalytical, finite element, finite volume, and thermal lattice Boltzmann-based computational fluid dynamics models for various electron and laser-based AM processes [40–45]. As the CA algorithm can simulate dendritic growth of any cubic crystal structure, it has been applied to austensitic stainless steel, β titanium, and Inconel solidification, and accurately reproduced experimentally observed grain size, morphology, and texture distributions in many cases.

The large thermal gradients in the melt pool tend to favor epitaxial growth of existing grains at the melt pool boundary. However, nucleation can still be important under certain processing conditions and/or in melt pools inoculated with additional particles, as nucleation of fine, relatively equiaxed grains has been linked to strength and ductility compared to alloy microstructures with more columnar grains [46]. While some CA simulations neglect nucleation entirely, focusing on conditions where nucleation's role on microstructure development can be safely ignored, many other CA models include this effect either at the melt pool surface, in the bulk undercooled liquid, or both [40,45,47–51]. The nucleation densities in these studies vary several orders of magnitude, and only a few studies examine the effect of changing nucleation density values on the resulting grain structure prediction [48,52]. This is a notable source of uncertainty in CA simulations of AM microstructure, as details of heterogeneous grain nucleation during AM processing is still an area of active research.

In additional to temperature field and nucleation considerations, the grain structure of a layer's substrate (whether it is a baseplate, or partially remelted grains from a previously deposited layer) will also have an effect on the competition between epitaxial and nucleated grain growth [53]. After a large number of layers (as is the case in many AM parts, which may consist of thousands of deposited layers or more), the effect of the baseplate grain structure would be expected to be negligible, with impingement of grains from previous layers (from the baseplate and nucleation events) eventually reaching a quasisteady state in grain size as a function of build height. The final layer of a build is also distinct from other regions as it does not remelt, and will also have a different microstructure than the layers near the substrate and in the "bulk" part [54,55]. It has been observed that different alloy phases, grain shapes, textures, hardness, and yield strengths were present in regions of microstructure taken from varied distances from the top build surface [56,57]. While a grain structure used for calculation of constitutive properties should be representative of a bulk part's grain structure, and not skewed by the grain structure of the baseplate nor the transient microstructure of the final layer, the number of layers needed to achieve this representative region of microstructure is likely scan-pattern- and alloy-dependent. For example, the aluminum alloy grain structure analyzed by [58] and the titanium alloy prior-β grain structure analyzed by [59] both appeared to show an initial transient microstructure in approximately the first 0.5 to 0.7 mm of build above the baseplate surface, before becoming relatively static further along in the build. The CA calculations and EBSD measurements of [44] for powder bed fusion of Inconel 718 show that cross-sections taken from 1 and 10 mm above the baseplate have significantly different grain areas and texture. Meanwhile, the CA model of [42] used to simulate selective laser melting of Ti-6Al-4V showed that the number of grains present as a function of simulated layer leveled off after about 15 layers, and the CA model of [60] used to simulate selective electron beam melting of Inconel 718 showed that the average grain diameter became static after 4 to 5 mm of build.

While many models of AM have focused on the connection between process and microstructure, others have expanded to include modeling of part properties using the simulated microstructure. Modeling of material stress–strain response requires information regarding multiple microstructural features, including grain-scale microstructural information as well as phase and subgrain data [36]. Mechanical property modeling has been performed using generated grain-scale microstructure data from phase field models [61],

Voronoi tessellations [62], procedurally generated grain structures designed to resemble those from AM processing [63], and CA grain structures [60,64,65]. Given the dependence of constitutive part properties on microstructure, and the dependence of microstructure on processing conditions, frameworks developed to combine process, grain structure, and property models fill an important multiscale materials modeling need [64]. The present work represents the results of such a framework, coupling a process model (OpenFOAM), a grain structure model (ExaCA), and a crystal plasticity finite element model (ExaConstit) as part of the ExaAM workflow. By doing so, we aim to answer questions regarding the development of grain size and texture as a function of the number of layers simulated, as well as how this development is influenced by heterogeneous nucleation. Whether or not the initial condition (substrate grain size) or the boundary condition (heterogeneous nucleation density) dominates grain structure development, and the downstream impact of each of these sources of uncertainty on ExaConstit-simulated properties, will be addressed.

3. Materials and Methods

3.1. Workflow Description

An example of the workflow used to generate microstructure results is shown in Figure 2. Temperature input data were generated using a custom application developed using the OpenFOAM computational fluid dynamic platform to simulate heat and mass transfer in welding and additive manufacturing; more details on this application are given in [24]. This data were generated for two raster scan patterns: an "odd" layer, with each pass by the heat source alternating between the positive and negative X directions, and an "even" layer, with each pass by the heat source alternating between the positive and negative Y directions. The scan and material parameters for the relevant material, Inconel 625, are given in Table 1; the result for each layer is a series of melt tracks approximately 30 to 40 μm deep and 150 to 175 μm wide. As shown in Figure 2a, these temperature data are a "sparse" form of the full time-temperature history as previously described by [66], consisting of a list of Cartesian coordinates from the OpenFOAM simulation that went above the alloy liquidus temperature. These coordinates are given on a regular grid, alongside the times at which each coordinate went below the liquidus and solidus temperatures for the final time. A trilinear interpolation scheme is used to interpolate this liquidus and solidus time from the OpenFOAM grid spacing to the CA cell size. By using temperature data of this form, the CA implicitly makes the assumption that solidification from remelted boundaries has a negligible effect on microstructure, as only the final time that a given CA cell undergoes solidification is calculated. This assumption previously appeared valid for shallow melt pools and raster scan patterns resembling those used in this present work [66]. In addition to the interpolated liquidus and solidus time data, input values are required for the number of layers simulated, the ordering of the layers, and the offset in the Z (build) direction between layers. Unless otherwise mentioned, in the present work, we simulate 65 layers of alternating even and odd scan patterns, offset by 20 μm in the build direction. Using the temperature data and multilayer input values, a multilayer field of liquidus and solidus time values can be assembled. This assembly is schematically illustrated in Figure 2b. Temperature data from older layers are overwritten with data from newer layers, as only the final time that a given CA cell undergoes solidification is calculated. A field of "layer ID" values lists the layer during which each cell solidifies; since liquidus and solidus time values start at zero for each layer, linking the solidus and liquidus data with the appropriate layer is necessary.

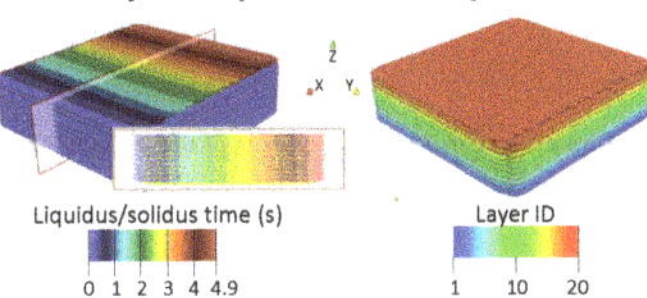

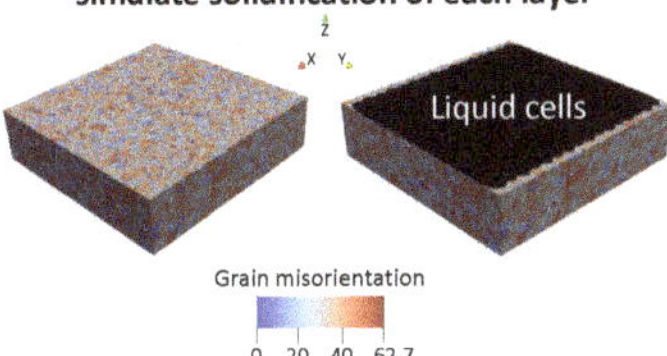

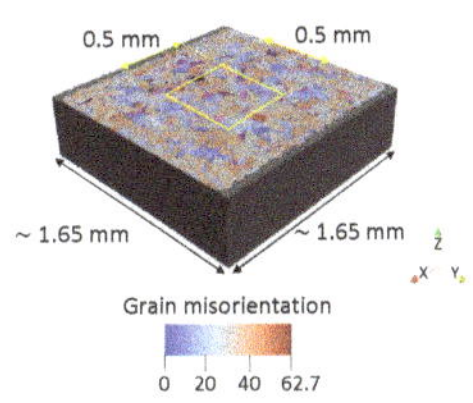

Figure 2. Schematic of the CA model workflow. Liquidus and solidus time data is first interpolated from the OpenFOAM regular grid to the CA grid for each layer pattern (**a**), and multilayer data fields are constructed from the interpolated data and input file instructions (**b**). The final times that a given cell goes below the liquidus and solidus temperatures, and the layer associated with the solidification event is associated with, are tracked. These temperature data are used alongside an initial substrate grain structure and solidification parameters to initialize a CA simulation (**c**) and, in turn, generate as-solidified grain structures (**d**). Subsections of simulated microstructure are extracted from a "representative" area for grain area, texture, and property analysis.

Table 1. Material and scan parameters used in the OpenFOAM model for liquid and solid Inconel 625. The temperature-dependent parameters were considered as constants beyond 2000 K.

Parameter	Value
Laser efficiency	31%
Laser power	195 W
Laser depth	4.899×10^{-5} m
Laser radius (Gaussian beam shape)	1.041×10^{-4} m
Laser velocity	0.8 m/s
Laser passes per layer	16
Laser line offset	100 μm
Liquidus temperature	1620 K
Solidus temperature	1410 K
Thermal expansion coefficient for liquid	1.25×10^{-4} K^{-1}
Dendrite arm spacing for drag force	1 μm
Thermocapillary coefficient	-3.08×10^{-4} N/(m · K)
Solid and liquid density (function of T)	8603.74–0.68278·T kg/m^3
Solid heat capacity	428.38 + 0.23638·T J/(kg·K)
Solid thermal conductivity	8.9164 + 0.014743·T W/(m·K)
Liquid heat capacity	725.74 J/(kg·K)
Liquid thermal conductivity	8.9164 + 0.014743·T W/(m·K)
Latent heat of fusion	217,540 J/kg
Dynamic viscosity	0.003032 kg/(m·s)
Reference density for buoyancy	7569.92 kg/m^3
Reference temperature for buoyancy	1620 K
Spatial resolution of generated liquidus/solidus time data	5 μm

Once the OpenFOAM data has been read and initialized by ExaCA, a substrate grain structure must be generated. This was performed using ExaCA itself as a preprocessing step to generate a "baseplate" microstructure as shown in Figure 2c: initializing an appropriately sized simulation domain of liquid cells, calculating the probability that a given cell of size Δx will be the center of a baseplate grain based on an input mean substrate grain diameter S_0, and generating a random number $R_{substrate}$ between 0 and 1 for each cell. For cells where the condition

$$R_{substrate} < \frac{\Delta x^3}{S_0^3} \tag{1}$$

is satisfied, randomly oriented grain seeds (each grain has one of 10,000 possible random orientations) are placed at these sites. The model described in Section 3.2 is then used to let these grains growth at a constant rate until all liquid cells are claimed by a grain.

After reading in the appropriate OpenFOAM and substrate data for a given run, ExaCA will also read an input data file containing cell size, time step, and parameters governing nucleation and growth; these parameters and their use will be discussed in the following section. Finally, after the completion of an ExaCA simulation, a region 0.5 mm in the X and Y directions is taken from the center of the resulting microstructure for use in grain statistics calculation (shown schematically in Figure 2d and described further in Section 3.3) and properties analysis (described further in Section 3.4).

3.2. CA Model Description

The computational domain is divided into a regular grid of cells, with spacing Δx. A time step, Δt, is selected and used to convert the "liquidus time" data into values of "critical time step", i.e., the time step at which a cell goes below the liquidus temperature for the final time. The difference between the liquidus and solidus times and temperatures is used together with Δt to define an "undercooling change" value, or the rate at which the magnitude of the undercooling increases each time step following the "critical time step", at which the cell has zero undercooling. Cells that do not have liquidus and solidus time data are part of the substrate, and are initialized as solid type. Cells that have liquidus and solidus time data are part of the melt pool and initialized as liquid type, with the exception of those that border solid cells that are initialized as active type. The cell undercooling is only tracked for active and liquid type cells, and only proceeding each cell's critical time step (at which the undercooling is set to 0, and decremented by the cell's undercooling change value each time step). Cells initialized as solid or active type have a grain ID value taken from the substrate input file, where each grain ID corresponds to an orientation taken from a list of 10,000 possible random orientations for the FCC Inconel 625 crystal structure.

The CA algorithm uses a series of rules associated with specific cell types to advance the system state each time step. The main set of rules relates to tracking the solidification front inside of undercooled active cells, which represent the solid–liquid interface, and the "capture" of adjacent liquid cells by grains associated with active cells. The decentered octahedron method, originally published by Gandin and Rappaz [39] and briefly summarized here, is used to model these grain growth and cell capture processes. The decentered octahedron method ensures that grain growth, regardless of grain orientation, is independent of the Cartesian grid directions. All active cells are initialized with associated octahedra, with the six-half diagonal directions of each octahedron aligned with the $\langle 001 \rangle$ directions of each cell's grain orientation. These octahedra represent the grain envelopes, containing cellular or dendritic morphologies that the present model does not explicitly simulate. All octahedra are tracked independently from one another, even those belonging to the same parent grain ID. When the time step exceeds an active cell's critical time step, the dimensionless size of said active cell's octahedron, L, is increased by an increment

$$\Delta L = \frac{\Delta t}{\Delta x}\left(A(\Delta T)^3 + B(\Delta T)^2 + C(\Delta T)\right). \tag{2}$$

each time step based on the local undercooling in the cell, ΔT. This interfacial response function links the local cell undercooling to the dimensionless dendrite growth velocity (in turn dependent on subscale phenomena such as solute diffusion and interfacial energy) and the expansion of the grain envelope. The polynomial form of the interfacial response function shown in Equation (2) and the associated constants A, B, and C are taken from a previously modeled nickel-based alloy [41], which in turn was chosen based on the interfacial response function calculated for pure nickel [67] and adjusted to match experimental grain structure data. As each active cell's octahedron grows, any liquid cell adjacent to the active cell may become "captured" when the octahedron engulfs the liquid cell center. The liquid cell then becomes an active cell, becomes a member of the original active cell's grain, and is given its own octahedron with the same orientation as the original active cell's octahedron. The placement of octahedra centers, the initial size of new octahedra, and the geometry calculations involved in the cell capture process are provided in the original work on the decentered octahedron algorithm [39]. The max size of ΔL in Equation (2) is capped at 0.05 to avoid octahedron growth so rapid that the grain growth becomes dependent on the code's iteration order through active cells. When an active cell no longer has any liquid neighbors, it becomes a solid cell, and its octahedron is no longer tracked. This process is repeated until all liquid cells associated with a given layer ID have been converted to active or solid cells, after which simulation of the next layer ID begins.

Heterogeneous nucleation in the CA model is governed by a nucleation density N_0, a mean nucleation undercooling ΔT_N, and a standard deviation of the nucleation undercooling ΔT_σ. This is performed in a slightly simplified manner as in the literature due to the simplifying assumptions of [66]: it is known a priori which cells will undergo solidification, and that each of these cells will only solidify once (intermediate remelting and solidification events are neglected). At the beginning of the simulation, a random number $R_{nucleation}$ between 0 and 1 is generated for each liquid cell. If the condition

$$R_{nucleation} < N_0 (\Delta x)^3 \tag{3}$$

is satisfied, the liquid cell is considered a potential nucleus site. Each potential nucleus is randomly assigned a nucleation undercooling value, taken from a Gaussian distribution with a mean of ΔT_N and standard deviation of ΔT_σ. The potential nucleus will either become extinct if the liquid cell is captured by an adjacent active cell prior to reaching its assigned nucleation undercooling, or successfully become a nucleus if it remains liquid type upon reaching its assigned nucleation undercooling. If the nucleation event is successful, the cell becomes a new active cell, becomes part of a new grain, and is assigned an octahedron which will grow via the previously discussed decentered octahedron algorithm.

The values used for all CA model parameters are given in Table 2. These are used unless otherwise mentioned—notably, the heterogeneous nucleation density N_0 and mean substrate grain diameter S_0 will later be varied from their default values to examine their roles on microstructure development. As a brief aside, the CA cell size of 1.666 μm is one third of the OpenFOAM data resolution of 5 μm. A previous study using OpenFOAM data at 5 μm resolution and various ExaCA Δx using similar scan parameters found that this was the largest cell size able to achieve reasonably converged grain shape results [66]. For a general set of scan parameters, CA cell size and temperature field resolution required to fully resolve texture are unknown; however, cell sizes on the order of microns generally appear able to capture texture evolution in AM melt pool solidification [68]. Cell sizes on the order of microns are also more than an order of magnitude smaller than most AM part feature sizes (for example, the thin open cellular structure shown by [2] has internal support bars on the order of 10 s of microns), which should allow future work modeling grain structure in fine geometries in addition to bulk parts.

Table 2. Default values for CA model parameters considered in the present work.

Parameter	Symbol	Value
Cell size	Δx	1.666 μm
Time step	Δt	0.083 μs
Interfacial response fitting parameter 3rd order	A	-1.0302×10^{-7} m/(s · K^3)
Interfacial response fitting parameter 2nd order	B	1.0533×10^{-4} m/(s · K^2)
Interfacial response fitting parameter 1st order	C	2.2196×10^{-3} m/(s · K)
Heterogeneous nucleation density	N_0	10^{13}/m^3
Mean nucleation undercooling	ΔT_N	5 K
Standard deviation of nucleation undercooling	ΔT_σ	0.5 K
Mean substrate grain diameter	S_0	45 μm
Offset in build direction for new layers	None	20 μm
Number of layers simulated	None	65

3.3. CA Data Analysis

To quantify the change in grain structure as a function of distance from the domain's bottom surface (i.e., how the grain structure changes as a function of deposited layer), we introduce two parameters: a mean grain cross-sectional area ς, and a mean weighted grain cross-sectional area $\bar{\varsigma}$. Using the entire domain cross-section for calculation of these parameters would skew the means via inclusion of substrate grains at and beyond the lateral bounds of the melted region, so we instead define an area in X and Y that is expected to be representative of the grain structure far from the scan edges. In the case of the even and odd layer scan patterns simulated through OpenFOAM, we select the 0.5 by 0.5 mm region (area of 0.25 mm^2, see yellow outline in Figure 2d) furthest from the boundaries for analysis. With a cell size of 1.666 μm, this region is 300 cells wide in the X and Y directions. At a given Z coordinate within this representative area, the mean grain cross-sectional area is given by

$$\varsigma = \frac{0.25 \text{ mm}^2}{N_G},$$ (4)

where N_G is the number of unique grain ID values represented at the Z coordinate of interest. As many AM microstructures are dominated by a small number of grains with large cross-sectional areas, using a simple average of grain cross-sectional areas can skew small if many small grains are present. We introduce a weighted mean grain cross-sectional area for this reason, defined as

$$\bar{\varsigma} = \frac{1}{0.25 \text{ mm}^2} \sum_{i=1}^{i=N_G} A_i{}^2.$$ (5)

In Equation (5), the "weights" are the grain areas themselves, as A_i is the cross-sectional area of a grain with grain ID "i". Multiplying each A_i by itself in Equation (5) in turn weights the average toward the areas of the largest grains. Paraview images of the resulting grain structure provide a qualitative way to examine ExaCA results. Two different colormaps are used in these Paraview images, one for grains that were originally part of the substrate, and one for grains that formed via heterogeneous nucleation events. The colormaps describe grains' misorientation relative to the positive Z (build) direction, that is, the difference in angle between the positive Z direction and the closest aligned <001> direction for the grain orientation, which can range from 0 to 62.8 degrees based on crystal geometry [69].

Regions of grain structure are printed for use by ExaConstit in property calculations, in a format listing cell X, Y, Z, and grain ID values (with a corresponding file mapping each grain ID to one of the possible 10,000 random orientations). These representative microstructure regions are 300 by 300 by 300 CA cells in size, using the same 300 by 300 area in X and Y from the mean grain area calculations. Two separate 300 cell regions in the Z direction are selected, a "lower" region consisting of Z coordinates 170 through 469 (roughly corresponding to layers 15 through 39), and an "upper" region consisting of Z coordinates 470 through 769 (roughly corresponding to layers 40 through 64). The microstructure from the first 14 layers is excluded from these representative volumes as it is expected to skew towards the random texture and grain size of the substrate, and the final (65th) layer is excluded as the final layer of a build is well-known as being unrepresentative of the grain structure as a whole. In addition to comparing properties of grain structures with different S_0 and N_0, upper and lower regions are compared to quantify property variation in different spatial locations of the build.

3.4. Crystal Plasticity Model

In order to run the crystal plasticity simulations, we rely on three open source libraries. The crystal plasticity constitutive response is provided by the ExaCMech library [70] and then fed into ExaConstit to solve for the bodies response to an applied load. ExaConstit [26] is a general quasistatic nonlinear solid mechanics velocity-based finite element application built on the MFEM framework [71]. ExaConstit solves for the balance of linear momentum as given in Equation (6) using typical solid mechanics FEM discritizations of the weak form for nonlinear materials.

$$\nabla \cdot \sigma = 0 \tag{6}$$

ExaCMech provides a number of different elastoviscoplastic crystal plasticity models available to the user. These models all employ large strain kinematics with varying crystal-scale constitutive responses for the elastic and plastic response. A detailed description of these sorts of models can found in [72,73]. A brief description of the kinematics and constitutive response of the material will be described here. Starting from the velocity gradient, $\mathbf{L}$, we can decompose the kinematic response into the following:

$$\mathbf{L} = \dot{\mathbf{F}}\mathbf{F}^{-1} = \dot{\mathbf{V}}^e\mathbf{V}^{e-1} + \mathbf{V}^e\hat{\mathbf{L}}^p\mathbf{V}^{e-1} \tag{7}$$

where $\mathbf{F}$ is the deformation gradient, $\mathbf{V}^e$ is the elastic left stretch tensor, and $\hat{\mathbf{L}}^p$ is defined as:

$$\hat{\mathbf{L}}^p = \dot{\mathbf{R}}^*\mathbf{R}^{*T} + \mathbf{R}^*\dot{\mathbf{F}}^p\mathbf{F}^{p-1}\mathbf{R}^{*T} \tag{8}$$

where $\mathbf{R}^*$ is the lattice rotation and $\mathbf{F}^p$ is the plastic deformation gradient. The plastic velocity gradient, $\bar{\mathbf{L}}^p = \dot{\mathbf{F}}^p\mathbf{F}^{p-1}$, appearing in Equation (8) is directly related to the combined macroscopic shearing of the slip systems:

$$\bar{\mathbf{L}}^p = \sum_{\alpha=1}^{n} \dot{\gamma}^\alpha \bar{\mathbf{s}}^\alpha \otimes \bar{\mathbf{m}}^\alpha \tag{9}$$

where $\dot{\gamma}^\alpha$ is the plastic shearing rate on the α slip system and $\bar{\mathbf{s}}^\alpha \otimes \bar{\mathbf{m}}^\alpha$ is the Schmid tensor defined in the crystal reference frame. Our plastic shearing rate is described by our choice of models for our slip kinetics and hardening models. For this study, we make use of a power law slip kinetics formulation with a Voce hardening model. Our slip kinetics takes on the following form:

$$\dot{\gamma}^\alpha = \dot{\gamma}_0 \left(\frac{\tau^\alpha}{g^\alpha}\right)^{\frac{1}{m}} \mathrm{sgn}(\tau^\alpha) \tag{10}$$

where $\dot{\gamma}_0$ is a reference shearing rate, τ^α is a resolved shear stress on the α slip system, g^α is the critically resolved shear strength on the α slip system, and m is the rate sensitivity of slip. The resolved shear stress can be computed as the following:

$$\tau^\alpha = \sigma : (\hat{\mathbf{s}}^\alpha \otimes \hat{\mathbf{m}}^\alpha)_{sym} \tag{11}$$

where σ is the Cauchy stress. We describe g^α using a Voce hardening model as defined below:

$$\dot{g}^{\alpha} = h_0 \left(\frac{g_{sat} - g^{\alpha}}{g_{sat} - g_0^{\alpha}} \right)^{n_{slip}} \sum_{\beta=1} \dot{\gamma}_{\beta} \tag{12}$$

where h_0 is the strength hardening coefficient, g_0 is the initial critically resolved shear strength, and g_{sat} is the saturation critically resolved shear strength.

The crystal elastic moduli for the Inconel 625 used in the simulations are listed in Table 3 along with the slip system parameters. The elastic moduli were taken from [74]. The choice of elastoviscoplastic model parameters are based upon stress–strain data obtained from the stress-relieved AM Inconel 625 in [75].

Table 3. Single crystal elastic constants (Inconel 625) employed in simulation, along with the elasto-viscoplastic model constants used for Inconel 625.

Parameter	Value
C_{11}	243.3 GPa
C_{12}	156.7 GPa
C_{44}	117.8 GPa
h_0	791.0 MPa
g_0	328.5 MPa
g_{sat}	788.0 MPa
m	0.03
m'	0.0
$\dot{\gamma}_0$	$1.0\,\mathrm{s}^{-1}$
$\dot{\gamma}_{s0}$	$5 \times 10^{10}\,\mathrm{s}^{-1}$

4. Results

4.1. Baseline Uncertainty in CA Microstructures

To quantify the variation in grain structure as a function of the number of deposited layers, a baseline for the spread in grain area statistics based on the two significant sources of "randomness" in ExaCA needs to be established. First, we set the mean substrate grain diameter S_0 and heterogeneous nucleation density N_0 to the default values from Table 2. The first source of randomness is the random number generator seed used to generate substrate grain orientations and locations, i.e., the seed used in generation of $R_{substrate}$ values used in the comparison from Equation (1). The second source of uncertainty is the random number generator seed used to generate heterogeneous nuclei locations, i.e., $R_{nucleation}$ values used in the comparison from Equation (3), and their associated activation undercooling values. Three different random number generator seeds are used for each of these two sources of uncertainty, generating three statistically equivalent substrates and three statistically equivalent sets of nucleation sites. The nine possible permutations of these seeds are used in simulation initialization, and the resulting mean grain cross-sectional area (ς) and weighed mean cross-sectional area ($\bar{\varsigma}_W$) as a function of build height are shown in Figures 3 and 4, respectively. As the curves in Figure 3 are similar and have a degree of noise, ς values are plotted as a moving average with a period of 25 data points (41.666 μm) for curve visibility.

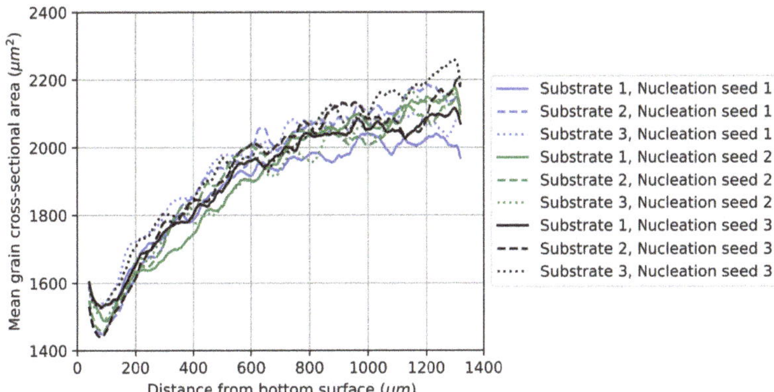

Figure 3. Mean grain cross-sectional area (ς) as a function of build height for the various permutations of random number generation seed for substrate and grain nucleation. ς values shown are moving averages with a period of 25 data points.

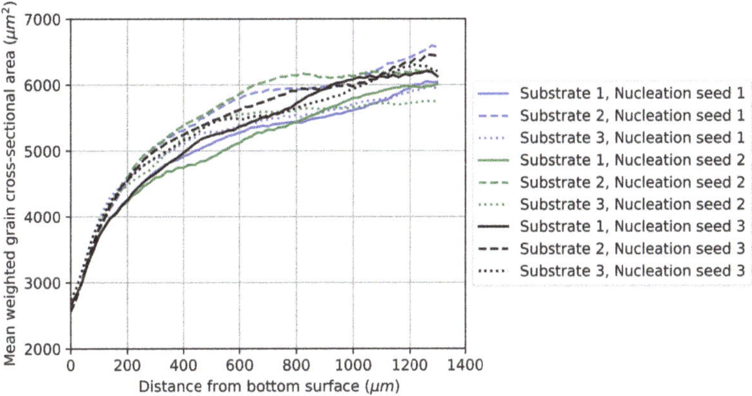

Figure 4. Mean weighted grain cross-sectional area ($\bar{\varsigma}$) as a function of build height for the various permutations of random number generation seed for substrate and grain nucleation.

For all substrates and nucleation seeds, three regimes are seen in Figure 3 as a function of distance from the bottom surface: a sharp drop in ς, a steady and slowly flattening increase in ς, and a final sharp drop in ς (partially obscured as ς is plotted as a moving average). The two drops are related to nucleation of grains during simulation of the first few layers and the final layer, while the increase is related to the quasisteady-state nucleation and impingement of less favorably oriented grains occurring during the successive simulation of each new layer. Unlike the plot of ς, the plot of $\bar{\varsigma}$ in Figure 4 does not show the initial transient nor final transient decrease—by weighting the largest grains more heavily in the average, a smoother, less noisy description of the dominant microstructural features is obtained. The increases in ς and $\bar{\varsigma}$ flattened off significantly after the first 600 μm (30 layers or so), and even more after the first 1000 μm, though it is not clear that the grain areas have reached a steady state with respect to build height even after all 65 layers. Additionally, the increase in $\bar{\varsigma}$ toward the steady state in Figure 4 appears to be happening more slowly than the corresponding $\varsigma(z)$ curves in Figure 3.

The spread in ς due to random substrate and nucleation seed increases as a function of build height through the first 250 μm of solidification, beyond which there is approximately a ± 150 μm^2 or $\pm 7\%$ variation from the mean across all substrates and nucleation seeds.

The spread in ξ due to random substrate and nucleation seed increases as a function of build height through the first 500 µm of solidification, beyond which there is approximately a ± 500 µm^2 or around $\pm 8\%$ variation from the mean value. Neither nucleation seed nor substrate seems to be the dominant factor in the randomness of the grain area data: we do not see clustering of colors (which would indicate that substrate controls the scatter), nor do we see clustering of line types (which would indicate that nucleation seed controls the scatter).

4.2. CA Predictions with Varied Mean Substrate Grain Diameter

Now that the development of ς and ξ as a function of build height and random seeds have been quantified, we examine the effect of varied mean substrate grain size S_0 on microstructure far from the substrate. We generate five substrates with S_0 values ranging from 25 to 65 µm as reasonable guesses for AM baseplate grain sizes. ς and ξ as a function of distance from the bottom build surface starting from these varied initial conditions are shown in Figures 5 and 6, respectively.

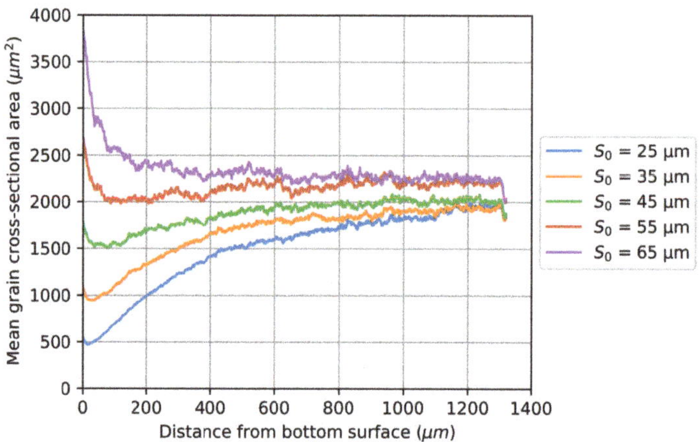

Figure 5. Mean grain cross-sectional area (ς) as a function of build height with various mean substrate grain diameters S_0.

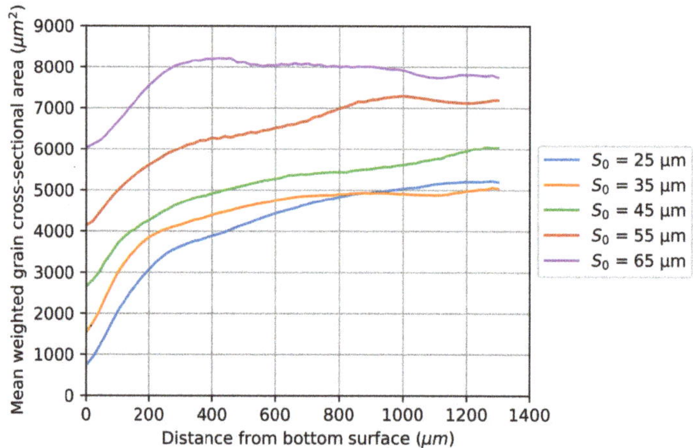

Figure 6. Mean weighted grain cross-sectional area (ξ) as a function of build height with various mean substrate grain diameters S_0.

As shown in Figure 5, after approximately 600 μm (20–25 layers), ς for all substrates is within 20% of its "final" (after all 65 layers) value. After 65 layers of simulation for all substrates, there is a mean grain area spread of around 250 μm^2, a spread similar to the scatter in ς induced through random number variation in the model from Section 4.1. For the three smaller S_0 values, ς shows more of a true convergence, as the curves for $\varsigma(z)$ with $S_0 < 55$ μm are mostly indistinguishable after 65 layers. However, the $S_0 = 55$ and 65 μm curves remain distinct from the $S_0 = 25$, 35, and 45 μm curves. In Figure 6, the $\bar{\varsigma}$ curves show a less clear trend in convergence as a function of distance from the bottom surface, for all substrate grain diameters but particularly $S_0 = 55$ and 65 μm. After 65 layers of simulation, a $\bar{\varsigma}$ spread of around 3000 μm^2 is present as a function of S_0—a spread more significant than that induced through statistical variation in substrate/nucleation random seed alone (which was closer to 500 μm^2 as shown in Section 4.1).

The difference in mean grain area and mean weighted grain area development among small and large S_0 simulations can be rationalized as a result of the two types of competition: that among the baseplate grains, and that between baseplate and nucleated grains. For small S_0, favorably oriented grains impinge less favorably oriented baseplate grains relatively quickly during the initial transient layers. However, with large S_0, there are fewer baseplate grains and impingement is less likely. For large S_0, grains unfavorably oriented with the local thermal gradient direction are therefore more likely to advance due to this lack of competition. In this case, nucleation and growth of new grains (some of which will have more favorable orientations than adjacent epitaxial grains) is the only other mechanism for impinging the less favorably oriented grains. This process is more likely with large S_0, as advancing baseplate grains with unfavorable orientation relative to the local thermal gradient direction require larger undercooling to advance in said direction. This larger undercooling provides the opportunity for nucleated grains to grow despite the large thermal gradient magnitude that would normally suppress nucleated grain growth. However, this process of baseplate grain impingement via growing nucleated grains takes many more layers to occur as heterogeneous nucleation of more favorably oriented grains is relatively rare, and the large magnitude of G present at the melt pool boundary still tends to suppress growth of nucleated grains. This explains why the decrease in $\varsigma(z)$ and $\bar{\varsigma}(z)$ for $S_0 = 55$ and 65 μm in Figures 5 and 6 toward the other curves is much more gradual than the initial transient increase in $\varsigma(z)$ and $\bar{\varsigma}(z)$ seen for the $S_0 = 25$, 35, and 45 μm curves.

This conclusion is supported by Figure 7, which plots the interior region (away from the scan edges) of the simulated grain structure colored by grain source (baseplate or nucleation) and grain misorientation. It is shown that while high aspect ratio columnar grains of relatively similar area are present for $S_0 = 25$ and 65 μm simulations, the larger S_0 simulation volume appears to have more nucleated grains (shades of red), particularly toward the top surface. Figure 8 also supports this conclusion; it shows the 001 pole figures using data from the top halves of the volumes from Figure 7 plotted using the MTEX toolbox. For all S_0, a reasonably strong (2–3x a uniform random distribution) 001 texture is observed, along with a secondary near-110 texture. However, these also show that the 001 texture is slightly weaker as S_0 is increased (lighter shades of green/yellow), and the regions of the pole figure corresponding to unfavorable orientations show slightly increased intensity (lighter shades of gray). This likely corresponds to both unfavorably oriented baseplate grains as well as nucleated grains, as unfavorably oriented baseplate grains as well as nucleated grains (which will have random orientations, though the more favorably oriented nucleated grains would be expected to grow to the largest sizes) lead to a slight decrease in the overall texture. We also note that the tall grains extending through many layers observed in Figure 7 and the 001 textures observed in Figure 8 are commonly seen in simulated and experimental AM of Inconel alloys [18,21,44,60]. While the material and scan parameters used in the present simulation were not designed to reproduce a specific result from experiment, the similarity of the simulated result with other simulations and experimental grain structures validates the present set of simulated process and solidification model parameters as relevant to real AM processing.

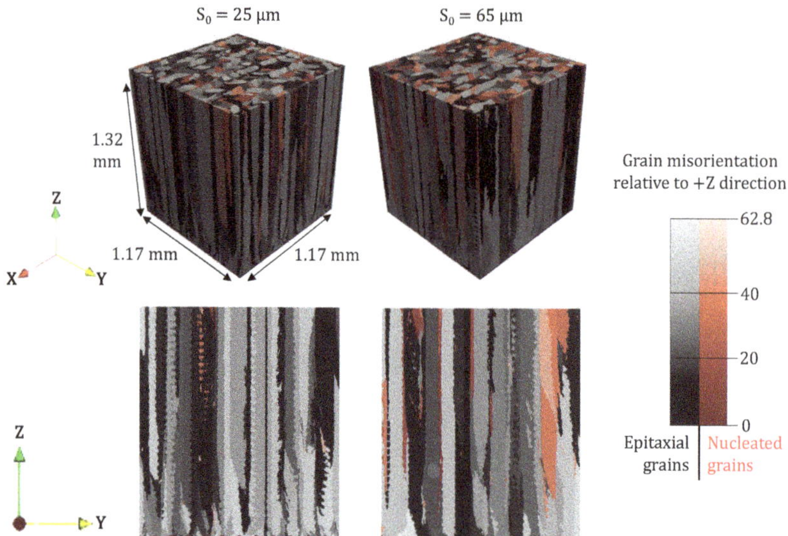

Figure 7. Grain structure variation from a rectilinear volume far from the scan edges using the lower and upper extremes in mean baseplate grain diameter (S_0 = 25 and 65 μm), colored by grain misorientation relative to the build direction and grain source (epitaxial grains from the baseplate versus nucleated grains).

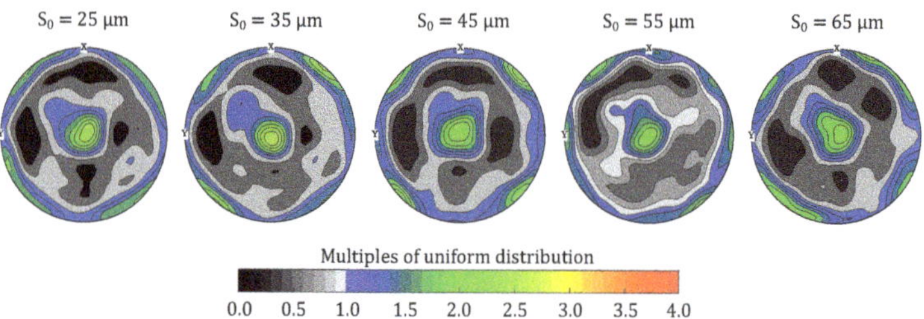

Figure 8. 001 pole figures from the upper halves of the simulation volumes from Figure 7, with varied mean substrate grain diameter S_0.

4.3. CA Predictions with Varied Nucleation Density

The development of grain structure induced through varying heterogeneous nucleation density N_0 was calculated at multiple S_0 values to compare the relative importance of N_0 and S_0. For S_0 values of 25, 45, and 65 μm, $\xi(z)$ was calculated using $N_0 = 10^{12}$ m^{-3} and 10^{14} m^{-3}, and compared to a the $N_0 = 10^{13}$ m^{-3} results from Section 4.2. This is plotted in Figure 9, where different line colors are used for different N_0 and different line markers are used for different S_0. Purely epitaxial growth ($N_0 = 0$) for the intermediate substrate grain size $S_0 = 45$ μm is also plotted alongside results for the other permutations of N_0 and S_0.

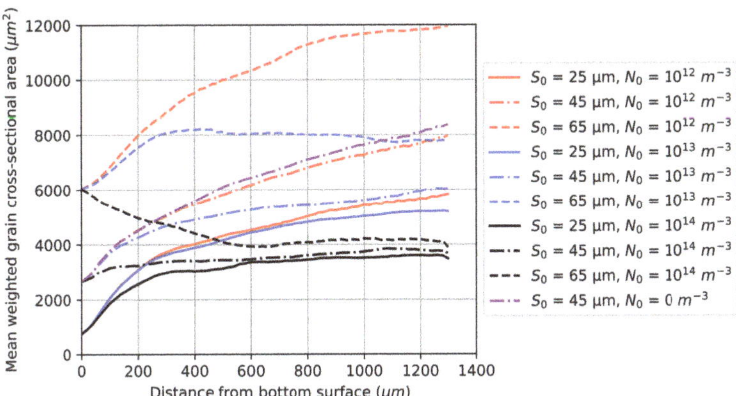

Figure 9. Mean weighted grain cross-sectional area ($\bar{\varsigma}$) as a function of build height with varied nucleation density N_0 (including a baseline with no nucleation) and mean substrate grain diameter S_0.

Convergence of $\bar{\varsigma}$ curves with varied S_0 is strongly dependent on N_0. The largest N_0 in Figure 9 (black) shows $\bar{\varsigma}$ with different substrates to be within 1000 μm^2 after 500 μm of build, while the smallest N_0 (red) shows little to no convergence of $\bar{\varsigma}$ (each substrate grain diameter's curve "flattens out", but at different values $\bar{\varsigma}$ values). The curve corresponding to purely epitaxial growth (purple dashes) is similar to the curve corresponding to the smallest N_0 of 10^{12} m^{-3} (red dashes). Given that nearly entirely epitaxial solidification is unlikely during AM processing (otherwise, we would see grains spanning the entirety of parts), this nucleation density is likely on the small side. The slower convergence of $\bar{\varsigma}$ curves at smaller N_0 and faster convergence of $\bar{\varsigma}$ curves at larger N_0 is expected, as reaching a value of $\bar{\varsigma}$ independent of distance from the bottom surface depends on the balance of nucleation and impingement. Impingement is a slow process of baseplate grains blocking other baseplate grains as additional layers are deposited, while nucleation occurs in every layer and introduces new grains. The more new grains are introduced, the fewer layers are needed to reach the quasisteady-state balance of nucleation and impingement. The extreme condition with zero nucleation would be expected to take a large number of layers to reach the point where all impingement has occurred and $\bar{\varsigma}$ is independent of additional layer deposition, while the extreme condition with infinite nucleation would be expected to reach this point after a single layer, as all cells would be home to their own grain. As might be expected, $\bar{\varsigma}$ as a function of build height has a stronger dependence on the substrate grain structure when N_0 is small, as impingement of substrate grains via nucleation becomes less common.

Figure 10 shows sections of the grain structures for each N_0 value at the fixed S_0 of 45 μm. As expected based on the minimal difference in Figure 9 between purely epitaxial growth (purple dashes) and the nucleation density of 10^{12} m^{-3} (red dashes) for the 45 μm substrate diameter, the 10^{12} m^{-3} grain structure is almost entirely epitaxial grains, with a small handful of large nucleated grains. The 10^{14} m^{-3} grain structure, after the first 0.5 mm or so of the build, is around 50% nucleated grains. Figure 11 shows the 001 and 110 pole figures for the upper halves of the simulation volumes shown in Figure 10, showing that while the overall 001 texture is present in approximately equal magnitudes for all N_0, a stronger competing 110 texture is present in the $N_0 = 10^{14}$ m^{-3} case. This is confirmed by Figure 10, as the largest nucleated grains tended to have misorientation angles relative to the build direction of around 45 degrees (i.e., 110 grain orientations); these 110 grains have tall, columnar shapes similar to the baseplate grains. It is possible that the 110 orientations favored by the largest nucleated grains are related to the local thermal gradient direction in regions of the melt pool favorable for nucleation and growth. The overall grain shape appeared relatively independent on N_0 as it did with S_0, with more narrow grain areas the primary difference as N_0 was increased.

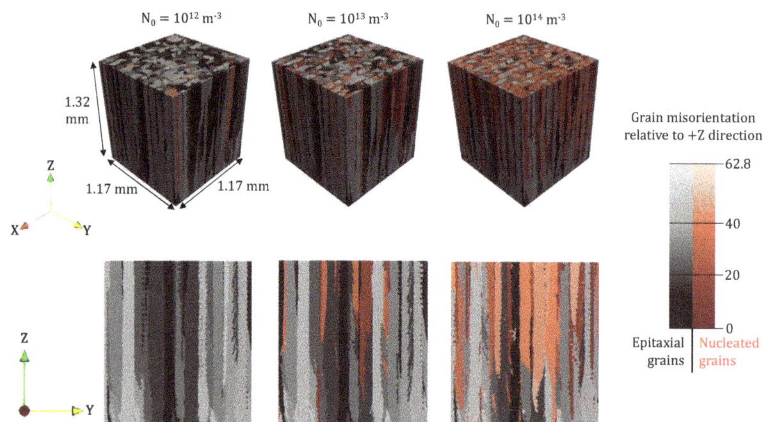

Figure 10. Grain structure variation from a representative area (far from the scan volume edges) using a fixed mean baseplate grain diameter $S_0 = 45$ μm, and three values for heterogeneous nucleation density N_0. Grains are colored by grain misorientation relative to the build direction and grain source (epitaxial grains from the baseplate versus nucleated grains).

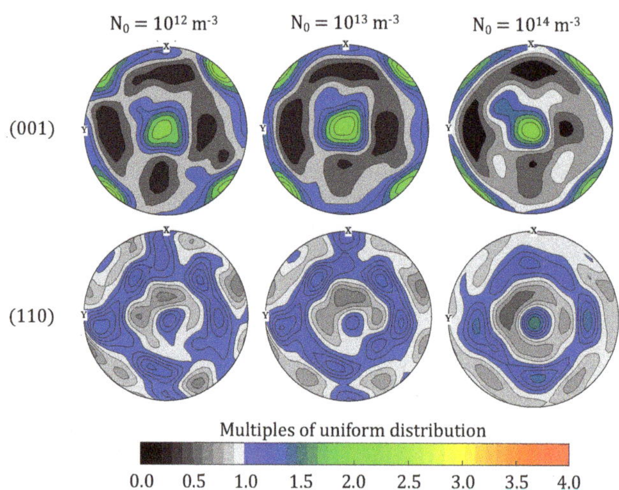

Figure 11. 001 and 110 pole figures from the upper halves of the simulation volumes from Figure 10, with varied heterogeneous nucleation density N_0.

4.4. ExaConstit Model of Constitutive Properties with Varied Mean Substrate Grain Diameter and Nucleation Density

Representative volume elements (RVEs) from the ExaCA simulations using $N_0 = 10^{12}$ and 10^{13} m^{-3} and $S_0 = 25$, 45, and 65 μm were passed to ExaConstit to model their macroscopic behaviors for part scale predictions. The macroscopic stress–strain response and distribution of triaxiality will be examined in this study as they are of use in fitting macroscopic models for part scale predictions [76–82]. As discussed in Section 3.3, these RVEs consisted of cubes with 300 cells (0.5 μm of material) per side, and two RVEs were extracted from each ExaCA simulation: an "upper" RVE corresponding to the top 24 layers of deposited material (excluding the final layer), and a "lower" RVE corresponding to the 24 layers of material below the upper sample. Given that ς and ζ evolution clearly depended on N_0 and S_0, crystallographic texture development as a function of build height may not be the same. This difference in texture would be expected to result in differences in ExaConstit's macroscopic property predictions, particularly when comparing the lower (more influenced by the baseplate) and upper (less influenced by the baseplate) RVEs.

Simulations were carried out under uniaxial tension at $1 \times 10^{-3}\ \mathrm{s}^{-1}$ strain rates out to 5% strain. Figure 12 shows ExaConstit's prediction for the stress–strain behavior of the various permutations of S_0, N_0, and RVE location in the build. As expected, the elastic response and plastic response of these RVEs are largely similar given similar grain morphologies and textures of each RVE. However, we do see differences in the yield stress and later responses in the fully-developed plastic flow areas of the curve. For example, the upper $S_0 = 25\ \mu\mathrm{m}$, $N_0 = 10^{12}\ \mathrm{m}^{-3}$ curve has the lowest yield stress and macroscopic stress response of the various RVEs. The lower $S_0 = 65\ \mu\mathrm{m}$, $N_0 = 10^{12}\ \mathrm{m}^{-3}$ RVE is on the other end of things and has the highest yield stress. To better compare differences in the response, we make use of a heat map to compare the relative percent difference between yield stresses of all samples. The percent differences in yield stress are plotted in Figure 13, and this figure shows that most samples are similar to one another. The yield stress for the upper $S_0 = 25\ \mu\mathrm{m}$, $N_0 = 10^{12}\ \mathrm{m}^{-3}$ is the biggest outlier, with a greater than a 5% difference in the response between this RVE and those using $S_0 = 65\ \mu\mathrm{m}$. For the most part, these differences are largely due to nucleation or baseplate structures having different textures, and the differences are relatively minor. Other limiting factors on the material response, such as defects or pores in the solidified material or statistical variation in builds could play a larger role in an application needs.

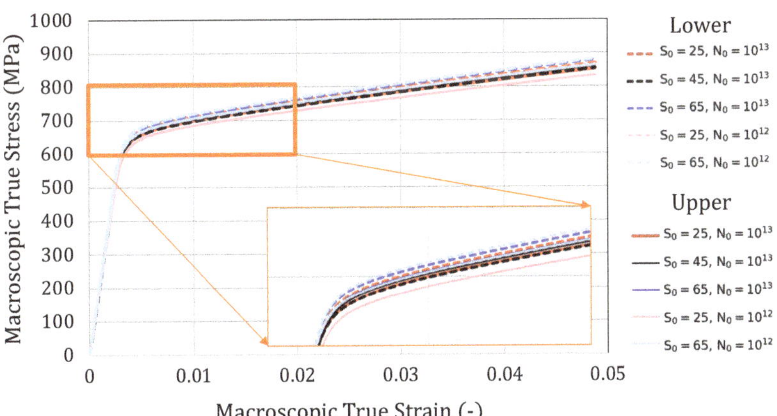

Figure 12. Macroscopic stress-strain curves as a function of applied load in the Z (build) direction for various ExaCA-simulated RVEs.

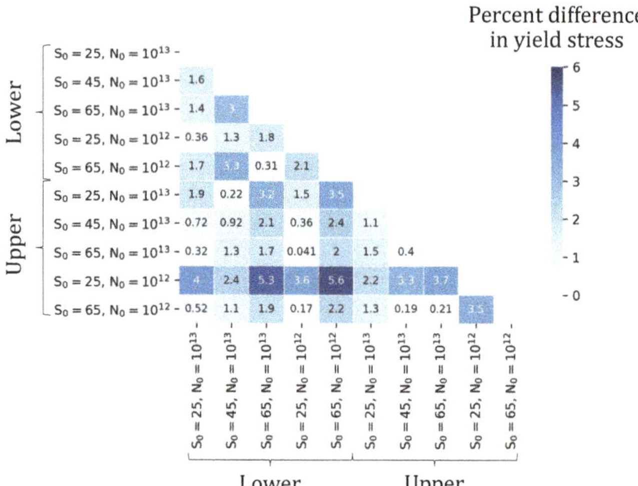

Figure 13. Relative percent differences in yield stress between various ExaCA-simulated RVEs.

To drive porosity-models at the part scale, distributions of the hydrostatic stress can be used to instantiate these models [83]. Rather than examining the hydrostatic stress, we will examine the stress triaxiality, which is defined as

$$Triaxiality = \frac{\sigma_h}{\sigma_{vm}}, \tag{13}$$

where σ_h is the hydrostatic stress and σ_{vm} is the von Mises stress. Figure 14 shows the stress triaxiality distributions for the lower and upper RVEs of varied N_0 and S_0. These distributions are often used in the mechanics community to investigate potential areas with an increased likelihood of fracture. The width and maximum height of these distributions are nearly identical for all lower and upper RVEs, though it is noted that the lower $S_0 = 65$ μm, $N_0 = 10^{12}$ m^{-3} and upper $S_0 = 65$ μm, $N_0 = 10^{13}$ m^{-3} distributions have long tails at small triaxiality. While these tails look significant on the histograms, we note that the y-axis counts are given on a log scale, and that the number of elements represented in these tails is relatively insignificant compared to the distribution peaks, which are similar for all RVEs. These tails likely consist of one or two anomalous grains, either from the baseplate or from a specific heterogeneous nucleation event.

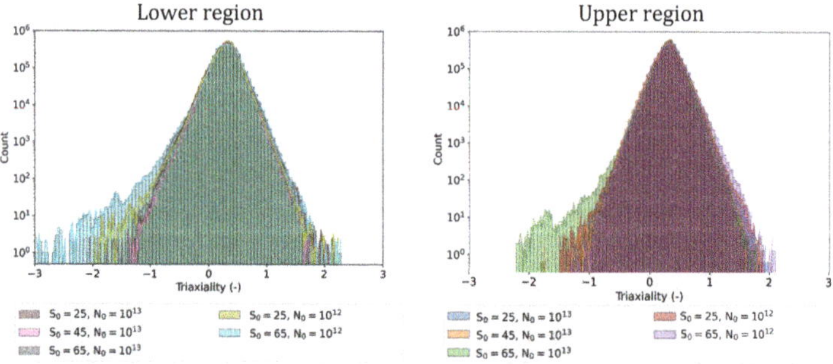

Figure 14. Stress triaxiality distributions for the lower (**left**) and upper (**right**) RVEs with varied N_0 and S_0.

4.5. CA Predictions Using Extremes in Substrate Initial Conditions

We can also examine grain structure development starting from the largest possible "extremes" in initial substrate. For the first extreme, we use a single grain as the substrate, selecting the orientation from the list of 10,000 with all axes as closely aligned with the <001> directions as possible. For the second extreme, a single grain is again used as the substrate, but the orientation with growth directions as misaligned as possible from the positive Z direction is selected from the list of 10,000 possible orientations. For the final extreme, each CA cell that is part of the substrate is assigned a different grain ID and orientation, essentially simulating a substrate grain size equivalent to the CA cell size.

Figure 15 shows the resulting grain structures after 120 layers of CA calculations. The top surfaces of the simulation domains show the grain shapes and sizes beginning to resemble those starting from the standard equiaxed substrates from Figures 7 and 10. Starting from the single crystal substrates, nucleated grains partially block regions of the substrate from advancing and growth through multiple layers, though regions of the single crystal substrates are still visible at the top surface after 120 layers. The orientation of the single substrate grain plays a role in layerwise grain structure development as well, with more nucleated grains appearing and advancing after fewer layers with the large misorientation substrate, relative to the small misorientation substrate. This difference is due to the fact that the low misorientation substrate tends to be closer aligned with the thermal gradients in more regions of the melt pool, and therefore its growth is more competitive with nucleated grains (which are less likely to be as closely aligned with the local thermal gradients). The high misorientation grain is less likely to be closely aligned with the thermal gradient direction (as evidenced by the fact that misorientations greater than 45 degrees were rare in Figures 7 and 10), and this high misorientation grain is less competitive with nucleated grains, which are therefore able to form and grow more readily. Large area nucleated grains extending through multiple layers are not as readily seen when starting from the extreme condition where each cell is its own grain. This is likely because with more substrate grains present, regions of the melt pool favorable for nucleation through large misalignment between the thermal gradient direction and substrate grain orientation are less common.

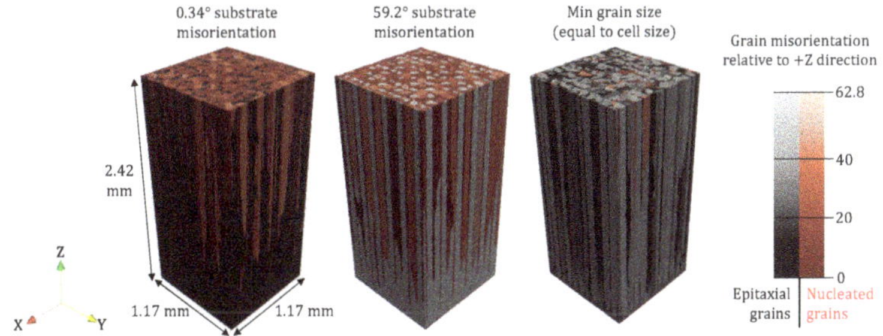

Figure 15. Images of grain misorientation from a representative region of the simulation volume starting from extremes in possible initial substrate.

The textures of these simulations are also interesting to note; Figure 16 shows the 001 pole figures using the top 30 layers of the volumes shown in Figure 15. The $S_0 = 45$ μm pole figure, reproduced from Figure 8, is also plotted for reference. For the low misorientation single grain substrate, the 001 texture is stubborn (likely due to the fact that it is strongly favored by the thermal conditions), though a weak competing 110 texture does appear through the growth of nucleated grains. For the high misorientation single crystal substrate, the 111 texture from the substrate is still visible in the pole figure, but reasonably strong 001 and 110 textures have also developed. Finally, when starting from the smallest possible grain size, 001 and 110 textures are visible with approximately the same intensity as the $S_0 = 45$ μm pole figure, but other orientations such as 210 and 111 remain present as well.

These results show that despite the relative insensitivity of texture to S_0 after 15+ layers observed in Section 4.2, reasonable initial conditions are still required to obtain a reasonable texture without the need to simulate several millimeters of build. However, we note that if these simulations were continued for thousands of layers, it is expected that even with disparate initial conditions, the microstructures far enough from the initial substrate would still be expected to become indistinguishable.

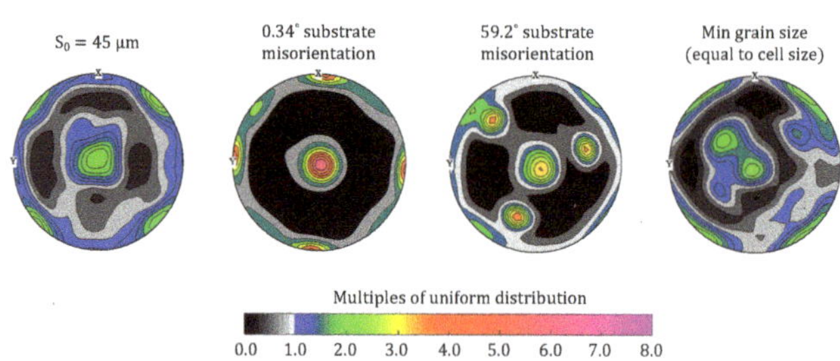

Figure 16. 001 pole figures from the top 30 layers of the simulation volumes from Figure 15, alongside the previously generated pole figure using $S_0 = 45$ μm.

5. Discussion

The coupled modeling of process, microstructure, and properties were investigated using three codes that serve as part of the ExaAM framework for self-consistent modeling of AM processes. AM raster pattern time–temperature history from the computational fluid dynamics software OpenFOAM served as input to multilayer grain structure simulations with the cellular-automata-based ExaCA code. A series of ExaCA simulations were performed, with the main takeaways as follows:

- Mean grain area (ς) as a function of build height (z) was plotted for fixed nucleation density and substrate, while varying the random number generator seeds used to generate statistically equivalent sets of nucleation and substrate data. The resulting $\varsigma(z)$ curves slowly became independent of additional layer deposition (i.e., z), with spreads in the steady-state $\varsigma(z)$ values of 7%–8% from statistical error due to the random number generation processes. The same was true for mean weighted grain area ($\bar{\varsigma}$).

- Statistical error resulting from substrate generation and nucleation data generation appeared to contribute to the spread in predicted ς and $\bar{\varsigma}$ equally. The average of the ς curves was within 10% of its 65 layer value, and the average of the $\bar{\varsigma}$ curves was within 15% of its 65 layer value, after around 500 μm of build (or about 23 layers).

- Examining ς and $\bar{\varsigma}$ curves with ±20 μm from the default mean substrate grain diameter $S_0 = 45$ μm, it was found that curves with smaller S_0 tended to converge more quickly than those at larger S_0, and reach steady-state values more quickly than those at larger S_0. The spread of these curves with $S_0 > 45$ μm after 1.2 mm of simulated build remained significant relative to the uncertainty in $\varsigma(z)$ and $\bar{\varsigma}(z)$ due to random number generation alone.

- Steady-state values of $\bar{\varsigma}$ as a function of z were much more readily reached at large N_0, regardless of S_0; at $N_0 = 10^{14}$ m^{-3}, this steady state appeared to be reached after 0.6 mm of simulated microstructure. As N_0 was reduced to 10^{12} m^{-3}, S_0 had a much larger impact on the $\bar{\varsigma}(z)$ curves, and $\bar{\varsigma}$ was still increasing as a function of z after 1.2 mm of simulated microstructure.

- While $\bar{\varsigma}$ differences due to N_0 and S_0 were seen throughout the simulated builds (simulations with smaller S_0 and larger N_0 tending to have smaller grains and reach the steady state more quickly than those with larger S_0 and smaller N_0), the simulated microstructures were qualitatively similar. The strengths of the 001 and 110 textures, as well as the tall and narrow grain shapes, were similar across all simulations, though it was noted that nucleated grains tended to be more likely than epitaxial grains to have 110 textures.

- Using RVEs from different regions of these simulated microstructures as input to ExaConstit calculations, it was found that similar stress–strain behavior and stress triaxiality distributions resulted from RVEs using various permutations of N_0 and S_0, and yield stress values were within $\pm 6\%$ of each other. Only minor differences were noted in macroscopic mechanical responses between RVEs taken from layers 15 through 39 and RVEs taken from layers 40 through 64, despite differences in ζ of up to 3x between the RVEs with the smallest and largest grain areas.

We conclude that the initial substrate during simulation of multilayer AM grain growth has a significant effect on grain size development during the first 2 mm of builds for the simulated scan pattern, particularly for small N_0 and large S_0. In turn, a CA simulation looking to obtain a microstructure representative of a structure far from the baseplate for conditions that favor epitaxial growth should err on the side of smaller S_0, particularly when N_0 is small. However, reasonable estimations for constitutive mechanical properties can be obtained from simulated grain structures with a range of possible S_0 and N_0 values. For all S_0 and N_0 values used, the grain structures and textures were sufficiently similar after 15 simulated layers such that any differences in predicted properties were on the order of the differences due to S_0 and N_0. This is despite the fact that the microstructures showed non-negligible differences in grain area for different S_0, N_0, and location in the build, suggesting that the texture and distribution of orientations in space are relatively agnostic to these parameters for the conditions tested. However, we note that when using extremes in S_0, such as an extremely small initial grain size or a single grain baseplate, this texture convergence does not readily appear, and that while texture after 15 layers may be relatively insensitive to S_0 and N_0 for the given material and scan pattern, reasonable input values are still required.

    A few caveats apply to the scope of these conclusions. The present study is limited to modeling constitutive property sensitivity to microstructure input, and the sensitivity of microstructure to various uncertain input parameters. In real AM processing, additional physics not accounted for in the present models could complicate the relationships among substrate, nucleation, microstructure as a function of build height, and location-dependent properties. Experimental verification for a given material and scan parameters would be needed to determine to what extent the conclusions drawn regarding microstructure and properties modeling are valid in practice.

    In addition to experimental validation of location-specific microstructure, future work using ExaCA includes revising the sparse data assumption from [66], allowing ExaCA analysis using time–temperature history resulting from more complex scan patterns. Generating multiple OpenFOAM input data sets to gauge the relative significance of time–temperature history uncertainty (due to energy absorption, etc) on ExaCA output, relative to the uncertainty in N_0 and S_0 examined in this study, is also a relevant next step toward consistent prediction of constitutive properties. Future science capability within ExaConstit includes varying the critically resolved shear strength as a function of distance from grain boundaries; a nonuniform resistance to plastic slip would likely yield property predictions that are more sensitive to the varying grain areas observed in the RVEs. Simulation of additional layers within ExaCA to determine the amount of material needed for grain area convergence with small N_0 and large S_0 may be needed to obtain far from baseplate grain structures, to be fed to an updated ExaConstit that is more sensitive to grain area.

Author Contributions: Conceptualization, M.R.; methodology, M.R. and R.C.; software, M.R. and R.C.; validation, M.R. and R.C.; formal analysis, M.R. and R.C.; investigation, M.R. and R.C.; resources, J.B.; data curation, M.R. and R.C.; writing—original draft preparation, M.R. and R.C.; writing—review and editing, M.R. and R.C.; visualization, M.R. and R.C.; supervision, J.B.; project administration, J.B.; funding acquisition, J.B. All authors have read and agreed to the published version of the manuscript.

Funding: This research was supported by the Exascale Computing Project (17-SC-20-SC), a collaborative effort of the U.S. DOE Office of Science and the NNSA. This manuscript has been authored by UT-Battelle, LLC under Contract No. DE-AC05-00OR22725 with the U.S. Department of Energy. The United States Government retains and the publisher, by accepting the article for publication, acknowledges that the United States Government retains a nonexclusive, paid-up, irrevocable, world-wide license to publish or reproduce the published form of this manuscript, or allow others to do so, for United States Government purposes. Work was performed in part under the auspices of Lawrence Livermore National Laboratory under contract DE-AC52-07NA27344, and supported in part by the Exascale Computing Project (17-SC-20-SC), a collaborative effort of the U.S. DOE Office of Science and the NNSA.

Institutional Review Board Statement: Not applicable.

Informed Consent Statement: Not applicable.

Data Availability Statement: The Department of Energy will provide public access to these results of federally sponsored research in accordance with the DOE Public Access Plan: http://energy.gov/downloads/doe-public-access-plan (accessed on 7 February 2022).

Acknowledgments: The authors would like to thank Gerry Knapp for generating the OpenFOAM thermal data used to drive the ExaCA grain structure simulations, as well as Adam Creuziger for his help using MTEX to generate the pole figures shown in the present work.

Conflicts of Interest: The authors declare no conflict of interest.

References

1. Zhang, W.; Lei, Y.; Meng, W.; Ma, Q.; Yin, X.; Guo, L. Effect of deposition sequence on microstructure and properties of 316L and Inconel 625 bimetallic structure by wire arc additive manufacturing. *J. Mater. Eng. Perform.* **2021**, *30*, 8972–8983. [CrossRef]
2. Calleja-Ochoa, A.; Barrio-Gonzalez, H.; López de Lacalle, N.; Martínez, S.; Albizuri, J.; Lamikiz, A. A new approach in the design of microstructured ultralight components to achieve maximum functional performance. *Materials* **2021**, *14*, 1588. [CrossRef] [PubMed]
3. Collins, P.; Brice, D.; Samimi, P.; Ghamarian, I.; Fraser, H. Microstructural Control of Additively Manufactured Metallic Materials. *Annu. Rev. Mater. Res.* **2016**, *46*, 63–91. [CrossRef]
4. DebRoy, T.; Wei, H.; Zuback, J.; Mukherjee, T.; Elmer, J.; Milewski, J.; Bese, A.M.; Wilson-Heid, A.; De, A.; Zhang, W. Additive manufacturing of metallic components—Process, structure and properties. *Prog. Mater. Sci.* **2018**, *92*, 112–224. [CrossRef]
5. Zhang, X.; Yocom, C.J.; Mao, B.; Liao, Y. Microstructure evolution during selective laser melting of metallic materials: A review. *J. Laser Appl.* **2019**, *31*, 031201. [CrossRef]
6. Oliveira, J.; LaLonde, A.; Ma, J. Processing parameters in laser powder bed fusion metal additive manufacturing. *Mater. Des.* **2020**, *193*, 108762. [CrossRef]
7. Sing, S.; Huang, S.; Goh, G.; Goh, G.; Tey, C.; Tan, J.; Yeong, W. Emerging metallic systems for additive manufacturing: In-situ alloying and multi-metal processing in laser powder bed fusion. *Prog. Mater. Sci.* **2021**, *119*, 100795. [CrossRef]
8. Wan, H.; Zhou, Z.; Li, C.; Chen, G.; Zhang, G. Effect of scanning strategy on grain structure and crystallographic texture of Inconel 718 processed by selective laser melting. *J. Mater. Sci. Technol.* **2018**, *34*, 1799–1804. [CrossRef]
9. Roehling, T.T.; Shi, R.; Khairallah, S.A.; Roehling, J.D.; Guss, G.M.; McKeown, J.T.; Matthews, M.J. Controlling grain nucleation and morphology by laser beam shaping in metal additive manufacturing. *Mater. Des.* **2020**, *195*, 109071. [CrossRef]
10. Lu, Y.; Wu, S.; Gan, Y.; Huang, T.; Yang, C.; Junjie, L.; Lin, J. Study on the microstruture, mechanical property and residual stress of SLM Inconel-718 alloy manufactured by different island scanning strategy. *Opt. Laser Technol.* **2015**, *75*, 197–206. [CrossRef]
11. Raghavan, N.; Simunovic, S.; Dehoff, R.; Plotkowski, A.; Turner, J.; Kirka, M.; Babu, S. Localized melt-scan strategy for site specific control of grain size and primary dendrite arm spacing in electron beam additive manufacturing. *Acta Mater.* **2017**, *140*, 375–387. [CrossRef]
12. Knapp, G.; Raghavan, N.; Plotkowski, A.; DebRoy, T. Experiments and simulations on solidification microstructure for Inconel 718 in powder bed fusion electron beam additive manufacturing. *Addit. Manuf.* **2019**, *25*, 511–521. [CrossRef]
13. Ghayoor, M.; Lee, K.; He, Y.; Chang, C.h.; Paul, B.K.; Pasebani, S. Selective laser melting of 304L stainless steel: Role of volumetric energy density on the microstructure, texture and mechanical properties. *Addit. Manuf.* **2020**, *32*, 101011. [CrossRef]
14. Cheng, M.; Xiao, X.; Luo, G.; Song, L. Integrating control of molten pool technology and solidification texture by adjusting pulse duration in laser additive manufacturing of Inconel 718. *Opt. Laser Technol.* **2021**, *142*, 107137. [CrossRef]
15. Wang, Y.; Chen, X.; Shen, Q.; Su, C.; Zhang, Y.; Jayalakshmi, S.; Singh, R.A. Effect of magnetic field on the microstructure and mechanical properties of inconel 625 superalloy fabricated by wire arc additive manufacturing. *J. Manuf. Process.* **2021**, *64*, 10–19. [CrossRef]
16. Todaro, C.; Easton, M.; Qiu, D.; Zhang, D.; Bermingham, M.; Lui, E.; Brandt, M.; StJohn, D.; Qian, M. Grain structure control during metal 3D printing by high-intensity ultrasound. *Nat. Commun.* **2020**, *11*, 142. [CrossRef]

17. Ma, Q.; Chen, H.; Ren, N.; Zhang, Y.; Hu, L.; Meng, W.; Yin, X. Effects of ultrasonic vibration on microstructure, mechanical properties, and fracture mode of Inconel 625 parts fabricated by cold metal transfer arc additive manufacturing. *J. Mater. Eng. Perform.* **2021**, *30*, 6808–6820. [CrossRef]
18. Pérez-Ruiz, J.D.; Martin, F.; Martínez, S.; Lamikiz, A.; Urbaikain, G.; López de Lacalle, L.N. Stiffening near-net-shape functional parts of Inconel 718 LPBF considering material anisotropy and subsequent machining issues. *Mech. Syst. Process.* **2022**, *168*, 108675. [CrossRef]
19. Köhen, P.; Ewald, S.; Schleifenbaum, J.H.; Belyakov, A.; Haase, C. Controlling microstructure and mechanical properties of additively manufactured high-strength steels by tailored solidification. *Addit. Manuf.* **2020**, *35*, 101389.
20. Bermingham, M.; StJohn, D.; Krynen, J.; Tedman-Jones, S.; Dargusch, M. Promoting the columnar to equiaxed transtion and grain refinement of titanium alloys during additive manufacturing. *Acta Mater.* **2019**, *168*, 261–274. [CrossRef]
21. Tang, Y.T.; Panwisawas, C.; Ghoussoub, J.N.; Gong, Y.; Clark, J.W.; Nemeth, A.A.; McCartney, D.G.; Reed, R.C. Alloys-by-design: Application to new superalloys for additive manufacturing. *Acta Mater.* **2021**, *202*, 417–436. [CrossRef]
22. Turner, J.A.; Belak, J.; Barton, N.; Bement, M.; Carlson, N.; Carson, R.; DeWitt, S.; Fattebert, J.L.; Hodge, N.; Jibben, Z.; et al. ExaAM: Metal additive manufacturing simulation at the fidelity of the microstructure. *Int. J. High Perform. Comput. Appl.* **2022**, *36*, 13–39. [CrossRef]
23. OpenFOAM. 2021. Available online: https://github.com/OpenFOAM (accessed on 7 February 2022).
24. Coleman, J.; Plotkowski, A.; Stump, B.; Raghavan, N.; Sabau, A.; Krane, M.; Heigel, J.; Ricker, R.; Levine, L.; Babu, S. Sensitivity of Thermal Predictions to Uncertain Surface Tension Data in Laser Additive Manufacturing. *J. Heat Transf.* **2020**, *142*, 122201. [CrossRef]
25. Rolchigo, M.; Reeve, S.; Stump, J. ExaCA. 2021. Available online: https://github.com/LLNL/ExaCA (accessed on 7 February 2022).
26. Carson, R.A.; Wopschall, S.R.; Bramwell, J.A. ExaConstit. 2019. Available online: https://www.osti.gov/doecode/biblio/31691 (accessed on 7 February 2022).
27. Basak, A.; Das, S. Epitaxy and Microstructure Evolution in Metal Additive Manufacturing. *Annu. Rev. Mater. Res.* **2016**, *46*, 125–149. [CrossRef]
28. Wang, D.; Song, C.; Yang, Y.; Bai, Y. Investigation of crystal growth mechanism during selectiv laser melting and mechanical property characterization of 316L stainless steel parts. *Mater. Des.* **2016**, *100*, 291–299. [CrossRef]
29. Bai, L.; Le, G.; Liu, X.; Li, J.; Xia, S.; Li, X. Grain morphologies and microstructures of laser melting depositied V-5Cr-5Ti alloys. *J. Alloys Compd.* **2018**, *745*, 716–724. [CrossRef]
30. Wei, H.; Mazumder, J.; DebRoy, T. Evolution of solidification texture during additive manufacturing. *Sci. Rep. Nat.* **2015**, *5*, 16446. [CrossRef]
31. Pham, M.S.; Dovgyy, B.; Hooper, P.A.; Gourlay, C.M.; Piglione, A. The role of side-branching in microstructure development in laser powder-bed fusion. *Nat. Commun.* **2020**, *11*, 749. [CrossRef]
32. Kok, Y.; Tan, X.; Wang, P.; Nai, M.; Loh, N.; Liu, E.; Tor, S. Anisotropy and heterogeneity of microstructure and mechanical properties in metal additive manufacturing: A critical review. *Mater. Des.* **2018**, *139*, 565–586. [CrossRef]
33. Shamsaei, N.; Yadollahi, A.; Bian, L.; Thompson, S.M. An overview of Direct Laser Deposition for additive manufacturing; Part II: Mechanical behavior, process parameter optimization and control. *Addit. Manuf.* **2015**, *8*, 12–35. [CrossRef]
34. Tian, Z.; Zhang, C.; Wang, D.; Liu, W.; Fang, X.; Wellman, D.; Zhao, Y.; Tian, Y. A review on laser powder bed fusion of Inconel 625 nickel-based alloy. *Appl. Sci.* **2020**, *10*, 81. [CrossRef]
35. Li, J.; Zhou, X.; Brochu, M.; Provatas, N.; Zhao, Y.F. Solidification microstructure simulation of Ti-6Al-4V in metal additive manufacturing: A review. *Addit. Manuf.* **2020**, *31*, 100989. [CrossRef]
36. Gatsos, T.; Elsayed, K.A.; Zhai, Y.; Lados, D.A. Review on Computational Modeling of Process-Microstructure-Property Relatonships in Metal Additive Manufacturing. *JOM* **2020**, *72*, 403–419. [CrossRef]
37. Rappaz, M.; Gandin, C.A. Probabilistic modeling of microstructure formation in solidification processes. *Acta Metall. Et Mater.* **1993**, *41*, 345–360. [CrossRef]
38. Gandin, C.A.; Rappaz, M. A coupled finite-element cellular-automaton model for the prediction of dendritic grain structures in solidification processes. *Acta Metall. Mater.* **1994**, *42*, 2233–2246. [CrossRef]
39. Gandin, C.A.; Rappaz, M. A 3D cellular automaton algorithm for the prediction of dendritic grain growth. *Acta Mater.* **1997**, *45*, 2187–2195. [CrossRef]
40. Zinoviev, A.; Zinovieva, O.; Ploshikhin, V.; Romanova, V.; Balokhonov, R. Evolution of grain structure during laser additive manufacturing. Simulation by a cellular automata method. *Mater. Des.* **2016**, *106*, 321–329. [CrossRef]
41. Rai, A.; Markl, M.; Korner, C. A coupled Cellular Automaton-Lattice Boltzmann model for grain structure simulation during additive manufacturing. *Comput. Mater. Sci.* **2016**, *124*, 37–48. [CrossRef]
42. Zinovieva, O.; Zinoviev, A.; Ploshikhin, V. Three-dimensional modeling of the microstructure evolution during metal additive manufacturing. *Comput. Mater. Sci.* **2018**, *141*, 207–220. [CrossRef]
43. Akram, J.; Chalavadi, P.; Pal, D.; Stucker, B. Understanding grain evolution in additive manufacturing through modeling. *Addit. Manuf.* **2018**, *21*, 255–268. [CrossRef]
44. Koepf, J.A.; Goterbarm, M.R.; Markl, M.; Körner, C. 3D multi-layer grain structure simulation of powder bed fusion additive manufacturing. *Acta Mater.* **2018**, *152*, 119–126. [CrossRef]

45. Liu, D.R.; Wang, S.; Yan, W. Grain structure evolution in transition-mode melting in direct energy deposition. *Mater. Des.* **2020**, *194*, 108919. [CrossRef]

46. Wang, Z.; Lin, X.; Kang, N.; Hu, Y.; Chen, J.; Huang, W. Strength-ductility synergy of selective laser melted Al-Mg-Sc-Zr alloy with a heterogenous grain structure. *Addit. Manuf.* **2020**, *34*, 101260.

47. Yang, J.; Yu, H.; Yang, H.; Li, F.; Wang, Z.; Zeng, X. Prediction of microstructure in selective laser melted Ti-6Al-4V alloy by cellular automaton. *J. Alloys Compd.* **2018**, *748*, 281–290. [CrossRef]

48. Li, M.; Li, J.M.; Zheng, Q.; Wang, G.; Zhang, M.X. Effect of solutes on grain refinement of as-cast Fe-4Si alloy. *Metall. Mater. Trans. A* **2018**, *49A*, 2235–2247. [CrossRef]

49. Lian, Y.; Gan, Z.; Yu, C.; Kats, D.; Liu, W.K.; Wagner, G.J. A cellular automaton finite volume method for microstructure evolution during additive manufacturing. *Mater. Des.* **2019**, *169*, 107672. [CrossRef]

50. Mohebbi, M.S.; Ploshikhin, V. Implementation of nucleation in cellular automaton simulation of microstructural evolution during additive manufacturing of Al alloys. *Addit. Manuf.* **2020**, *36*, 101726. [CrossRef]

51. Dezfoli, A.R.A.; Lo, Y.L.; Raza, M.M. Prediction of epitaxial grain growth in single-track laser melting of IN718 using integrated finite element and cellular automaton approach. *Materials* **2021**, *14*, 5202. [CrossRef]

52. Xiong, F.; Huang, C.; Kafka, O.L.; Lian, Y.; Yan, W.; Chen, M.; Fang, D. Grain growth prediction in selective electron beam melting of Ti-6Al-4V with a cellular automaton method. *Mater. Des.* **2021**, *199*, 109410. [CrossRef]

53. Anderson, T.D.; DuPont, J.N.; DebRoy, T. Origin of stray grain formation in single-crystal superalloy weld pools from heat transfer and fluid flow modeling. *Acta Mater.* **2010**, *58*, 1441–1454. [CrossRef]

54. Wang, T.; Zhu, Y.; Zhang, S.; Tang, H.; Wang, H. Grain morphology evolution during behavior of titanium alloy components during laser melting deposition additive manufacturing. *J. Alloys Compd.* **2015**, *632*, 505–513. [CrossRef]

55. Sabau, A.S.; Yuan, L.; Raghavan, N.; Bement, M.; Simunovic, S.; Turner, J.A.; Gupta, V.K. Fluid dynamics effects on microstructure predition in single-laser tracks for additive manuacturing of IN625. *Metall. Trans. B* **2020**, *51*, 1263–1281. [CrossRef]

56. Liu, Y.; Liu, Z.; Jiang, Y.; Wang, G.; Yang, Y.; Zhang, L. Gradient in microstructure and mechanical property of selective laser melted AlSi10Mg. *J. Alloys Compd.* **2018**, *735*, 1414–1421. [CrossRef]

57. Li, Z.; Chen, J.; Sui, S.; Zhong, C.; Lu, X.; Lin, X. The microstructure evolution and tensile properties of Inconel 718 fabricated by high-deposition-rate laser directed energy deposition. *Addit. Manuf.* **2020**, *31*, 100941. [CrossRef]

58. Dinda, G.; Dasgupta, A.; Mazumder, J. Evolution of microstructure in laser depositied Al-11.28%Si alloy. *Surf. Coat. Technol.* **2012**, *206*, 2152–2160. [CrossRef]

59. Al-Bermani, S.S.; Blackmore, M.L.; Zhang, W.; Todd, I. The Origin of Microstructural Diversity, Texture, and Mechanical Properties in Electron Beam Melted Ti-6Al-4V. *Metall. Mater. Trans. A-Phys. Metall. Mater. Sci.* **2010**, *41*, 3422–3434. [CrossRef]

60. Kergassner, A.; Koepf, J.A.; Markl, M.; Korner, C.; Mergheim, J.; Steinmann, P. A novel approach to predict the process-induced mechanical behavior of additively manufactured materials. *J. Mater. Eng. Perform.* **2021**, *30*, 5235–5246. [CrossRef]

61. Lim, H.; Abdeljawad, F.; Owen, S.J.; Hanks, B.W.; Foulk, J.W.; Battaile, C.C. Incorporating physically-based microstructures in materials modeling: Bridging phase field and crystal plasticity frameworks. *Model. Simul. Mater. Sci. Eng.* **2016**, *24*, 045016. [CrossRef]

62. Kergassner, A.; Mergheim, J.; Steinmann, P. Modeling of additively manufactured materials using gradient-enhances crystal plasticity. *Comput. Math. Appl.* **2019**, *78*, 2338–2350. [CrossRef]

63. Romanova, V.; Balokhonov, R.; Zinovieva, O.; Emelianova, E.; Dymnich, E.; Pisarev, M.; Zinoviev, A. Micromechanical simulations of additive manufactured aluminum alloys. *Comput. Struct.* **2021**, *244*, 106412. [CrossRef]

64. Yan, W.; Lian, Y.; Yu, C.; Kafka, O.L.; Liu, Z.; Liu, W.K.; Wagner, G.J. An integrated process-structure-property modeling framework for additive manufacturing. *Comput. Methods Appl. Mech. Eng.* **2018**, *339*, 2184–204. [CrossRef]

65. Heo, T.W.; Khairallah, S.A.; Shi, R.; Berry, J.; Perron, A.; Calta, N.P.; Martin, A.A.; Barton, N.A.; Roehling, J.; Roehling, T.; et al. A mesoscopic digital twin that bridges length and time scales for control of additively manufactured metal microstructures. *J. Phys. Mater.* **2021**, *4*, 034012. [CrossRef]

66. Rolchigo, M.; Stump, B.; Belak, J.; Plotkowski, A. Sparse thermal data for cellular automata modeling of grain structure in additive manufacturing. *Model. Simul. Mater. Sci. Eng.* **2020**, *28*, 065003. [CrossRef]

67. Bragard, J.; Karma, A.; Lee, Y. Linking Phase-Field and Atomistic Simulations to Model Dendritic Solidification in Highly Undercooled Melts. *Interface Sci.* **2002**, *10*, 121–126. [CrossRef]

68. Teferra, K.; Rowenhorst, D.J. Optimizing the cellular automata finite element model for additive manufacturing to simulate large microstructures. *Acta Mater.* **2021**, *213*, 116930. [CrossRef]

69. MacKenzie, J. Second paper on statistics associated with the random disorientation of cubes. *Biometrika* **1958**, *45*, 229–240. [CrossRef]

70. Barton, N.R.; Carson, R.A.; Wopschall, S.R. ECMech. 2018. Available online: https://www.osti.gov/doecode/biblio/28720 (accessed on 7 February 2022).

71. MFEM: Modular Finite Element Methods Library. Available online: https://www.osti.gov/doecode/biblio/35738 (accessed on 7 February 2022).

72. Marin, E.; Dawson, P. On modelling the elasto-viscoplastic response of metals using polycrystal plasticity. *Comput. Methods Appl. Mech. Eng.* **1998**, *165*, 1–21. [CrossRef]

73. Moore, J.A.; Barton, N.R.; Florando, J.; Mulay, R.; Kumar, M. Crystal plasticity modeling of β phase deformation in Ti-6Al-4V. *Model. Simul. Mater. Sci. Eng.* **2017**, *25*, 075007. [CrossRef]
74. Wang, Z.; Stoica, A.D.; Ma, D.; Beese, A.M. Diffraction and single-crystal elastic constants of Inconel 625 at room and elevated temperatures determined by neutron diffraction. *Mater. Sci. Eng. A* **2016**, *674*, 406–412. [CrossRef]
75. Groeber, M.; Schwalbach, E.; Donegan, S.; Uchic, M.; Chapman, M.; Shade, P.; Musinski, W.; Miller, J.; Turner, T.; Sparkman, D.; et al. AFRL AM Modeling Challenge Series: Challenge 3 Data Package. 2018. Available online: https://acdc.alcf.anl.gov/mdf/detail/groebermichael_afrl_am_package_v1.1/ (accessed on 7 February 2022).
76. Barlat, F.; Aretz, H.; Yoon, J.; Karabin, M.; Brem, J.; Dick, R. Linear transfomation-based anisotropic yield functions. *Int. J. Plast.* **2005**, *21*, 1009–1039. [CrossRef]
77. Barton, N.R.; Knap, J.; Arsenlis, A.; Becker, R.; Hornung, R.D.; Jefferson, D.R. Embedded polycrystal plasticity and adaptive sampling. *Int. J. Plast.* **2008**, *24*, 242–266. [CrossRef]
78. Knap, J.; Barton, N.R.; Hornung, R.D.; Arsenlis, A.; Becker, R.; Jefferson, D.R. Adaptive sampling in hierarchical simulation. *Int. J. Numer. Methods Eng.* **2008**, *76*, 572–600. [CrossRef]
79. Van Houtte, P.; Yerra, S.K.; Van Bael, A. The Facet method: A hierarchical multilevel modelling scheme for anisotropic convex plastic potentials. *Int. J. Plast.* **2009**, *25*, 332–360. [CrossRef]
80. Segurado, J.; Lebensohn, R.A.; LLorca, J.; Tomé, C.N. Multiscale modeling of plasticity based on embedding the viscoplastic self-consistent formulation in implicit finite elements. *Int. J. Plast.* **2012**, *28*, 124–140. [CrossRef]
81. Gawad, J.; Banabic, D.; Van Bael, A.; Comsa, D.S.; Gologanu, M.; Eyckens, P.; Van Houtte, P.; Roose, D. An evolving plane stress yield criterion based on crystal plasticity virtual experiments. *Int. J. Plast.* **2015**, *75*, 141–169. [CrossRef]
82. Zhang, H.; Diehl, M.; Roters, F.; Raabe, D. A virtual laboratory using high resolution crystal plasticity simulations to determine the initial yield surface for sheet metal forming operations. *Int. J. Plast.* **2016**, *80*, 111–138. [CrossRef]
83. Chu, C.C.; Needleman, A. Void Nucleation Effects in Biaxially Stretched Sheets. *J. Eng. Mater. Technol.* **1980**, *102*, 249–256. [CrossRef]

Article

Dynamic Simulations of Manufacturing Processes: Hybrid-Evolving Technique

Amir M. Horr * and Johannes Kronsteiner

LKR Light Metals Technologies, Austrian Institute of Technology, 1210 Vienna, Austria;
Johannes.Kronsteiner@ait.ac.at
* Correspondence: amir.horr@ait.ac.at; Tel.: +43-(0)50550-6918

Abstract: Hybrid physical-data-driven modeling techniques have steadily been developed to address the multi-scale and multi-physical aspects of dynamic process simulations. The analytical and computational features of a new hybrid-evolving technique for these processes are elaborated herein and its industrial applications are highlighted. The authentication of this multi-physical and multi-scale framework is carried out by developing an integrated simulation environment where multiple solver technologies are employed to create a reliable industrial-oriented simulation framework. The goal of this integrated simulation framework is to increase the predictive power of material and process simulations at the industrial scale.

Keywords: hybrid modeling; dynamic material processes; evolving domains; data-driven modeling; genetic algorithm symbolic regression; cooling process

Citation: Horr, A.M.; Kronsteiner, J. Dynamic Simulations of Manufacturing Processes: Hybrid-Evolving Technique. *Metals* **2021**, *11*, 1884. https://doi.org/10.3390/met11121884

Academic Editor: Alain Pasturel

Received: 1 November 2021
Accepted: 17 November 2021
Published: 23 November 2021

Publisher's Note: MDPI stays neutral with regard to jurisdictional claims in published maps and institutional affiliations.

1. Introduction

The predictive power of numerical simulations for modeling and optimizing dynamic material processes have rigorously been examined by scientists and engineers to avoid extensive and costly experimental and repeated trial series. The most vital issue related to these numerical simulation schemes is to generate reliable and accurate results in a reasonable computational time while handling this process modeling with reasonable details and realistic conditions. The conventional Finite Element (FE) and Computational Fluid Dynamic (CFD) solvers have broadly been employed as pillars of the computing strategy for solving multi-physical phenomena during the considered processes. However, due to the multi-phase and multi-scale natures of these processes, more sophisticated simulation frameworks are required to handle the modeling of different phases of material at different length scales. Hence, in recent years, some multi-solver simulation frameworks along with extensive interfacing and bridging technologies were proposed to handle the simulation of material processes at different phases and length scales.

To establish this integrated simulation framework, reliable interfacing and coupling procedures are required to implement the multi-solver scheme which is a pillar of the proposed Through Process Simulations (TPS) of materials which might include casting process to the forming of final parts [1]. To help establish a ground for such an integrated simulation framework, new and innovative hybrid techniques have widely been proposed [2–4], which help to improve analytical and numerical aspects of simulations. These upgraded and sophisticated analytical and numerical simulations are founded on hybrid physical-data-driven schemes where data processing and handling techniques are used to improve the analytical and numerical modeling. The center stone of the idea is to combine the power of Artificial Intelligent (AI) and Machine Learning (ML) techniques with various existing capabilities of analytical and numerical techniques [5]. Hence, the key here is to integrate sound physical and analytical techniques with emerging numerical simulation technologies and supplement the integration process with benefits of the hybrid modeling scheme.

In the work herein, the ongoing research on new evolving and dynamic mesh numerical concepts along with hybrid physical-data driven modeling and the integration of AI-computational framework (using Genetic Algorithm Symbolic Regression, GASR) will be introduced [6]. The framework would address some of the long-standing problems of speed, accuracy, and reliability of numerical tools for dynamic material processes. The hybrid-evolving numerical solution proposed herein is based on a computational concept which institutes the sound physical\mathematical models and does not only rely on the improved algorithm or solver technology for the material and process simulations. Hence, some fundamental numerical performances of the simulation technique are described. Additionally, some outcomes of a simple case study are shown which have been utilized using the proposed hybrid framework. By this means, the aspects of hybrid physical-data-driven modeling are interrogated using the developed GASR tool. One of the main contributions of this paper is to show how new computational technologies combined with hybrid modeling and suitable AI scheme can transform the traditional material and process simulation techniques.

2. Dynamic Processes

The optimization of different industrial processes including, casting, extrusion, forging, and rolling have long been considered, and many technical\economic advantages on the one hand and also difficult technical challenges, on the other hand, have been examined. The use of numerical simulations for optimizing these processes has undoubtedly gained popularity in the last couple of decades and new and innovative methods have been developed to be a part of design-optimize-verify processes or even used as a part of real-time monitoring and automation in the production lines [7–11]. The FE and CFD solvers have traditionally been utilized for these material process simulations where multi-physical\phase and multi-scale aspects of the material evolutions have been considered.

The conventional meshing and spatial discretization schemes have initially been utilized in these numerical techniques to define a fixed number of simpler subspaces\elements with approximate or exact functions (e.g., shape functions). These discretized numerical domains are then formulated into an algebraic system of equations which can be solved with the FE and CFD solvers. However, many material processes, like casting, extrusion or even Additive Manufacturing (AM) processes are characterized by their gradual production in the time domain. These continuously growing domains for the dynamic processes are principally described with the rate of material and thermal energy insertion during the process (e.g., casting speed, AM deposition rate). The main challenges for the simulation of these processes are the dynamic nature of these numerical domains, their continuous material insertion, and the evolution of thermal energy characteristics during these processes.

Although today's meshing technologies have become very efficient during simulations (e.g., adaptive re-meshing), even for the complex geometries and also bodies with large deformations, the issues of dynamic changes of the domain during simulations and their evolving discretization scheme have not been systematically addressed. For the dynamic processes like the casting process and AM, the traditional numerical discretization strategy was based on the domain discretization for the whole final part\billet. The generated elements in these domains are then deactivated at the start of the transient simulation to be later activated piece-wise based on the casting speed (or deposition rate) during the process simulation.

Although this "Block deactivation\activation Technique" (BT) can successfully be applied to some industrial casting applications like vertical and horizontal casting processes (shown in Figure 1), the size of the numerical matrices and system of equations is large for the duration of the transient analysis (deactivation of elements at the start of the process would not remove their matrices from the equation solver). To mitigate this problem, there have been some attempts to introduce innovative discretization techniques [12–14] for dynamic systems including mixed and merge/splitting element and cell division techniques. In these techniques, the numerical domain can be divided into Lagrangian, Eulerian, and Arbitrary Lagrangian-Eulerian (ALE) zones. For the casting and extrusion (and also AM) applications, the transient extension of the part during the simulation can numerically be modeled using the splitting element layers at interfaces. Although the approach can alleviate the existence of a large system of equations from the start of the simulation, it has limitations in terms of interface geometries and boundaries for the more complex dynamical systems.

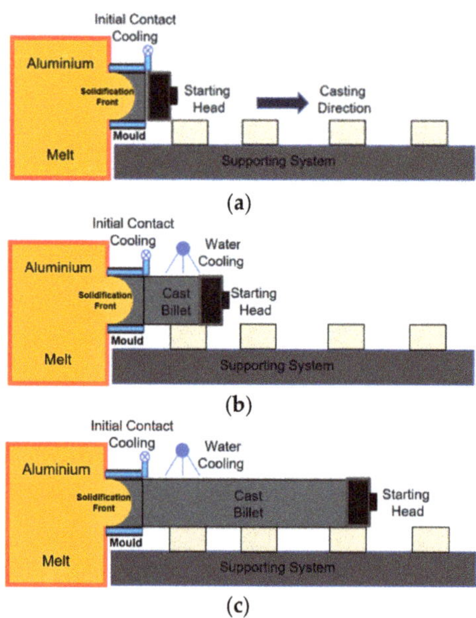

Figure 1. Schematic horizontal casting process at; (**a**) initial condition; (**b**) end of transient condition and; (**c**) steady-state condition.

For the process simulation of the continuous and semi-continuous casting and its complicated fluid-mushy-solid coupled analyses (e.g., multi-phase analyses), it is essential to use an advanced numerical approach, capable of dealing with melt flow, thermal energy transfer, solidification, and also solid-state conditions. The melt fluid flows into the mould and its low turbulence/laminar characteristics, mushy and solidification zone (with its metallurgical aspects), and also the final solid zones need to be simulated accurately to achieve desired results. These sophisticated numerical simulations have already found their way into the mainstream industrial tools [15–17] to enable the simulation of chain processes for production lines. Although the early multi-phase simulations of these processes are an important part of the framework, the details of these simulations are not discussed here in this paper. To limit the in-depth technical discussions within a single manuscript, only the thermal-mechanical simulations of dynamic material processes are presented herein. More technical discussions about the conventional material processes can be found in [18–20].

3. Evolving Domain Technique

The increasing computational power during the last couple of decades has greatly affected the way the traditional material process simulations were conducted and more advanced schemes like coupled multi-physical and multi-phase material modeling have quickly emerged. Furthermore, the pace of design for more energy-efficient and quality-oriented industrial processes has significantly picked up in the last two decades where more innovative and smart virtual tools are utilized to accelerate these processes.

The new concept of evolving domain and its Dynamic Mesh Technique (DMT) has been developed recently [20–23] to overcome numerical problems related to the continuous evolution and transient generation of numerical domains. It treats the extended parts of the domain as a dynamic zone (e.g., mesh block) which can be appended\evolved in a predefined or calculated manner and ultimately be attached to the main domain through a mapping-boundary concept. As the newly generated meshes are attached to the original domain, a mapping procedure would be performed to handle the new material\energy input.

For the mapping at the interfaces between the existing discretized domain with a newly appended mesh block and to overlap grids with multi-resolution (e.g., overlapping meshes with different sizes) some mathematical representations have already been developed. The Partition of Unity Finite Element Method (PUFEM), Overlapping Sphere Elements (OSE) and a few other techniques have already been proposed to handle the grid-inconsistency [24–26]. For overlapping meshes at the boundary of a numerical domain, let us assume two interfacing grids with overlapping conditions at the boundaries of $\Omega 1$ and $\Omega 2$ numerical domains. For linear finite element grids with base functions $\{v_i\}_{i\in I}$ and $\{w_j\}_{j\in J}$, the overlapping set can be written as [24–26];

$$\left\{ \varnothing_1 v_i \, , \, \varnothing_2 w_j \right\}_{i\in I, \, j\in J} \tag{1}$$

If the overlapping meshes are represented with a linear independent relation as;

$$\varnothing_1 \sum_i \alpha_i v_i + \varnothing_2 \sum_j \beta_j \, w_j = 0 \tag{2}$$

and if it is assumed that $\varnothing_2 = 1 - \varnothing_1$ then it can be shown that;

$$\varnothing_1 \left(\sum_i \alpha_i v_i - \sum_j \beta_j \, w_j \right) = -\sum_j \beta_j \, w_j \tag{3}$$

$$\sum_i \alpha_i v_i - \sum_j \beta_j \, w_j = Const. \tag{4}$$

To form a stiffness matrix for the interfacing\overlapping meshes, an assembly can firstly be formed for non-overlapping elements (as a conventional assembly process) and secondly, overlapping elements' terms can be entered with multiple entries. More comprehensive discussions about the multi-resolution interfacing\overlapping grids can be found in the literature [25,26].

Figure 2 shows a schematic workflow and dynamic mesh generation for the evolving domain framework where mesh blocks are inserted during the transient simulation at subsequent time steps. For the continuous casting and extrusion simulations, a directional vector (horizontal or vertical) can be predefined to start the directional boundary insertion and mapping scheme for the generated mesh blocks\layers to resemble the transient extension of the billet\part. The dynamically generated mesh blocks or element rows can be appended to the original numerical domain between restarts depending on the speed and requirement of the numerical model for the industrial process. As the numerical solution is performed on a full parallel-processing machine, decomposition of the extending domain for multiple instances of the solver at each restart shall independently be carried out for the transient solution. The generated element blocks (with the same or different sizes) are integrated into the original domain matrices by a reassembly process at restarts points (and become a part of the main domain) as the billet extension step with pre-defined casting speed comes to the end.

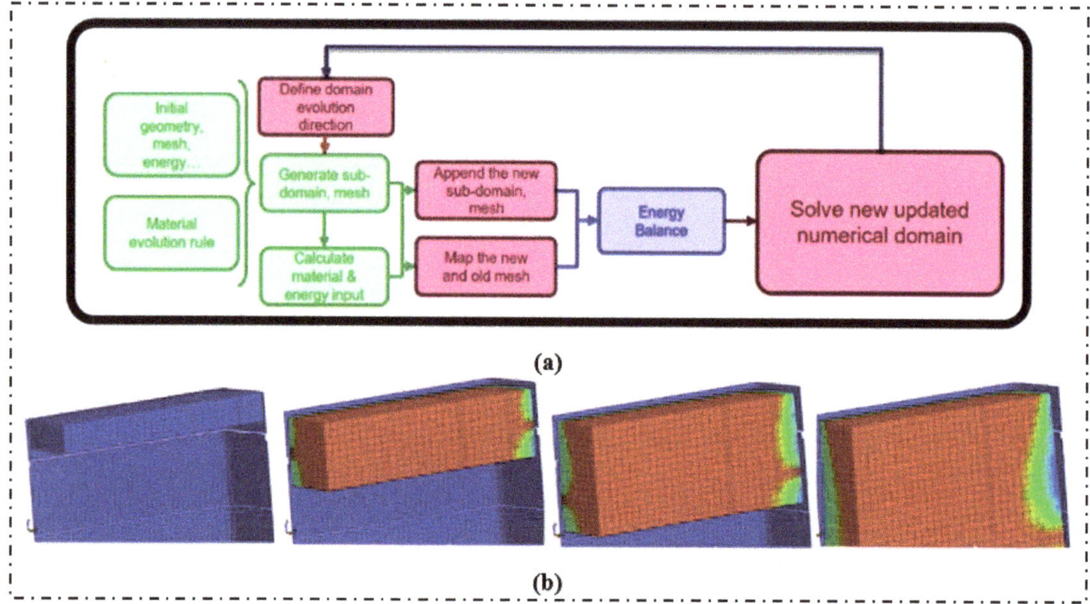

Figure 2. (a) Schematic workflow for evolving domain framework; (b) meshing initiation step and its generation and insertion during transient simulation at subsequent restarts [11].

Additionally, for the estimation of the thermal energy evolution and cooling during the transient process, HTC calculations were carried out to calculate the cooling rates and thermal gradients. For continuous and semi-continuous casting processes, one of the popular cooling strategies is to use water spray cooling. The numerical modelings of cooling processes are also essential to develop a deep understanding of the thermal evolution and the volatile temperature fields during casting processes. In many casting processes, the cast billet is impinged by a water jet to extract the thermal energy from inside of the billet. During the cooling process, the billet undergoes large temperature fluctuations, especially near its surface. Hence, it is essential to investigate and control the thermal event to minimize the occurrence of defects on the surface or sub-surface (hot tearing, cold cracking...).

4. Computational Performance

For the continuous casting application, the issues of long billet sizes and time-dependent generation scheme based on a pre-defined casting speed are encouraging a new approach for simulating such processes. One of the main burdens of these new techniques is the computational time and resources. For dynamic systems, it is essential to have a strategy to evolve the numerical domain according to the real pace of the manufacturing process which ultimately delivers a dynamic and growing computational domain. The proposed DMT technique can deliver a method where accurate results can be obtained using growing and evolving domains [4,9,11]. The computational resources for this technique are commissioned gradually as the size of the real billet and its simulated counterpart are growing during the process. Restarting solvers at discrete time steps comes at additional costs for the repeated matrices' assemblies and extra Input\Output (IO) activities.

The practical applications of the proposed DMT for real industrial cases can only be feasible when it can be shown that the technique is capable of competing with the best available techniques (e.g., BT) in terms of computational performance. To assess the computational efficiency of the DMT technique against the conventional BT a comprehensive numerical investigation has been carried out to study an industrial size vertical semi-continuous casting process. For the BT simulation, element blocks are predefined and the whole model is pre-loaded at the start of the simulation while the contacts between the blocks are activated consecutively. Table 1 shows the casting process parameters for both simulation techniques, while Table 2 shows a scenario table designed to assess the computational time and resources of both techniques for all scenarios on a single computing node. Additionally, Figure 3a shows the BT and DMT numerical models for the vertical semi-continuous casting process and their meshing strategies. Due to the double symmetric condition of the rectangular billet, only a quarter of the billet is modeled and meshed for the transient simulation to save computational time and resources. As it is shown in Figure 3a, different meshing and contact strategies were used to carry out the casting simulations for DMT and BT. While both techniques have employed contact elements to simulate the cooling boundaries (air and water cooling), for BT, the contacts had to be set up for the full length of the billet. Alternatively, for DMT, as new mesh layers are generated and attached to the main domain at solver restarts, contact setups needed to be updated [4,9,11].

Table 1. Defined parameters for casting process.

Melt Temperature [°C]	Billet Width [m]	Billet Thickness [m]	Casting Speed [m s^{-1}]	Cooling Water Temperature [°C]	HTC- Air Cooling [kW m^{-2} K^{-1}]	HTC- Water Cooling [kW m^{-2} K^{-1}]
630	1.24	0.3	0.01	20	1.5 (average)	11 (average)

Table 2. Scenarios Table for DMT and BT case studies using a single computing node with 16 cores.

Scenario No.	Billet Length [m]	No. of Elements	CPU Time DMT [s]	CPU Time BT [s]	IO Time DMT [s]	IO Time BT [s]	CPU Ratio DMT/BT
S1	0.5	27,189	16,345	20,581	78.62	6.90	79%
S2	1	41,357	45,577	77,387	179.18	7.02	59%
S3	1.4	59,573	84,545	163,063	271.84	13.77	52%

The computational efficiency of DMT in terms of CPU time and IO resources should be investigated using industrial-scale case studies on a parallel computing platform. Hence, a table for numerical simulations of industrial-scale scenarios has been prepared for estimation of computational efficiency. Table 2 shows the simulated scenarios for both DMT and BT models on different numbers of computing nodes. For the purpose of comparison, numerical models with three different geometries (for the billet lengths 0.5 m, 1 m, and 1.4 m) were created and their corresponding structured meshes were generated accordingly. These models were simulated on a parallel computing platform with the technical specifications shown in Table 3.

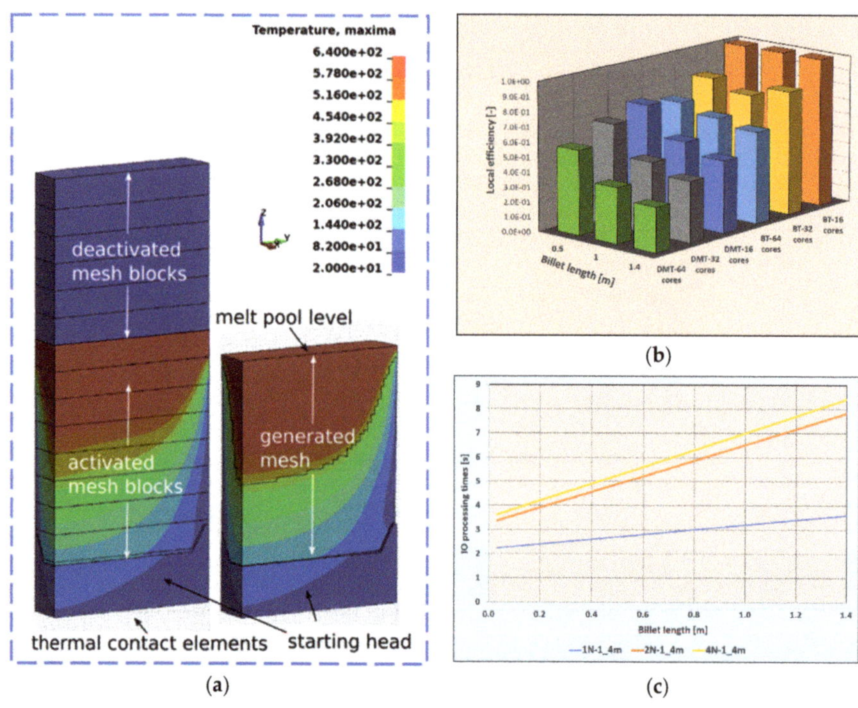

Figure 3. (**a**) BT and DMT models with their meshing strategies and overlaying temperature contours; (**b**) 3D plot of computational efficiency for various billet lengths and number of cores; (**c**) filtered IO times for DMT at each restart point (for 1.4 m length billet) using one, two and four nodes.

Table 3. Technical specifications and version of operating software for parallel computing nodes.

CPU Name	No. of Sockets	Cores per Socket	Total Memory [MB]	Communication Between Nodes	Parallelization Scheme	LS-DYNA Release	Accuracy
Intel Xeon E5-2687W v4	2	8	65,536	InfiniBand	Platform MPI 08.02.00.00 [10060]	MPP R8.1.0	Double precision

To compare the simulation results, the computational time-history results were postprocessed to calculate the CPU and IO computational times. The three-dimensional plot of local computational efficiency against three billet lengths and the number of computational cores is shown in Figure 3b. The local efficiency can be defined as the ratio of the total simulation time of a base scenario for the specified billet length (i.e., wall clock time) divided by the CPU time for each scenario. The most efficient scenario using 16 cores for the DMT technique was used as a base scenario for each billet length. Accordingly, the higher the efficiency values shown in Figure 3, the lower the total simulation times for the casting scenarios.

The results of the comparative study presented in Figure 3 show that concerning numerical efficiency, DMT can outperform BT almost in every scenario. It can also be clearly seen how the billet length and the parallelization influence the computational efficiency where DMT can benefit more from heavy parallelization than BT. Only in the case of 64 cores and only for the short billet length of 0.5 m, BT can slightly beat DMT using 16 cores [4]. Furthermore, Figure 3c shows the increase in IO processing times for DMT at each restart point (for 1.4 m length billet) using different numbers of computational nodes (1, 2, and 4 number of nodes). Due to the increased number of elements and thus history variables, the time and effort for reading and writing data, decomposing the mesh for parallelization, and initializing the matrices increase steadily over the whole transient simulation. It is also clear that using more than one computational node leads to a significant increase in input/output time while the difference between using 2 nodes (16 cores) and 4 nodes (64 cores), on the other hand, is not significant. Despite the higher IO time for the DMT, the total simulation wall clock time is still significantly lower than the conventional technique due to the fact that the CPU computational time is generally more than two orders of magnitude larger than IO times for real-size casting applications.

5. Conventional Cooling Models

Different aspects of numerical simulations of material processes are founded on complicated phenomena governed by the multi-physical and multi-phase interactions including thermal evolution and heating\cooling modeling. The cooling modelings during these material processes and their numerical simulations are traditionally based on a methodology of using theoretical and analytical knowledge along with some practical experiences. As conventional water impingement cooling is still popular in many material processes, the numerical modelings of these processes are essential to developing a deep understanding of thermal evolutions and their effect on final product quality.

During these processes (e.g., casting, rolling, extrusion, etc.), billet\parts are impinged with a water jet to extract thermal energy and to cool down to room temperature. Although a lot of research has been conducted in the last twenty years to numerically calculate\estimate the thermal evolution and heat transfer phenomena during industrial material processes, the complex multi-physical\phase nature of these events could not fully be described. Hence, simpler analytical estimations of Heat Transfer Coefficient (HTC) during these processes have broadly been used to accelerate the modeling process. The mathematical representation of the cooling process on hot surfaces can be expressed with the simplified governing equations for the water spray cooling system as [27–29];

$$\nabla.U = 0 \quad \frac{\partial U}{\partial t} = -\nabla\left(\frac{P}{\rho}\right) + \frac{1}{Re}\nabla^2 U \tag{5}$$

where U, P, ρ and Re are velocity vector, pressure, density, and Reynolds number, respectively. The spray wet area can also be defined as;

$$A = f(x,y,z)\frac{Q}{V_s} \tag{6}$$

where $f(x,y,z)$, vs. and Q are coordinate distribution function, surface flow rate, and water flow rate, respectively. Various parameters would affect the thermal energy dissipation during water spray cooling, namely: temperature difference between billet surface and water, water flow rate, angle of impingement, water quality, surface roughness, and some other minor parameters. Since, the type of boiling regime on the hot surface has major effects on the heat dissipation rate, the complex interaction between surface/steam/water and air would determine the real cooling HTC. Figure 4 shows a schematic bubble generation and departure on a horizontal hot surface during a water spray action.

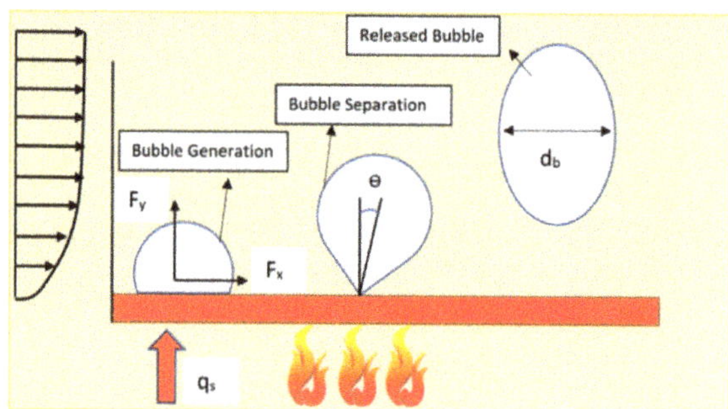

Figure 4. Schematic bubble dynamics stages on hot surface. Adapted with permission from ref. [22]. 2021 Elsevier.

Although there have been many attempts in recent years to carry out interfacial HTC estimations based on the inverse heat transfer calculations using available experimental data (with fitting techniques), the generality and accuracy of these estimation techniques have not been verified [30,31]. To make the HTC estimations simpler, the thermal energy transfer during cooling can be assumed as the summation of different interchanging mechanisms between a hot surface and the water as;

$$q_T = q_1 + q_2 + q_3 \tag{7}$$

where q_1, q_2 and q_3 are single-phase convective, evaporation, and quenching heat flux densities, respectively. With some mathematical manipulation, for the case of continuous liquid phase (water) and dispersive gases (water vapor and air), the three-phase continuity equation for phase k can be written as [32,33]:

$$\frac{\partial \alpha_k}{\partial t} + \nabla(\alpha_k \vec{U}_k) = \frac{\Gamma_{ki} - \Gamma_{ik}}{\rho_k} \tag{8}$$

where k denotes the phase and can be either liquid or gas and i is the non-k phase. The α, ρ, and U represent the phase volume fraction, density, and velocity vector respectively. More detailed discussions about the conventional spray-cooling modeling can be found in [27–33].

6. Hybrid Cooling Models

The use of hybrid physical-data-driven modeling and its application in material science and engineering have increasingly been promoted over recent years and the implementation of ML and AI technologies within alloy designs and processes have gradually been utilized. These emerging techniques have a great potential for the future of material process engineering and industrial manufacturing where it has already been proven that their potential to generate values in various material process applications and material modeling domains are significant [34,35]. For the material process simulations, ML and AI mean algorithm-based and data-driven schemes which enable better representations of phenomena with more accurate predictions and better continuous learning and improvement. The hybrid models are categorically grouped into different classes, namely; auxiliary, augmented, full, and dynamic hybrid models where different applications can be handled using an appropriate type of hybrid model. The description of these models can briefly be presented as;

- ➢ For auxiliary hybrid models, parameters in the physical/empirical models are function-fitted using ML and AI tools
- ➢ In augmented hybrid models, physical models are augmented with terms derived by function-fitting features of ML and AI tools

> Fully hybrid models where data trends from ML and AI are used along with physical laws to derive the hybrid model.
> Trained (or dynamic) hybrid models where the existing hybrid model is a subject of ML and AI tools for improvement

As these techniques are becoming ostensible for many materials and process modeling applications including dynamic casting and forming applications, the need for a greater understanding of technology utilization has raised within material and process simulation communities. The modeling capabilities of these techniques share the same fundamental intentions, namely; faster and more accurate modeling and the ability to dynamically change the rules and functions using new emerging data.

To ease the analytical overburden for the cooling modeling, the concepts of augmented hybrid modeling have willingly been employed here to estimate HTCs during the casting process. Since the mass conservation of the sum of the volume fractions in (8) should be unity ($\sum_k \alpha_k = 1$), the momentum terms for any of the phases in the equation should be conserved as [36–40];

$$\frac{\partial}{\partial t}(\alpha_k \rho_k U_k) + \nabla.(\alpha_k \rho_k U_k U_k) = -\alpha_k \nabla p + \alpha_k \rho_k g + \nabla.\alpha_k\left(\tau_k + \tau_k^t\right) + M_k \tag{9}$$

where p and g are pressure and gravitational acceleration. The M_k is the total interfacial force which can be defined as the rate of bubble generations and their movement relative to the hot surface during material processes as [36–40];

$$M_k = F_D + F_W + F_L + F_B + F_V + F_{TD} + \sum_{j=1}^{N}\left(\dot{m}_{jk}U_j - \dot{m}_{kj}U_k\right) \tag{10}$$

where F_D, F_W, F_L, F_B, F_V and F_{TD} are the drag, the wall lubrication (smoothness), lift, buoyancy, virtual mass, and turbulence drag forces (the Basset force is ignored herein due to negligible effects). The last term on the right-hand side of the equation is defined as momentum transfer associated with mass transfer. The energy conservation equation for each phase can also be written as [36–40];

$$\frac{\partial}{\partial t}(\alpha_k \rho_k h_k) + \nabla.(\alpha_k \rho_k U_k h_k) - \nabla.\left[\alpha_k\left(\lambda_k \nabla T_k + \frac{\mu_t}{\sigma_h}\nabla h_k\right)\right] = Q_k \tag{11}$$

where h, λ, T and Q are enthalpy, thermal conductivity, temperature, and interfacial heat transfer respectively. For the wall lubrication force (also known as Frank force), which is the effect that moving the bubbles away from the hot surface, the values can be estimated as [36–40];

$$F_w = \alpha_d \rho_c C_w\left\{(v_d - v_c) - [(v_d - v_c).n_w]n_w\right\}^2 n_w \quad C_W = C_{WE}\, max\left[0, \frac{1}{C_{WD}}\frac{1 - \frac{y_w}{C_{WC}d_b}}{y_W\left(\frac{y_W}{C_{WC}d_b}\right)^{p-1}}\right] \tag{12}$$

where n_w, C_w, C_{wE} are the normal unit vector for the hot surface, the wall lubrication coefficient, and bubble Bond coefficient. The p, $C_{WC,}$ and C_{WD} are correction coefficients and can be assumed as $p = 1.7$, $C_{WC} = 10$ and $C_{WD} = 6.8$ for the casting cooling spray system. The bubble Bond coefficient can also be estimated as,

$$C_{WE} = \begin{cases} e^{-0.933E + 0.179} & 1 \leq E \leq 5 \\ 0.007E + 0.04 & 5 \leq E \leq 33 \quad E = \frac{\Delta\rho\, g\, L^2}{\sigma} \\ 0.179 & 33 < E \end{cases} \tag{13}$$

where E is the bond number which is defined as a dimensionless number measuring the importance of surface tension forces compared to body forces and is used to characterize the shape of bubbles moving in a continuous fluid. The $\Delta\rho$, L and σ are density differences, characteristic length, and surface tension force per unit length. For the spray cooling during the casting process, the bond number can be written as [36–40];

$$E = \frac{g(\rho_c - \rho_d)d_b}{\sigma} \tag{14}$$

where d_b is the expected bubble diameter. Although the drag force F_D can approximately also be estimated using analytical calculations (with some simplified assumptions), the low accuracy of the results would hamper the HTC calculations for jet and spray cooling. Hence, an augmented hybrid model is used here to combine the analytical and data-driven techniques for the calculation of the interfacial force.

To start generating data for the calculation of the bubble drag force, a series of numerical simulations using detailed CFD modelings have been performed. A scenario table has then been defined for the drag force calculations based on varying bubble diameters and flow velocities. Table 4 shows the scenario table for the numerical simulations where different scenarios based on bubble diameters and water flow rate (e.g., water velocity at nozzles) are defined. For the data handling and postprocessing of the drag force data, the GASR technique has been employed. GASR is one of the interesting data processing techniques where measured and/or calculated data can be fitted by suitable mathematical formulae (from a different family of functions) using genetic and\or evolutionary algorithms. The technique has gradually been developed for computer implementation in recent years and some computer tools are already available based on GASR technologies. HeuristicLab [6] is one of the open-source academic software tools which have been developed to deal with a variety of data-driven modeling problems. It is prominently useful for problems where computational simulations are combined with optimization and design features within science and engineering research activities.

Table 4. Numerical simulation scenarios for different bubble diameters and water flow rates.

Dia. [m] \ Vel. [m/s]	0.1	0.2	0.5	0.8	1	1.5	2	2.5	3	4
0.001	×	×	×	×	×	×	×	×	×	×
0.002	×	×	×	×	×	×	×	×	×	×
0.005	×	×	×	×	×	×	×	×	×	×
0.02	×	×	×	×	×	×	×	×	×	×
0.05	×	×	×	×	×	×	×	×	×	×
0.1	×	×	×	×	×	×	×	×	×	×

The verification of drag force calculation for a single bubble has initially been carried out using simplified analytical and 2D detailed CFD techniques. One of the simple and popular analytical technique which can be used to approximate the bubble drag force can be written as [36–40];

$$F_D = C_D \frac{1}{8} \rho_c A_{int} |v_r| v_r{}^\alpha \tag{15}$$

$$C_D = \frac{24}{Re_b}\left(1 + 0.15 Re_b{}^{0.678}\right) \quad Re_b = \frac{\rho_c |v_r| d}{\mu_c} \tag{16}$$

where C_D, ρ_c, $A_{int} = \frac{\pi d^2}{4}$, d, α, and v_r are the drag coefficient, density of continuous phase (water), interfacial area, bubble diameter, correction factor, and the phase's relative velocity ($v_r = v_d - v_c$, the velocity difference between continuous and disperse phases), respectively. Figure 5 shows the comparison of drag forces for a single bubble using the simplified analytical (red) and the detailed CFD techniques (blue), assuming $d = 2$ mm, $\alpha = 1.5$, and $\rho_c = 1000$ kg/m^3.

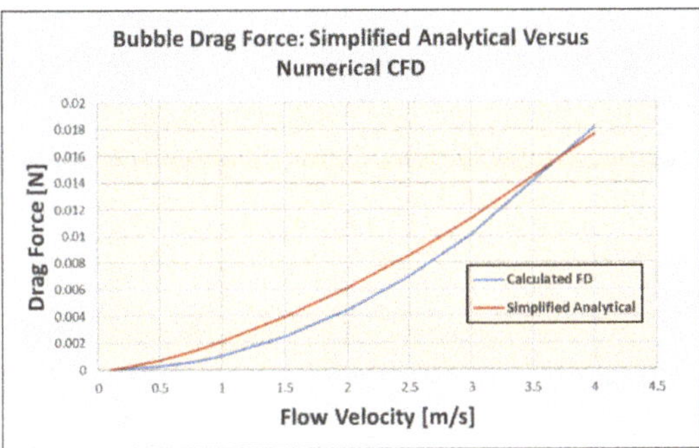

Figure 5. Comparison of bubble drag forces for simplified analytical and CFD technique.

The drag force data for the GASR multi-dimension regression analyses have then been generated using the results of the detailed CFD simulations. For the initial velocity field near the hot surface, global transient CFD analyses have been performed to calculate the water velocity field near the wall where bubbles are generated. Figure 6a,b show the global CFD simulation for the water cooling impingement on the billet surface and the thermal boundary setup for the vertical casting process. While Figure 6c show the detailed CFD simulations for the bubble dynamics for two different bubble diameters.

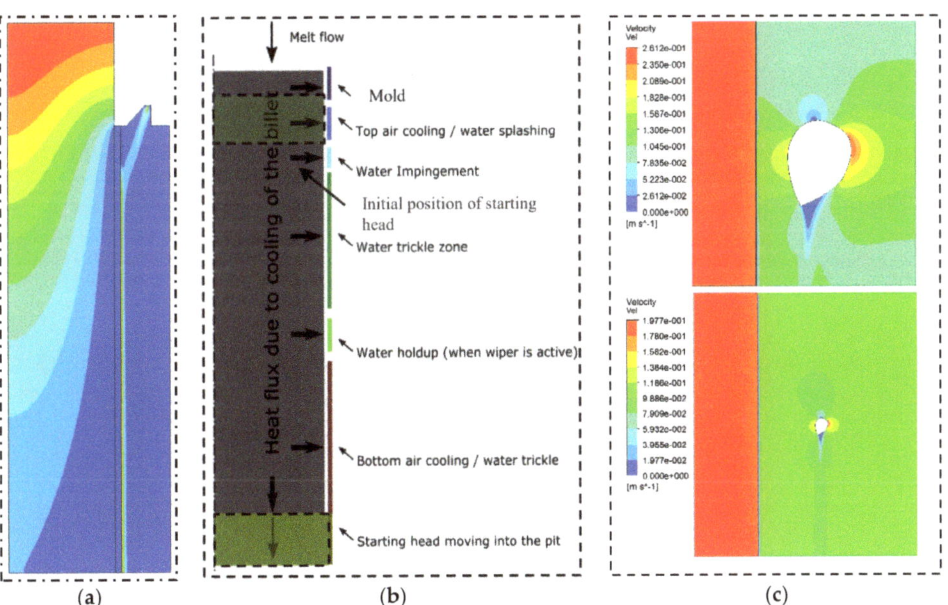

Figure 6. (a) Global CFD simulation for water impingement; (b) thermal boundary setup for vertical casting process; (c) local velocity results of detailed CFD bubble dynamics simulations for two different bubble diameters.

The calculation of bubble's drag forces using detailed CFD simulations has broadly been carried out for all planned scenarios and a small database has been generated for GASR training. In the work herein, the method of symbolic regression is freely used to obtain mathematical expressions of functions that can fit the cooling data based on simple rules and broad generalization. It differs from the conventional regression analyses (linear or nonlinear) where parameters are optimized in the mathematical model, rather it generates a routine to create models and their parameters for better a fitting scheme or for the purpose of getting better insights into the dataset.

Figure 7 shows the graphical presentation of the tree diagram for GASR function fitting along with resulting expressions for the bubble drag force with varying water velocities and the bubble diameters. The calculation of HTC values on the interface between the hot billet surface and water is generally governed by the classical thermal energy transfer law;

$$q = HTC \left(T_b - T_w \right) \tag{17}$$

where T_b, T_w are temperatures of billet surface and impinging water. As the water flow momentum\velocity and pressure vary over the billet surface during the cooling process, the HTC is not uniform over the whole billet surface. Additionally, the HTC calculations for the casting processes can be split into four distinct temperature ranges (see Table 5), namely;

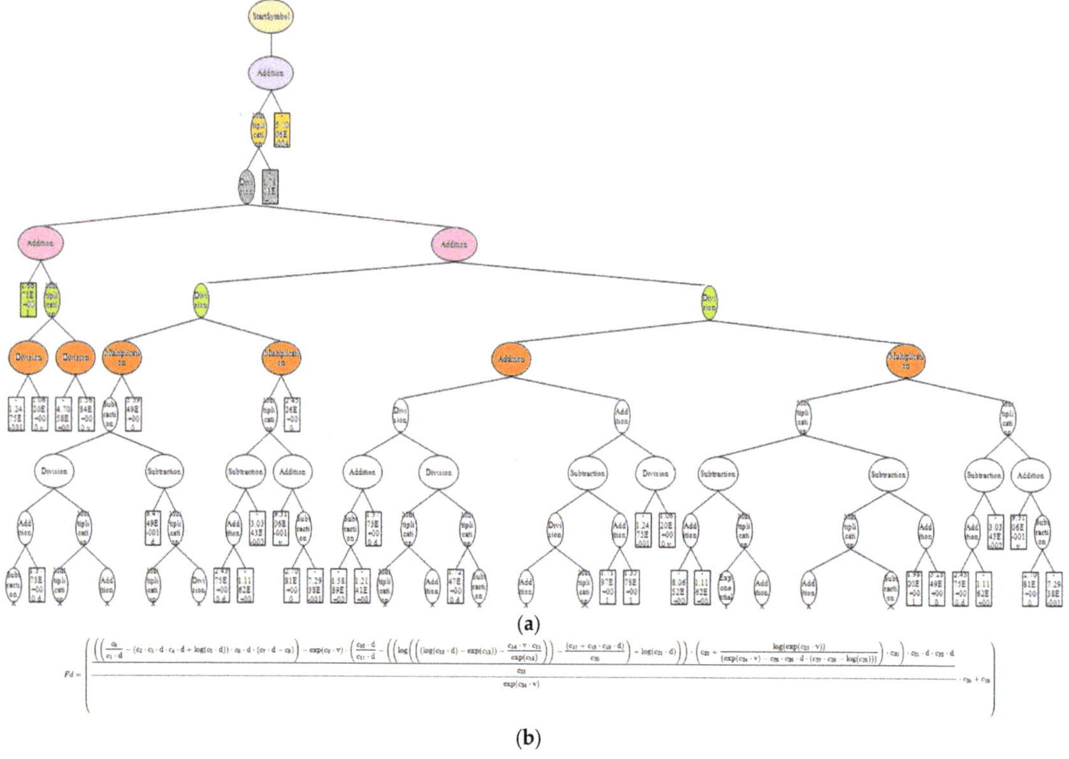

(a)

(b)

Figure 7. (**a**) Graphical representation of tree-diagram for GASR function fitting; (**b**) Resulting function for the bubble drag force with varying water velocities and the bubble diameters.

Table 5. New hybrid HTCs' equations for water impingement.

Zone Number	HTC Estimations
I–pre-boiling zone $40 < T_b < 100\,°C$	$HTC_{pre} = \alpha \left(\frac{Q}{T_b} \right)^{\beta} (7e2 * T_b - 1.67e3)$
II–nucleate boiling zone $100 < T_b < 235$	$HTC_{nu} = \left(\frac{\mu H_{fg}}{T_{sat}\, C_f^{\alpha}} \right) \left(\frac{C_p T_b}{H_{fg}} \right)^{\alpha} \left(\frac{\sigma \mu C_p}{K\, g\, (\rho_w - \rho_v)} \right)^{\beta}$
III–transition boiling zone $235 < T_b < 400$	$HTC_{tra} = HTC_{nu} \left(1 - \left(\frac{T_b - T_{CHF}}{T_b} \right)^{\gamma} \right)$
IV–film boiling zone $400 < T_b < 550$	$HTC_f = HTC_{tra} \left(1 - \left(\frac{T_b - T_{Le}}{T_b} \right) (H_1 + H_2)^{\lambda} \right)$ $H_1 = \frac{\gamma}{T_b} \times \left(\dfrac{k_w^3\, g\, \rho_w\, H_{fg}(\rho_w - \rho_v) \left(1 + \frac{c_p T_b}{2H_{fg}} \right)}{\mu_w\, T_b \left(\frac{\sigma}{g\,(\rho_w - \rho_v)} \right)^{\beta}} \right)^{\alpha}$ $H_2 = \frac{5.4 \times 10^4}{T_b} (1 + 0.055 T_b)\, v_s^{\kappa}$

➤ Pre-boiling: surface temperature below the water boiling temperature ($T < 100\,°C$) where single-phase calculation can be used to estimate the HTC values

➤ Nucleate boiling: surface temperature between water boiling temperature and Critical Heat Flux (CHF) temperature where nucleate the boiling regime is dominant. The formation of the discrete bubbles and their movement under drag, lift, and Frank lubrication forces (as discussed earlier) enhances the local fluid motion resulting in an increasing convective HTC. At higher billet surface temperatures near the CHF temperature, the discrete bubbles would coalesce into large size bubbles which further enhances the water flow (the so-called "fully developed nucleate boiling regime"). The rate of change for the surface HTC in the nucleate boiling regime is significant even with small surface temperature changes. For many industrial casting applications, the water-sprayed nucleate boiling regime is preferred since it is generating a high cooling rate.

➤ Transition boiling: surface temperature between CHF and Leidenfrost temperatures where transition boiling regime is taking place. The transition boiling regime and its modeling is one of the challenging issues for HTC calculations during industrial processes. In this type of boiling regime, the surface thermal energy exchange reduces as the billet surface temperature increases.

➤ Film boiling: surface temperature above the Leidenfrost temperature where vapor films are covering the surface of the hot billet. The modeling of cooling for this temperature range can be divided into several cooling regimes based on the different interface of fluid (water) and vapor as continuous, discrete, stable, and unstable.

The continuous HTC functions for the variation of billet surface temperatures have to be defined from hotter temperatures at the top of the billet to the cooler temperatures at the lower elevations to enable the simulation of the whole casting process at an industrial scale.

7. Hybrid-Evolving Technique: Case Study

To investigate the application of the evolving domain technique combined with the hybrid cooling models for industrial casting processes, a conventional vertical casting process has been simulated. As stated earlier, the conventional approach for dynamic systems like casting processes on FE solvers is to pre-mesh the whole cast domain (since the final shape of the domain is known) and to activate elements during the simulation. The CPU computational time and memory storage issues are the main drawbacks of such an approach which restricted the application of the technique for practical industrial casting simulations (especially for limited in-core computer storage space). Hence, for the continuous casting applications herein, a gradual extension of the numerical domain during the solution could avoid solving large matrices for all time steps and save computer resources during the calculation.

As part of the multi-physical and multi-phase simulation framework, the hybrid-evolving technique deals with the cooling phenomena at boundaries while the dynamic mesh module with its efficient parallel-processing scheme can perform fast and accurate FE analyses. For the casting process case study herein, the combination of hybrid physical-data driven technique and the dynamic generative computational scheme has been employed to promote the application of these new trends in the material science

7.1. HTC Estimations

For the vertical casting application, the thermal energy transfer during the process is complicated and cumbersome to accurately predict for the final FE simulations. The casting HTC values change rapidly during the startup condition where different cooling regimes add to the modeling complications. The HTCs generally rise rapidly with the increasing billet surface temperatures until a critical temperature is reached where the unstable and film boiling regimes would reduce the associated HTCs. As stated earlier, the estimation of CHF temperature point and its associated heat flux can help to understand and optimize the billet surface cooling rates and thermal-mechanical stress\strain conditions within the cast billet. The critical billet surface temperature for this study can be estimated at 220 °C to 275 °C (depending on surface roughness and water quality) where cooling on the hot billet surface is achieved by heat transfer from the billet to a free-falling film of water running down on the outside of the billet surface.

Using the analytical and hybrid physical-data-driven models presented in the previous section, a new set of HTC equations has been derived to model the heat energy evolution during the casting process. Appendix A present a brief description of the basic mathematical concepts and their parameter definitions while Table 5 and Figure 8 show the final derived HTC equations and their graphical representations. Although the use of a hybrid modeling technique results in long equations for the HTC estimations, the computer implementations of these equations would ease the calculation times.

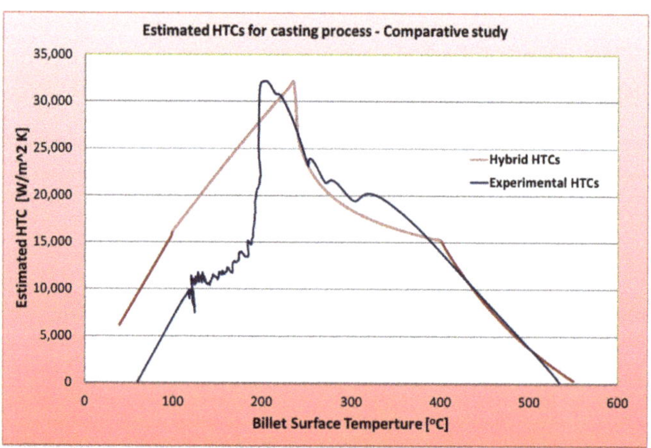

Figure 8. Comparison of calculated HTCs for hybrid and experimental methods.

As it can be seen in Figure 8, the HTC values are rising towards the CHF point where maximum cooling can be achieved for both experimental measurements and the hybrid model. Although the maximum thermal heat transfer for both models is similar the HTC measured and calculated values below CHF shows some inconsistency. At first glance, it seems that the hybrid model is overestimating the HTC values for billet surface temperatures between 60 °C to 230 °C. The experimental HTC graph is calculated based on the method in [41–43] where temperature measurements were obtained using a set of installed thermo-couples. Figure 9 shows the in-house testing apparatus for plate cooling tests along with the thermo-couple arrangements and the view of the water jet impingement. Using the apparatus, water can be impinged on a fixed and stationary plate using an appropriate size nozzle. Due to difficulties of measurements on the surface of the hot plate, the thermo-couples were mounted on the back of the plate inside a set of pre-drilled holes. The distance of these thermo-couples to the surface of the plate is a minimum 3 mm which makes the recorded temperatures within the bulk of the plate (not exactly on the surface). Hence, for these thermo-couples, the large thermal gradient forming at the very near of the plate surface cannot be captured accurately. Subsequently, the final HTC estimations using the experimental data seem to show underestimated values (e.g., showing zero HTC at about 60 °C) which may explain some of the inconsistencies between the recorded and calculated values at the lower temperatures.

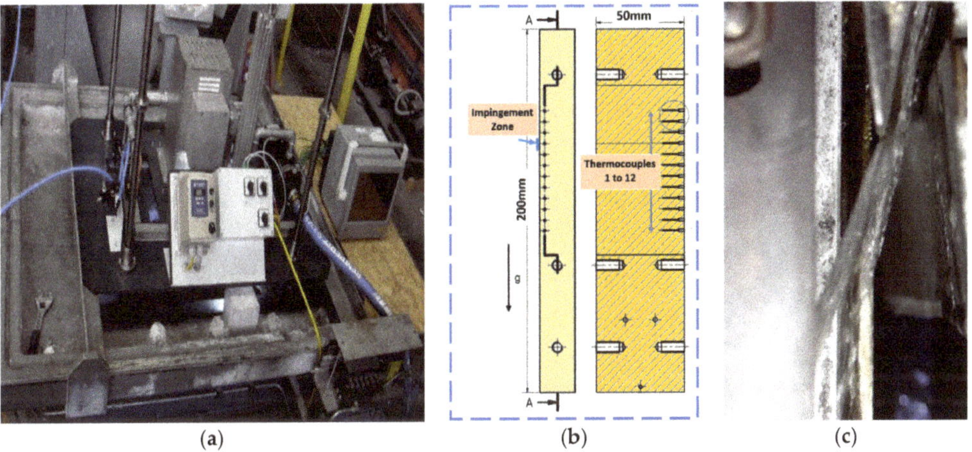

Figure 9. (**a**) Plate cooling test apparatus; (**b**) thermo-couple arrangements for the plate, (**c**) side view of water jet impingement.

7.2. Hybrid Evolving Simulation

In this case study, an industrial contact-less vertical casting process with rectangular cross-section has been modeled using the dynamic and evolving domain technique [4]. To avoid the complicated mold-filling and fluid-thermal-mechanical interaction simulations, a simple filling condition has been estimated and an initial state of the billet is assumed (e.g., after 50 mm of casting) along with initial thermal conditions. A preliminary structured mesh has been generated with thermal and displacement degrees of freedom. The convection and radiation of the melt (at the top of the mold) are taken into account using the results of free-convention CFD simulations (shown in Appendix B, Figure A1).

For the water-cooling HTCs, the values based on the hybrid modeling have been implemented and used for the rectangular quarter-model billet (using input parameters shown in Appendix B, Table A1). Figure 10 shows the FE simulation of the billet and the resulting cooling curves for the Hybrid-evolving model. Figure 10a,b show the generation of dynamic mesh layers with their temperature contours during the evolving scheme for the billet quarter model, while Figure 10c shows the temperature comparison for the measured and simulated billet surface temperatures below the water impingement zone. As it appears from the comparison of the cooling curves, although the hybrid cooling model can predict the temperature gradient at the start of the casting process relatively well, it shows some degree of instability in the transition boiling zone (temperature between 280 °C to 400 °C). Despite all the efforts to adjust the data on the simulated HTCs (results using the hybrid cooling model) to the phenomena at this temperature range, due to the difficulties in the modeling of the stochastic unstable film formation on the billet surface (film formation and breakage) the results show some degrees of discrepancy.

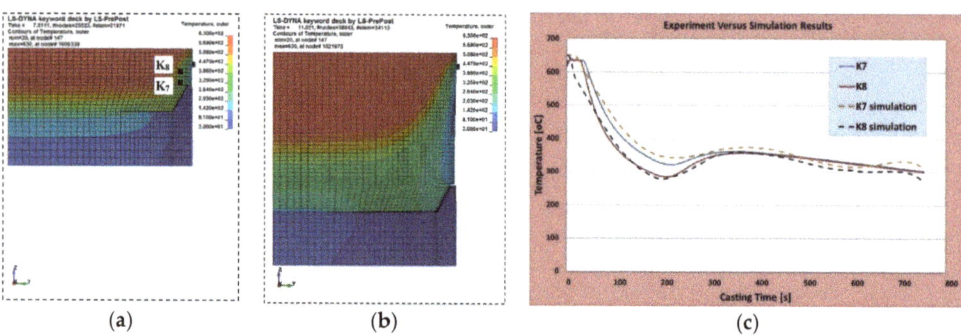

(a)   (b)   (c)

Figure 10. (a,b) Generation of dynamic mesh layers with overlaying temperature contours; (c) temperature curves for measured and simulated methods.

8. Conclusions

The hybrid-evolving approach introduced in this paper employs a new numerical evolving-domain technique combined with analytical-numerical-data driven hybrid models to simulate industrial casting processes. One of the advantages of this combined approach is to increase the accuracy and flexibilities of the process simulations while limiting the required computational time and resources. The technique relishes enough agility and flexibility to be implemented on mainstream commercial solvers for material processes. The framework is also easily adaptable to various dynamic material processes and provides a practical simulation routine for these processes at an industrial scale.

At the start of the manuscript, some technical aspects of the new evolving domain and dynamic mesh technique have shortly been presented and the computational performance of the technique is investigated. In the following parts of the paper, the contagious concepts of multi-phase cooling modeling combined with its augmented genetic-algorithm ML model have briefly been scrutinized and the potential application of the method for industrial material processes are highlighted using a case study for the vertical casting application.

At the first glance, it can be presumed that for the evolving domain technique the mandatory repeated initializations and solver matrices' re-assembly processes at every restart can introduce some disadvantages in terms of computational efficiency. However, detailed computational investigations have proved that processing smaller size matrices used throughout the simulations in this technique can outperform the conventional simulation techniques, especially on the parallel computational nodes. Furthermore, there is a possibility to shorten IO time for these initializations using embedded mesh data within the active computer memory of the FE solver. Thereby, the required input data could be passed to the solver directly instead of the conventional way via external input files.

As a final statement, the intention herein is to promote and encourage the use of a combination of hybrid modeling techniques with the latest developments on dynamic process simulations to upgrade computational material science to the next level. The opportunity to follow the trend on the new hybrid modeling schemes for further material processes would be taken and this would be the subject for the extension of the research work presented herein.

Author Contributions: The simulation of test scenarios along with partial writing of the manuscript has been carried out by J.K. where number of simulations on different size numerical models have been performed. He has also completed the post-processing of the numerical simulation results for presentation within the manuscript. The conceptual and fundamental technical arguments of the manuscript have been contributed by A.M.H., where basic theory and design of numerical framework have been elaborated. The draft preparation and editing of the manuscript have been shared by both authors. All authors have read and agreed to the published version of the manuscript.

Funding: This research was funded by the Federal State of Upper Austria in the FD Framework (within PSHeRo:ER Project OÖ Fin-010104/187).

Institutional Review Board Statement: Not applicable.

Informed Consent Statement: Not applicable.

Data Availability Statement: The sharing of raw-data required to reproduce the case studies are considered upon request by readers.

Acknowledgments: Authors would like to thank the Austrian Federal Ministry for Climate Action, Environment, Energy, Mobility, Innovation and Technology (in German, BMK), the Federal State of Upper Austria in the FD Framework (within PSHeRo:ER project OÖ Fin-010104/187) and also the Austrian Institute of Technology (AIT) for the technical/financial support in this research work. Authors would also like to thank DI Stefan Scheiblhofer, a former employee of the LKR Light Metals Technologies, for his technical input and contributions into the manuscript. The final review and proofing of the manuscript have been carried out internally by the LKR.

Conflicts of Interest: The authors declare no conflict of interest.

Appendix A

The conventional analytical approach for the calculation of HTC values in different temperature ranges can briefly be summarized for quick estimations. For the pre-boil temperature range (0 to 100 °C), a single-phase liquid (water) spray impingement cooling regime can be assumed where the HTC estimation is generally governed by the classical thermal energy transfer law;

$$q = HTC \left(T_b - T_w \right) \tag{A1}$$

where T_b, T_w are temperatures of the billet surface and impinging water. As the water flow momentum\velocity and pressure vary over the billet surface during cooling processes, HTCs are not uniform over the whole billet surface. If it is assumed that the spray flow is laminar (low velocity) and the surface of the billet is smooth and uniform, the following correlation functions can be derived for the Nusselt number distribution (Nusselt number is the ratio of convective to conductive heat transfer normal to the billet surface) at impingement point as [36–40];

$$N_u = 0.7212 \, Re^{0.5} Pr_L^{0.4} \quad 0.7 \le Pr \le 3 \tag{A2}$$

$$N_u = 0.7212 \, Re^{0.5} Pr_L^{0.37} \quad 3 \le Pr \le 10 \tag{A3}$$

where Re and Pr_L are Reynolds and Prandtl numbers. For the water flowing parallel to the billet surface below the impingement zone (trailing water), an approximate correlation function can be defined as;

$$N_u = 1.5874 \, Re^{0.33} Pr_L^{0.33} \left(25.735 \frac{\bar{r}^3}{Re} + 0.8566 \right)^{-0.67} \tag{A4}$$

where $\bar{r}$ is a non-dimensional radius (ratio of nozzle to impingement area radius). For HTCs during the nucleate boiling regime (temperature range between boiling and CHF), the simple empirical relationship for the thermal flux has been proposed by [44] as;

$$\frac{q_{nb}}{\mu.h_{fg}}\left(\frac{\sigma}{g(\rho_f - \rho_v)}\right)^{0.5} = \left(\frac{1}{C_{sf}}\left(\frac{c_{pf}\Delta T}{h_{fg}}\right)Pr_f^{-1.7}\right)^{1.95} \tag{A5}$$

where ρ_f, ρ_v, σ, μ, h_{fg}, c_{pf} and Pr_f are the density of water and vapor, surface tension, water dynamic viscosity, latent heat of evaporation from water to vapor, specific heat of water and water spray Prandtl number, respectively. The C_{sf} is a constant which depends on the interaction of the water spray with the billet surface and can be assumed as 3.07×10^{-3}. The use of this empirical equation is limited to certain ranges of water flow rates and temperature differences. The better empirical relationship which takes into account the water flow rates (spray velocity) and wider density ratios is proposed as [45,46];

$$\frac{q_c}{\rho_v.h_{fg}v_s} = 0.221 \times \left(\frac{\rho_w}{\rho_v}\right)^{0.645}\left(1+\frac{L}{l}\right)^{-0.364}\left(\frac{2\sigma}{\rho_l v_j^2(L-l)}\right)^{0.343} \tag{A6}$$

where q_c, ρ_w, v_s, L and l are critical heat flux (maximum heat flux), water density, velocity of water spray and characteristic length of heated surface and subcooled zones, respectively. This empirical function is valid for wide ranges of fluid to vapor density ratios (up to 1600) and Weber numbers (up to 5×10^6). The Weber number can be defined as $e = \frac{v_s^2 \rho_w L}{\sigma}$. For the film boiling regime, a simplified practical formulation can be adopted to estimate the maximum film boiling heat flux using simplified principals as [47];

$$q = 0.425 \times \left(\frac{k_f^3\, g\, \rho_f\, h_{pf}\left(\rho_w - \rho_f\right)\left(1 + \frac{c_{pf}\Delta T_{sat}}{2h_{pf}}\right)}{\left(\mu_w\, \Delta T_{sat}\left(\frac{\sigma}{g\,(\rho_w-\rho_f)}\right)^{0.5}\right)}\right)^{0.25} \tag{A7}$$

where ρ_w, ρ_f, k_f, μ_w, h_{pf}, c_{pf} and σ are density of water and vapor film, film thermal conductivity, water dynamic viscosity, latent heat of evaporation from water to vapor film, specific heat of water and surface tension, respectively.

Appendix B

The estimation of HTCs for the air cooling and water impingement boiling regimes is essential for any industrial material processes. The modeling of the convection, conduction, and radiation phenomena during cooling processes and their resulting thermal evolution within billets and parts is a cumbersome task. The convective HTC for the air cooling (for free convection) can be estimated using [36–40];

$$h_c = \frac{kN_u}{\lambda} \tag{A8}$$

where N_u is Nusselt number, k is the air thermal conductivity and λ is the characteristic length (for casting application it can be assumed as a distance from the melt-pool top surface to the water impinging point). The radiation HTC for air cooling can be assumed as;

$$h_r = \varepsilon\, \sigma\, T^4 / C^{\alpha T} \tag{A9}$$

where C and α are correction factors, σ is a constant (Stefan–Boltzmann constant) which can be assumed as 5.672×10^{-8} W/m^2-K^4 for casting applications and ε is the billet surface emissivity (for aluminum billets it can be assumed as 0.19 W/m^2). The C and α are correction factors which can be estimated using an initial free-convection CFD simulation of the casting process. Figure A1 shows the variation of radiation HTCs with billet surface temperature during the semi-continuous casting process of aluminum alloys

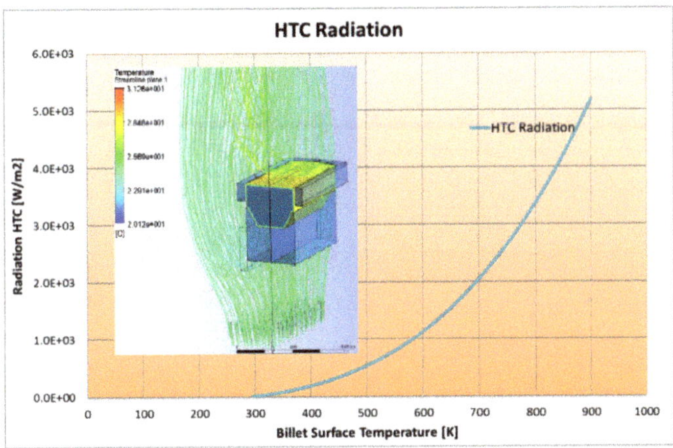

Figure A1. Radiation HTC for a range of billet surface temperatures with overlaying image of CFD free convection-radiation simulation.

For the hybrid-evolving framework used for the simulation of the casting process in this study, the water impingement process and its HTC curves are employed to extract thermal energy and to cool down the billet in a controlled manner. This would generally give the opportunity to control the microstructure formation, reducing thermal residual stresses and minimizing defects on the final billet. The transient fluctuations of HTCs during the process and its effects on solidification and microstructure formation of the cast billets are the most challenging parts of hybrid analytical-numerical simulation for cooling processes.

Hence, the proposed hybrid technique herein, has been focused on combining sound and simple analytical, efficient data-driven training techniques and a set of sequential detailed simulations in a way to achieve more accurate results in a reasonable computational time. Moreover, the overall planned simulation framework includes modules for the cooling and solidification which would be formulated and integrated into a commercial-based software platform as;

- HTC calculator using the hybrid cooling models for the water spray process (using augmenting hybrid method)
- Solidification module [48] for incorporating the change of phase into melt flow during casting processes (not discussed in this paper)
- Controlling module for mixing the cooling and solidification modules using in-house coding

Additionally, in this research work, special attention has been devoted to the development of more practical and affordable models with proper engineering accuracy and limited computational time and resources. Table A1 describes the parameters for the new hybrid HTC functions which have been implemented for the casting case study herein.

Table A1. Definition of parameters for hybrid HTCs equations.

Zone Number	Parameters	Validity
I–pre-boiling zone	Q–water flow rate [m^3/s] L_b–billet perimeter [m] T_b –billet surface temp [°C] $\alpha = 1.35$ $\beta = 0.3$	$40 < T_b < 100$
II–nucleate boiling zone	$\mu = 0.000267$ [Ns/m^2] $H_{fg} = 2257e3$ [J/kg] $c_f = 0.016$ [J/s] $C_p = 4219$ [J/kg K] $K = 0.68$ [W/m K] $g = 9.81$ [m/s^2] $r_w = 995$ [kg/m^3] $r_v = 0.6$ [kg/m^3] $\sigma = 0.059$ [N/m] $\alpha = 0.8$ $\beta = -0.515$	$100 < T_b < 235$
III–transition boiling zone	$T_{CHF} = 235$ [°C] $g = 0.42$	$235 < T_b < 400$
IV–film boiling zone	$v_s = 1$ [m/s] $\alpha = 0.4$ $\beta = 0.5$ $\gamma = 0.5$ $k = 0.6$ $l = 0.15$	$400 < T_b < 550$

References

1. Horr, A.M. Multi-resolution simulations of material processes for e-mobility applications. In Proceedings of the 7th Transport Research Arena TRA 2018, Vienna, Austria, 16–19 April 2018. [CrossRef]
2. Farias A de de Almeida, S.L.R.; Delijaicov, S.; Seriacopi, V.; Bordinassi, E.C. Simple machine learning allied with data-driven methods for monitoring tool wear in machining processes. *Int. J. Adv. Manuf. Technol.* **2020**, *109*, 2491–2501. [CrossRef]
3. Tao, F.; Cheng, J.; Qi, Q.; Zhang, M.; Zhang, H.; Fangyuan, S. Digital twin-driven product design, manufacturing and service with big data. *Int. J. Adv. Manuf. Technol.* **2018**, *94*, 3563–3576. [CrossRef]
4. Horr, A.M.; Kronsteiner, J. On Numerical Simulation of Casting in New Foundries: Dynamic Process Simulations. *Metals* **2020**, *10*, 886. [CrossRef]
5. Horr, A.M. Notes on New Physical & Hybrid Modelling Trends for Material Process Simulations. *J. Phys. Conf. Ser.* **2020**, *1603*, 012008.
6. Wagner, S.; Affenzeller, M. HeuristicLab: A Generic and Extensible Optimization Environment. In Proceedings of the International Conference in Coimbra, Portugal, 18–19 July 2005; Adaptive and Natural Computing Algorithms; Springer: Berlin/Heidelberg, Germany, 2005; pp. 538–541.
7. Vijayaram, T.R.; Sulaiman, S.; Hamouda, A.M.S.; Ahmad, M.H.M. Numerical simulation of casting solidification in permanent metallic molds. *J. Mat. Pro. Technol.* **2006**, *178*, 29–33. [CrossRef]
8. Zhihong, G.; Hua, H.; Yuhong, Z.; Shuwei, Q. Numerical Simulation of Squeeze Casting of AZ91D Magnesium Alloy. In Proceedings of the 2010 International Conference on Digital Manufacturing & Automation, Changcha, China, 8–20 December 2010; pp. 30–33.
9. Brötz, S.; Horr, A.M. Framework for progressive adaption of FE mesh to simulate generative manufacturing processes. *J. Manuf. Lett.* **2020**, *24*, 52–55. [CrossRef]
10. Lindwall, A.L.J.; Lindgren, L.E. Thermal FE-simulation of PBF using adaptive meshing and time stepping. In Proceedings of the Simulation for Additive Manufacturing, Munich, Germany, 11–13 October 2017; pp. 62–63.
11. Horr, A.M. Computational Evolving Technique for Casting Process of Alloys. *J. Math. Probl. Eng.* **2019**, *2019*, 6164092. [CrossRef]
12. Claus, S.; Bigot, S.; Kerfriden, P. Cutfem Method for Stefan-Signorini Problems with Application in Pulsed Laser Ablation. *SIAM J. Sci.* **2018**, *40*, B1444–B1469. [CrossRef]
13. Bastian, P.; Engwer, C. An unfitted finite element method using discontinuous Galerkin. *J. Num. Meth. Eng.* **2009**, *79*, 1557–1576. [CrossRef]

14. Fausty, J.; Bozzolo, N.; Muñoz, D.; Bernacki, M. A novel level-set finite element formulation for grain growth with heterogeneous grain boundary energies. *J. Mat. Des.* **2018**, *160*, 578–590. [CrossRef]
15. Kleeberg, C. Latest Advancements in Modelling and Simulation for High Pressure Die Castings. In Proceedings of the ALUCAST, Chennai, India, 3–5 December 2010; Aluminium Casters' Association; pp. 1–18.
16. Haldenwanger, H.G.; Stich, A. Casting simulation as an innovation in the motor vehicle development process. In *Modeling of Casting, Welding and Advanced Solidification*; Process IX, SIM 2000; Sahm, R., Hansen, P.N., Conley, J.G., Eds.; United Engineering Foundation: Philadelphia, PA, USA, 2000.
17. Malik, I.; Anwar Sani, A.; Medi, A. Study on using Casting Simulation Software for Design and Analysis of Riser Shapes in a Solidifying Casting Component. *J. Phys. Conf. Ser.* **2020**, *1500*, 012036. [CrossRef]
18. Francis, L. *Materials Processing, A Unified Approach to Processing of Metals, Ceramics and Polymers*; Academic Press: Cambridge, MA, USA, 2015; pp. 10–13.
19. Zhang, L.; Allanore, A.; Wang, C.; Yurko, J.A.; Crapps, J. *Materials Processing Fundamentals*; John Wiley: Hoboken, NJ, USA, 2013; pp. 63–72.
20. Horr, A.M. Integrated Material Process Simulation for Lightweight Metal Products. In *Computational Methods and Experimental Measurements XVII*; WIT Transactions on Modelling and Simulation; WIT Press: Southampton, UK, 2015; Volume 1, pp. 201–212.
21. Horr, A.M.; Kronsteiner, J.; Mühlstätter, C. Recent Advances in Aluminium Casting Simulation: Evolving Domains & Dynamic Meshing. In Proceedings of the Congress ALUMINUM, Verona, Italy, 20–24 June 2017; pp. 20–24.
22. Horr, A.M. Thermal Energy Approach for Aluminium Process Simulations: Source-Capacity Technique. *J. Mater. Today Proc.* **2019**, *10*, 263–270. [CrossRef]
23. Horr, A.M. On Analytical Concepts of Novel Multi-Resolution Casting Simulations. *Proc. IOP Conf. Ser. Mater. Sci. Eng.* **2019**, *529*, 012055. [CrossRef]
24. Liu, F.; Nashed, Z.; N'Guerekata, G.; Pokrajac, D.; Qiao, Z.; Shi, X.; Xia, X. *Advances in Applied and Computational Mathematics*; Nova Science Publishers: New York, NY, USA, 2006; pp. 38–46.
25. Cai, X.C.; Dryja, M.; Sarkis, M. Overlapping Nonmatching Grid Mortar Element Methods for Elliptic Problems. *SIAM J. Num Anal.* **1999**, *36*, 581–606. [CrossRef]
26. Mingyan, H.; Ziping, H.; Cheng, W.; Pengtao, S.; Shuang, Z. An overlapping domain decomposition method for a polymer exchange membrane fuel cell model. *Proc. Comp. Sci.* **2011**, *4*, 1343–1352.
27. Bellet, M.; Salazar-Betancourt, L.; Jaouen, O.; Costes, F. Modelling of Water Spray Cooling. Impact on Thermomechanics of Solid Shell and Automatic Monitoring to Keep Metallurgical Length Constant. In Proceedings of the 8th European Continuous Casting Conference ECCC, Graz, Austria, 23–26 June 2014; pp. 1202–1210.
28. Wendelstorf, J.; Spitzer, K.H.; Wendelstorf, R. Spray water cooling heat transfer at high temperatures and liquid mass fluxes. *Int. J. Heat Mass Transf.* **2008**, *51*, 4902–4910. [CrossRef]
29. Krause, F.; Schüttenberg, S.; Fritsching, U. Modelling and simulation of flow boiling heat transfer. *Int. J. Num. Meth. Heat Fluid Flow* **2010**, *20*, 312–331. [CrossRef]
30. Horr, A.M. Optimisation of Aluminium Casting Technology: Thermal Energy Approach. In Proceedings of the Euro Aluminium Congress, Dusseldorf, Germany, 25–26 November 2015; Session 2. pp. 1–13.
31. Horr, A.M.; Betz, A. Advanced fluid-thermal-mechanical casting simulation of lightweight alloys. In Proceedings of the International Symposium of Liquid Metal Processing & Casting, Leoben, Austria, 20–24 September 2015; pp. 490–497.
32. Slayzak, S.J.; Viskanta, R.; Incropera, F.P. Effects of interactions between adjoining rows of circular, free-surface jets on local heat transfer from the impingement surface. *ASME J. Heat Transf.* **1994**, *116*, 88–95. [CrossRef]
33. Lienhard, J.H. Liquid Jet Impingement. In *Annual Review of Heat Transfer*; Tien, C.L., Ed.; Begell House: New York, NY, USA, 1995; Chapter 4; Volume 6, pp. 199–270.
34. Montáns, F.J.; Chinesta, F.; Bombarelli, R.G.; Nathan Kutz, J. Data-driven modeling and learning in science and engineering. *C. R. Mécanique* **2019**, *347*, 845–855. [CrossRef]
35. Pan, Y.; Hu, M. A data-driven modeling approach for digital material additive manufacturing process planning. In Proceedings of the 2016 International Symposium on Flexible Automation (ISFA), Cleveland, OH, USA, 1–3 August 2016; pp. 223–228.
36. Wang, D.M.; Alajbegovič, A.; Su, X.M.; Jan, J. Numerical simulation of water quenching process of an engine cylinder head. In Proceedings of the 4th ASME/JSME Joint Fluids Engineering Conference, Honolulu, HI, USA, 6–10 July 2003; Volume 1, pp. 1571–1578.
37. Greif, D.; Kopun, R.; Kosir, N.; Zhang, D. Numerical Simulation Approach for Immersion Quenching of Aluminum and Steel Components. *Int. J. Auto. Eng.* **2017**, *8*, 45–49. [CrossRef]
38. Frank, T.; Shi, J.M.; Burns, A.D. Validation of Eulerian multiphase flow models for nuclear safety applications. In Proceedings of the 3rd International Symposium on Two-Phase Flow Modelling and Experimentation, Pisa, Italy, 22–25 September 2004.
39. Tomiyama, A. Struggle with computational bubble dynamics. In Proceedings of the ICMF '98, 3rd International Conference of Multiphase Flow, Lyon, France, 8–12 June 1998; pp. 1–18.
40. Ma, C.; Tian, Y.P.; Tian, Y.C.; Lei, D.H. Liquid jet impingement heat transfer with and without boiling. *J. Therm. Sci.* **1993**, *2*, 32–49. [CrossRef]
41. Kominek, J. Methodology of evaluation of heat transfer experiment on aluminum sample. In Proceedings of the 24th International Conference on Metallurgy and Materials, Brno, Czech Republic, 3–5 June 2015; pp. 1203–1208.

42. Raudensky, M. Heat transfer coefficient estimation by inverse conduction algorithm. *Int. J. Num Meth. Heat.* **1993**, *3*, 257–266. [CrossRef]
43. Pohanka, M.K.; Awoodbury, A. Downhill Simplex Method for Computation of Interfacial Heat Transfer Coefficients in Alloy Casting. *Inverse Probl. Eng.* **2003**, *11*, 409–424. [CrossRef]
44. Ruch, M.A.; Holman, J.P. Boiling Heat Transfer to a Freoon-113 Jet Impinging Upward onto a Flat Heated Surface. *Int. J. Heat Mass Transf.* **1975**, *18*, 51–60. [CrossRef]
45. Monde, M.; Kitajima, K.; Inoue, T.; Mitsutake, Y. Critical heat flux in a forced convective subcooled boiling with an impinging jet. *Inst. Chem. Eng. Sym. Ser. Heat Transf.* **1994**, *7*, 515–520.
46. Monde, M.; Okuma, Y. Critical heat flux in saturated forced convective boiling on a heated disk with an impinging jet–chf in l-regime. *Int. J. Heat Mass Transf.* **1985**, *28*, 547–552.
47. Berenson, P.J. Film-Boiling Heat Transfer from a Horizontal Surface. *J. Heat Transf.* **1961**, *83*, 351–356. [CrossRef]
48. Jäger, S.; Horr, A.M.; Scheiblhofer, S. Thermal evolution effects on solidification process during continuous casting. In Proceedings of the 6th Decennial International Conference on Solidification Processing, Old Windsor, UK, 25–28 July 2017; pp. 675–678.

Article

Mathematical Modelling of Isothermal Decomposition of Austenite in Steel

Božo Smoljan [1], Dario Iljkić [2],*, Sunčana Smokvina Hanza [2] and Krunoslav Hajdek [1]

[1] Department of Packaging, Recycling and Environmental Protection, University North, University Center Koprivnica, Trg Dr. Žarka Dolinara 1, 48000 Koprivnica, Croatia; bozo.smoljan@unin.hr (B.S.); krunoslav.hajdek@unin.hr (K.H.)

[2] Department of Materials Science and Engineering, Faculty of Engineering, University of Rijeka, Vukovarska 58, 51000 Rijeka, Croatia; suncana@riteh.hr

* Correspondence: darioi@riteh.hr; Tel.: +385-51-651-474

Abstract: The main goal of this paper is mathematical modelling and computer simulation of isothermal decomposition of austenite in steel. Mathematical modelling and computer simulation of isothermal decomposition of austenite nowadays is becoming an indispensable tool for the prediction of isothermal heat treatment results of steel. Besides that, the prediction of isothermal decomposition of austenite can be applied for understanding, optimization and control of microstructure composition and mechanical properties of steel. Isothermal decomposition of austenite is physically one of the most complex engineering processes. In this paper, methods for setting the kinetic expressions for prediction of isothermal decomposition of austenite into ferrite, pearlite or bainite were proposed. After that, based on the chemical composition of hypoeutectoid steels, the quantification of the parameters involved in kinetic expressions was performed. The established kinetic equations were applied in the prediction of microstructure composition of hypoeutectoid steels.

Keywords: mathematical modelling; computer simulation; austenite decomposition kinetics; microstructure transformations

Citation: Smoljan, B.; Iljkić, D.; Smokvina Hanza, S.; Hajdek, K. Mathematical Modelling of Isothermal Decomposition of Austenite in Steel. *Metals* **2021**, *11*, 1292. https://doi.org/10.3390/met11081292

Academic Editors: Jose Diaz and Henrik Saxen

Received: 15 June 2021
Accepted: 13 August 2021
Published: 16 August 2021

Publisher's Note: MDPI stays neutral with regard to jurisdictional claims in published maps and institutional affiliations.

1. Introduction

The research of the mathematical simulation of microstructure distribution in steel is one of the highest-priority research areas in the simulation of phenomena of the heat treatment of steel. By using the additivity rule and kinetic equations of isothermal decomposition of austenite, it is possible to calculate kinetics of austenite decomposition at continuous cooling of steel. The prediction of isothermal decomposition of austenite can be applied for understanding, optimization and control of microstructure composition and mechanical properties of steel [1–4].

The most common method of computer prediction of isothermal decomposition of austenite results is based on the chemical composition of steel by using time-temperature-transformation (TTT) diagrams [5].

Studies of the kinetics of isothermal decomposition of austenite have been intensified in the course of some pioneering studies on the isothermal decomposition of austenite [6–8].

The prediction of microstructure composition is usually based on semi-empirical methods derived from kinetic equations of microstructure transformation [9]. To describe the transformation kinetics by mathematical methods, a semi-empirical approach is employed using the Johnson–Mehl–Avrami–Kolmogorov (JMAK) equation together with additivity rule [10,11].

The phase transformations can be categorized into two categories: reconstructive phase transformations and displacive phase transformations. Decompositions of austenite into ferrite and pearlite in steels are typical examples of reconstructive phase transformations, while martensite, bainite, and Widmanstatten ferrite phase transformations can be recognized as displacive phase transformations [12].

The formation of ferrite occurs by nucleation at the austenite grain boundaries. After that the ferrite grows inside the austenite grains. The rate of volume fraction of the ferrite is a function of the nucleation rate and the velocity of the ferrite/austenite interface. The nucleation rate is primarily a function of the undercooling below the A_{e3} temperature and the grain size of austenite [13,14].

The nucleation mechanism of pearlite involves the formation of two phases, ferrite and cementite. The nucleation of cementite is a rate-limiting step in hypoeutectoid steels. The proeutectoid ferrite nucleates first and continues to grow with the same crystallographic orientation during the pearlite formation. For hypereutectoid steels, the role of the nucleation of ferrite is a limited process in comparison with the roles of the cementite nucleation. In eutectoid steel, the pearlite nucleation is assumed to occur at the austenite grain corners, edges, and boundaries.

Two different theories are proposed for the growth of pearlite. The Zener–Hillert theory assumes that the volume diffusion of carbon in the austenite is the rate-controlling mechanism [15,16]. In addition, Hillert theory assumes that grain boundary diffusion of the carbon atoms is the rate-controlling mechanism. The nucleation rate of pearlite follows the general nucleation theory [16].

The Johnson–Mehl–Avrami–Kolmogorov (JMAK) theory predicts the overall transformation rate on the basis of nucleation and growth rates. It is the most widely used model to describe the austenite–pearlite transformation kinetics [15].

Bainite was discovered nearly eight decades ago [17]. The research work carried out in the field of bainite is immense [17–20]. A qualitative theory to explain bainite formation still remains a subject of controversy [18,21]. One theory suggests a diffusion-controlled transformation where bainitic growth occurs by a diffusional ledge mechanism, while the other suggests that the bainite reaction is a displacive transformation [18]. Both theories have assumed models to predict the transformation kinetics [22,23]. The growth of bainite and Widmanstatten ferrite requires the partitioning of interstitial carbon. Because of this reason, their growth is controlled by diffusion of interstitial atoms of carbon [12].

Computer simulation of isothermal decomposition of austenite in steel is still a complex problem. The dependence of the physical quantities involved in the kinetic expressions of austenite decomposition has not yet been sufficiently defined in the literature. Efforts are being made to predict the dependence of physical quantities in the kinetic expressions of austenite decomposition on chemical composition.

This work proposes inversion methods for quantification of kinetic parameters and setting the kinetic expressions in the prediction of isothermal decomposition of austenite into ferrite, pearlite, or bainite.

The proposed method of setting kinetic relations can be used in the calculation of characteristic kinetic parameters for other groups of steel. The established model can be used for computer simulation of austenite decomposition in other steels with similar chemical compositions.

2. Materials and Methods

2.1. Methods for Estimation of Kinetic Parameters of Austenite Isothermal Decomposition

2.1.1. Kinetics Expressions of Austenite Decomposition in an Incremental Form

Kinetics of isothermal decomposition of austenite can be defined by Avrami's isothermal equation:

$$X = 1 - \exp(-kt^n) \tag{1}$$

where X is transformed part of the microstructure, t is time, and k and n are kinetic parameters. By extracting the time component, Equation (1) can be written as:

$$t = \frac{1}{k^{\frac{1}{n}}} \left(\ln\left(\frac{1}{1-X}\right) \right)^{\frac{1}{n}} \tag{2}$$

In computer-based mathematical analysis, it is convenient to define the kinetics of austenite decomposition in an incremental form. By differentiating Avrami's equation, it follows that:

$$\frac{dX}{dt} = \exp(-kt^n)nkt^{n-1} \tag{3}$$

After introducing Equation (2) in Equation (3) and a short rearrangement, it follows that:

$$\frac{dX}{dt} = nk^{\frac{1}{n}}\left(\ln\frac{1}{1-X}\right)^{1-\frac{1}{n}}(1-X) \tag{4}$$

Equation (4) can be written in an incremental form, and the volume fraction $\Delta X^{(N)}$ of austenite transformed in the time interval $\Delta t^{(N)}$ can be calculated as [3]:

$$\Delta X^{(N)} = nk^{\frac{1}{n}}\left(\ln\frac{1}{1-X^{(N-1)}}\right)^{1-\frac{1}{n}}\left(1-X^{(N-1)}\right)\Delta t^{(N)} \tag{5}$$

where $X^{(N-1)}$ is the volume fraction of austenite transformed in previous $N-1$ time intervals. Kinetic parameters k and n can be evaluated inversely by using data of time of isothermal transformation. The total volume fraction of austenite transformed during isothermal decomposition can be calculated as:

$$X = \sum_{i=1}^{N}\Delta X^{(N)} \tag{6}$$

2.1.2. Ferrite Transformation

The ferrite transformation takes place by the mechanism of nucleation and growth, with the following assumed kinetic parameters [13]:

$$n_F = 4 \tag{7}$$

$$k_F = \frac{\pi}{3}I_F S G_F^3 \tag{8}$$

where S is the surface of austenite grain suitable for nucleation, while I_F is the nucleation rate and G_F is the growth rate defined as:

$$I_F = T^{-\frac{1}{2}}D_0\exp\left(-\frac{Q_{dif}}{RT}\right)\exp\left(-\frac{k_1}{RT(\Delta T)^2}\right) \tag{9}$$

$$G_F = \frac{c_\gamma - c_0}{c_\gamma - c_\alpha}D_0\exp\left(-\frac{Q_{dif}}{RT}\right)\frac{1}{y^D} \tag{10}$$

In Equations (9) and (10), T is temperature, ΔT is the undercooling below the critical temperature A_{e3}, D_0 is the material constant, R is the universal gas constant, Q_{dif} is the diffusion activation energy, while c_0, c_α, and c_γ are the concentrations of steel, ferrite and austenite at the boundary with ferrite, respectively. The effective diffusion length is defined as [7,13]:

$$y^D = \frac{k_2}{(\Delta T)^{n_1}} \tag{11}$$

where k_1, k_2 and n_1 are the kinetic parameters dependent on chemical composition of steel. After introducing Equations (9)–(11) in Equation (8), and after some modification, Equation (8) can be rewritten as [14]:

$$k_F = SD_0^4\exp\left(\frac{-4Q_{dif}}{RT}\right)\exp\left(\frac{-k_1}{RT(A_{e3}-T)^2}\right)\left(\frac{c_\gamma - c_0}{c_\gamma - c_\alpha}\right)^3\left(\frac{(A_{e3}-T)^{n_1}}{k_2}\right)^3 \tag{12}$$

To determine the values of the constants k_1, k_2 and n_1, it is first necessary to determine the value of the coefficient k_F for three temperatures, and then to solve a system of three equations with three unknowns (k_1, k_2, n_1).

It was assumed that the ferrite transformation does not take place to the end, but to the maximum volume $V_{max} = V_{rF} \times V$, when the normalized volume fraction of ferrite can be defined as [18]:

$$\tilde{\zeta}_F = \frac{X_F}{V_{rF}} \tag{13}$$

In Equation (13), X_F is the volume fraction of ferrite and V_{rF} is the relative volume of ferrite. The linear temperature dependence of the volume V_{rF} can be evaluated using an Fe-Fe$_3$C diagram with the following assumptions: at temperatures A_{e3} and B_s, the volume V_{rF} is equal to 0, while at temperature A_{e1}, it takes the maximum value $V_{rF} = c_0/0.8$ (Figure 1).

$$V_{rF} = a_1 + a_2T, \text{ for } Ae_3 > T \geq Ae_1 \tag{14}$$

where:

$$a_1 = \frac{A_{e3}}{A_{e3} - A_{e1}} \frac{c_0}{0.8} \tag{15}$$

$$a_2 = \frac{-1}{A_{e3} - A_{e1}} \frac{c_0}{0.8} \tag{16}$$

$$V_{rF} = a_3 + a_4T, \text{ for } Ae_1 > T \geq T_F \tag{17}$$

where:

$$a_3 = \frac{-T_F}{A_{e1} - T_F} \frac{c_0}{0.8} \tag{18}$$

$$a_4 = \frac{1}{A_{e1} - T_F} \frac{c_0}{0.8} \tag{19}$$

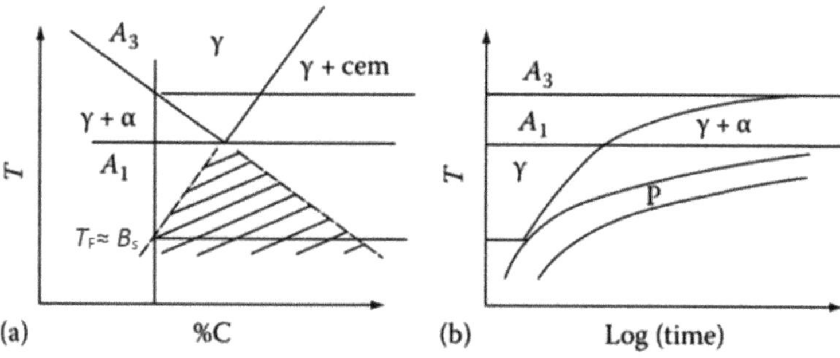

Figure 1. Effect of transformation temperature on the volume fraction of proeutectoid ferrite. (a) Scheme of an extension of the eutectoid field with a temperature of austenite decomposition in a Fe-C system; (b) scheme of a TTT diagram for hypoeutectoid steel.

The real volume of ferrite can be written as:

$$dV_F = \left(\frac{V_{max} - V_F}{V_{max}}\right) dV_{Fe} = \left(\frac{V_{rF}V - V_F}{V_{rF}V}\right) dV_{Fe} \tag{20}$$

$$dV_F = (1 - \tilde{\zeta}_F) dV_{Fe} \tag{21}$$

The extended volume of ferrite is defined as:

$$dV_{Fe} = \frac{4}{3}\pi I_F SG_F{}^3 V(t - t_i)^3 dt \tag{22}$$

where t_i is the incubation time. After introducing Equation (22) in Equation (21), it follows that:

$$V_{rF}\, d\zeta_F = (1 - \zeta_F)\frac{4}{3}\pi I_F SG_F{}^3(t - t_i)^3 dt \tag{23}$$

$$V_{rF}\, d\zeta_F = (1 - \zeta_F)4k_F(t - t_i)^3 dt \tag{24}$$

$$V_{rF}\,\frac{d\zeta_F}{(1 - \zeta_F)} = 4k_F(t - t_i)^3 dt \tag{25}$$

After integrating Equation (25), it follows that:

$$-\ln(1 - \zeta_F)V_{rF} = k_F t^4 \tag{26}$$

For small values of the normalized volume fraction of ferrite, Equation (26) can be written as:

$$-\ln(1 - X_F) = k_F t_i^4 \tag{27}$$

At any temperature, knowing an incubation time of ferrite transformation, found out from the IT diagram, the value of the kinetic parameter k_F can be written as:

$$k_F = \frac{-\ln\left(1 - X_{Ft_i}\right)}{(t_i)^4} = \frac{-\ln(1 - 0.01)}{(t_i)^4} \tag{28}$$

From Equation (26), follows the normalized volume fraction of ferrite which is:

$$\zeta_F = 1 - \exp\left(-\frac{k_F}{V_{rF}}t^4\right) \tag{29}$$

Equation (29) can be written in an incremental form, when the normalized volume fraction of ferrite formed by the mechanism of nucleation and growth in the time interval $\Delta t^{(N)}$ can be calculated as:

$$\Delta\zeta_F^{(N)} = 4\left(\frac{k_F}{V_{rF}}\right)^{\frac{1}{4}}\left(\ln\frac{1}{1 - \zeta_F^{(N-1)}}\right)^{\frac{3}{4}}\left(1 - \zeta_F^{(N-1)}\right)\Delta t^{(N)} \tag{30}$$

At any time of transformation, the real volume fraction of ferrite can be calculated as $X_F = \zeta_F V_{rF}$ (Equation (13)).

2.1.3. Bainite Transformation

The bainite transformation begins at a temperature B_s. Like ferrite transformation, it does not take place to the end, but to the maximum volume $V_{max} = V_{rB} \times V$, when the normalized volume fraction of bainite can be defined as [18]:

$$\zeta_B = \frac{X_B}{V_{rB}} \tag{31}$$

where V_{rB} is the relative volume of bainite, while X_B is the volume fraction of bainite defined as:

$$d\zeta_B V_{rB} = (1 - \zeta_B)I_B u\, dt \tag{32}$$

where u is the volume of the structural unit of bainite. The nucleation rate is defined as [14]:

$$I_B = T^{-\frac{1}{2}}D_0\exp\left(-\frac{Q_{dif} + k_6}{RT}\right)\exp\left(-\frac{k_7}{RT(\Delta T)^2}\right)\exp\left(-\frac{t_i}{t}\right) \tag{33}$$

where k_6 and k_7 are the kinetic parameters dependent on chemical composition of steel. After introducing Equation (33) to Equation (32), it follows that:

$$\frac{d\zeta_B \Delta V_{rB}}{(1 - \zeta_B)} = T^{-\frac{1}{2}} D_0 \exp\left(-\frac{Q_{dif} + k_6}{RT}\right) \exp\left(-\frac{k_7}{RT(\Delta T)^2}\right) \exp\left(-\frac{t_i}{t}\right) u\, dt \tag{34}$$

$$V_{rB} \int \frac{d\zeta_B}{(1 - \zeta_B)} = T^{-\frac{1}{2}} D_0 \exp\left(-\frac{Q_{dif} + k_6}{RT}\right) \exp\left(-\frac{k_7}{RT(\Delta T)^2}\right) u \int \exp\left(-\frac{t_i}{t}\right) dt \tag{35}$$

If the following is accepted [24]:

$$\int_0^t \exp\left(-\frac{t_i}{t}\right) dt \approx k t^n \tag{36}$$

Equation (36) can be rewritten as:

$$-\ln(1 - \zeta_B) V_{rB} = T^{-\frac{1}{2}} D_0 \exp\left(-\frac{Q_{dif} + k_6}{RT}\right) \exp\left(-\frac{k_7}{RT(\Delta T)^2}\right) k_8 t^{n_B} \tag{37}$$

where k_8 is the kinetic parameters. The time of bainite transformation can be expressed by:

$$t = \left[\frac{-\ln(1 - \zeta_B) V_{rB}}{T^{-\frac{1}{2}} D_0 \exp\left(-\frac{Q_{dif} + k_6}{RT}\right) \exp\left(-\frac{k_7}{RT(\Delta T)^2}\right) k_8}\right]^{\frac{1}{n_B}} \tag{38}$$

For small values of the normalized volume fraction of bainite it can be taken that $\ln(1 - \zeta_B) V_{rB} \approx \ln(1 - X_B)$; therefore, Equation (38) can be rewritten as:

$$t = \left[\frac{-\ln(1 - X_B)}{T^{-\frac{1}{2}} D_0 \exp\left(-\frac{Q_{dif} + k_6}{RT}\right) \exp\left(-\frac{k_7}{RT(\Delta T)^2}\right) k_8}\right]^{\frac{1}{n_B}} \tag{39}$$

As a rule, at low temperatures austenite is completely transformed into bainite, when it can be assumed that $V_{rB} \approx 1$ and $\zeta_B \approx X_B$. With this assumption, the kinetic parameter n_B can be defined as:

$$n_B \approx \frac{2.661}{\log(t_{0.99}) - \log(t_i)} \tag{40}$$

where t_i is the incubation time and $t_{0.99}$ is the finish time of the isothermal bainite transformation found out from the IT diagram. The denominator of Equation (39) is a function of temperature; therefore, for the incubation time and constant temperature, Equation (39) can be rewritten as:

$$t_i = t_{0.01} = \left[\frac{-\ln(1 - X_B)}{k_B}\right]^{\frac{1}{n_B}} \tag{41}$$

At any temperature, knowing an incubation time of bainite transformation, found out from the IT diagram, the value of the kinetic parameter k_B can be expressed by:

$$k_B = \frac{-\ln\left(1 - X_{Bt_i}\right)}{(t_i)^{n_B}} = \frac{-\ln(1 - 0.01)}{(t_i)^{n_B}} \tag{42}$$

For 99% of austenite transformed into bainite, Equation (38) can be written as:

$$t_{0.99} = \left[\frac{-\ln(1 - \zeta_B) V_{rB}}{k_B}\right]^{\frac{1}{n_B}} = \left[\frac{-\ln\left(1 - \frac{X_B}{V_{rB}}\right) V_{rB}}{k_B}\right]^{\frac{1}{n_B}} \tag{43}$$

where the linear temperature dependence of the volume V_{rB} is assumed:

$$V_{rB} = a_5 + a_6 T. \tag{44}$$

Coefficients a_5 and a_6 can be determined by corresponding values of the volume V_{rB} on two different temperatures in IT diagram. Based on Equations (39) and (41), the kinetic parameter k_B can be written as:

$$k_B = T^{-\frac{1}{2}} D_0 \exp\left(-\frac{Q_{dif} + k_6}{RT}\right) \exp\left(-\frac{k_7}{RT(\Delta T)^2}\right) k_8 \tag{45}$$

With the previously determined kinetic parameter n_B, the defined temperature dependence of the volume V_{rB} and with the known values of the constants k_6, k_7 and k_8, the kinetics of the bainite transformation is completely defined. To determine the values of the constants k_6, k_7 and k_8, it is first necessary to determine the value of the coefficient k_B for three temperatures, and then to solve a system of three equations with three unknowns.

Based on Equations (46) and (47), the volume fraction of bainite and the normalized volume fraction of bainite can be determined by:

$$X_B = V_{rB}\left(1 - \exp\left(-\frac{k_B}{V_{rB}} t^{n_B}\right)\right) \tag{46}$$

$$\zeta_B = \frac{X_B}{V_{rB}} = 1 - \exp\left(-\frac{k_B}{V_{rB}} t^{n_B}\right) \tag{47}$$

Equation (47) can be written in an incremental form, when the normalized volume fraction of bainite formed in the time interval $\Delta t^{(N)}$ can be calculated as:

$$\Delta \zeta_B^{(N)} = n_B \left(\frac{k_B}{V_{rB}}\right)^{\frac{1}{n_B}} \left(\ln \frac{1}{1 - \zeta_B^{(N-1)}}\right)^{1 - \frac{1}{n_B}} \left(1 - \zeta_B^{(N-1)}\right) \Delta t^{(N)} \tag{48}$$

At any time of transformation, the real volume fraction of bainite can be calculated as $X_B = \zeta_B V_{rB}$ (Equation (31)).

2.1.4. Pearlite Transformation

In the remaining undercooled austenite that has not transformed into ferrite or bainite, at temperatures lower than A_{e1}, the pearlite transformation takes place. The kinetics of pearlite transformation is independent of the kinetics of the previous ferrite or bainite transformation. At temperatures $A_{e1} > T \geq B_s$ the remaining volume available for pearlitic transformation is $V_{rP1} \cdot V$, while at temperatures $T < B_s$ the remaining volume is $V_{rP2} \times V$, where $V_{rP1} = 1 - V_{rF}$ and $V_{rP2} = 1 - V_{rB}$.

For pearlite transformation by the mechanism of nucleation and growth, the following kinetic parameters are assumed:

$$n_P = 4 \tag{49}$$

$$k_P = \frac{\pi}{3} I_P S G_P^{3} \tag{50}$$

where S is surface of austenite grain suitable for nucleation, while I_P is the nucleation rate and G_P is the growth rate defined as [14]:

$$I_P = T^{-\frac{1}{2}} D_0 \exp\left(-\frac{Q_{dif}}{RT}\right) \exp\left(-\frac{k_3}{RT(\Delta T)^2}\right) \tag{51}$$

$$G_P = \Delta T D_0 \exp\left(-\frac{Q_{dif}}{RT}\right) (c_{\gamma\alpha} - c_{\gamma Fe3C}) \tag{52}$$

In Equations (51) and (52), ΔT is the undercooling below the critical temperature A_{e3}, while $c_{\gamma\alpha}$ and $c_{\gamma Fe3C}$ are the concentrations of austenite at the boundary with ferrite and cementite, respectively. k_3 is the kinetic parameters dependent on chemical composition. After introducing Equations (51) and (52) into Equation (50) and after some modifications, it can be rewritten:

$$k_P = SD_0^4 \exp\left(\frac{-4(Q_{dif} + k_5)}{RT}\right) \exp\left(\frac{-k_3}{RT(A_{el} - T)^2}\right) (c_{\gamma\alpha} - c_{\gamma Fe3C})^3 (A_{el} - T)^3 k_4^{-4} \tag{53}$$

To determine the values of the constants k_3, k_4 and k_5, it is first necessary to determine the value of the coefficient k_P for three temperatures, and then to solve a system of three equations with three unknowns.

The real volume of pearlite can be written as:

$$dV_P = (1 - \zeta_P)dV_{Pe} \tag{54}$$

where the normalized volume fraction can be expressed by:

$$\zeta_P = \frac{X_P}{V_{rP1}} \tag{55}$$

The extended volume of pearlite is defined as:

$$dV_{Pe} = \frac{4}{3}\pi I_P S G_P^3 V (t - t_i)^3 dt \tag{56}$$

After introducing Equation (56) into Equation (55), it follows that:

$$V_{rP1} \, d\zeta_P = (1 - \zeta_P)\frac{4}{3}\pi I_P S G_P^3 (t - t_i)^3 dt \tag{57}$$

$$V_{rP1} \, d\zeta_P = (1 - \zeta_P)4k_P(t - t_i)^3 \Delta dt \tag{58}$$

$$V_{rP1} \frac{d\zeta_P}{(1 - \zeta_P)} = 4k_P(t - t_i)^3 dt \tag{59}$$

After integrating Equation (59), it follows that:

$$-\ln(1 - \zeta_P)V_{rP1} = k_P t^4 \tag{60}$$

For small values of the normalized volume fraction of perlite, Equation (60) can be written as:

$$-\ln(1 - X_P) = k_P t_i^4 \tag{61}$$

At any temperature, knowing the incubation time of pearlite transformation, found out from the IT diagram, the value of the kinetic parameter k_P can be written as:

$$k_P = \frac{-\ln\left(1 - X_{Pt_i}\right)}{(t_i)^4} = \frac{-\ln(1 - 0.01)}{(t_i)^4} \tag{62}$$

From Equation (60) follows the normalized volume fraction of pearlite, which is:

$$\zeta_P = 1 - \exp\left(-\frac{k_P}{V_{rP1}}t^4\right) \tag{63}$$

As for ferrite and bainite transformation, Equation (63) can be written in an incremental form. The normalized volume fraction of pearlite formed by the mechanism of nucleation and growth in the time interval $\Delta t^{(N)}$ can be calculated by:

$$\Delta \zeta_P^{(N)} = 4\left(\frac{k_P}{V_{rp1}}\right)^{\frac{1}{4}}\left(\ln\frac{1}{1-\zeta_P^{(N-1)}}\right)^{\frac{3}{4}}\left(1-\zeta_P^{(N-1)}\right)\Delta t^{(N)} \tag{64}$$

The presented method for estimation of k_P and ζ_P is also valid at temperatures lower than B_s. In that case, in the above equations, the relative volume V_{rP1} should be replaced by the relative volume V_{rP2}.

2.2. Materials

With the aim of qualitatively and quantitatively defining the influence of chemical composition on the isothermal decomposition of austenite, the values of kinetic parameters were investigated on a number of hypoeutectoid, low-alloy steels [25]. Their composition is shown in Table 1.

Table 1. Chemical composition of studied steels (balance Fe).

Designation (DIN)	Chemical Composition, wt. %									
	C	Si	Mn	P	S	Cr	Cu	Mo	Ni	V
42CrMo4	0.38	0.23	0.64	0.019	0.013	0.99	0.17	0.16	0.08	<0.01
Ck45	0.44	0.22	0.66	0.022	0.029	0.15	-	-	-	0.02
28NiCrMo74	0.30	0.24	0.46	0.030	0.025	1.44	0.20	0.37	2.06	<0.01
34Cr4	0.35	0.23	0.65	0.026	0.013	1.11	0.18	0.05	0.23	<0.01
25CrMo4	0.22	0.25	0.64	0.010	0.011	0.97	0.16	0.23	0.33	<0.01
36Cr6	0.36	0.25	0.49	0.021	0.020	1.54	0.16	0.03	0.21	<0.01
41Cr4	0.44	0.22	0.80	0.030	0.023	1.04	0.17	0.04	0.26	<0.01

3. Results

Section 2.1 presents methods for estimating kinetic parameters, which completely define the kinetics of austenite isothermal decomposition into ferrite, pearlite and bainite. The calculated values of the kinetic parameters depend on the chemical composition, i.e., they are valid only for one steel.

The critical temperatures of austenite decomposition were calculated based on Equations (65) and (66) [26], and Equation (67) [27].

$$\begin{aligned} A_{e3} = 883.49 \ & - 275.89\%C + 90.91(\%C)^2 - 12.26\%Cr + 16.45\%C\%Cr - 29.96\%Mn + 23.50\%C\%Mn \\ & + 8.49\%Mo - 10.80\%C\%Mo - 25.56\%Ni + 14.71\%C\%Ni + 1.45\%Mn\%Ni + 0.76(\%Ni)^2 \\ & + 13.53\%Si - 3.47\%Mn\%Si \end{aligned} \tag{65}$$

$$\begin{aligned} A_{e1} = 727.37 \ & + 13.40\%Cr - 1.03\%C\%Cr - 16.72\%Mn + 0.91\%C\%Mn + 6.18\%Cr\%Mn - 0.64(\%Mn)^2 \\ & + 3.14\%Mo + 1.86\%Cr\%Mo - 0.73\%Mn\%Mo - 13.66\%Ni + 0.53\%C\%Ni + 1.11\%Cr\%Ni \\ & - 2.28\%Mn\%Ni - 0.24(\%Ni)^2 + 6.34\%Si - 8.88\%Cr\%Si - 2.34\%Mn\%Si + 11.98(\%Si)^2 \end{aligned} \tag{66}$$

$$B_s = 830 - 270\%C - 90\%Mn - 37\%Ni - 70\%Cr - 83\%Mo \tag{67}$$

The dependence of kinetic parameters of ferrite, pearlite and bainite transformation on the content of carbon, chromium, molybdenum and nickel was estimated by regression analysis (Equations (68)–(78)). Because of the similar content of manganese and silicon in studied steels, these elements were not included in the regression analysis. Based on the proposed equations, the kinetic parameters involved in mathematical model of ferrite, pearlite and bainite transformation can be calculated for any other chemical composition of hypoeutectoid, low-alloy steels (Table 2).

$$k_2 = \exp(-4.02 - 11.11\%\text{C} - 1.99\%\text{Cr} + 20.76\%\text{Mo} - 40.99\%\text{Ni}) \tag{68}$$

$$n_1 = 6.23 - 2.25\%\text{C} - 0.81\%\text{Cr} + 5.17\%\text{Mo} - 8.83\%\text{Ni} \tag{69}$$

$$k_3 = 106003024.20 - 51200507.88\%\text{C} - 19751485.43\%\text{Cr} - 256581910.00\%\text{Mo} + 25605277.23\%\text{Ni} \tag{70}$$

$$k_4 = \exp(-33.63 + 61.93\%\text{C} - 15.96\%\text{Cr} + 6.20\%\text{Mo} - 3.60\%\text{Ni}) \tag{71}$$

$$k_5 = 97606.14 - 488262.55\%\text{C} + 134064.43\%\text{Cr} + 28612.33\%\text{Mo} + 29497.32\%\text{Ni} \tag{72}$$

$$k_6 = 190361.38 - 288009.61\%\text{C} - 76052.87\%\text{Cr} - 693123.59\%\text{Mo} + 78021.58\%\text{Ni} \tag{73}$$

$$k_7 = \exp(10.37 + 24.69\%\text{C} + 1.24\%\text{Cr} - 9.98\%\text{Mo} + 2.63\%\text{Ni}) \tag{74}$$

$$k_8 = \exp(81.35 - 73.62\%\text{C} - 16.43\%\text{Cr} - 127.88\%\text{Mo} + 14.57\%\text{Ni}) \tag{75}$$

$$n_\text{B} = 1.57 + 1.74\%\text{C} - 0.47\%\text{Cr} - 1.79\%\text{Mo} + 0.16\%\text{Ni} \tag{76}$$

$$a_5 = -1.0130 + 3.7800\%\text{C} + 1.8340\%\text{Cr} + 4.2012\%\text{Mo} + 0.4655\%\text{Ni} \tag{77}$$

$$a_6 = 0.003525 - 0.006847\%\text{C} - 0.002777\%\text{Cr} - 0.006653\%\text{Mo} - 0.001100\%\text{Ni} \tag{78}$$

Table 2. Kinetic parameters of austenite isothermal decomposition.

Transformation	Constant	Steel Designation (DIN)						
		42CrMo4	Ck45	28NiCrMo74	34Cr4	25CrMo4	36Cr6	41Cr4
Ferrite	k_1	2.5×10^5	2.5×10^5	-	2.5×10^5	2.5×10^5	2.5×10^5	2.5×10^5
	k_2	3.8321×10^{-5}	1.7441×10^{-4}	-	9.0625×10^{-10}	6.3183×10^{-8}	8.8213×10^{-9}	1.1941×10^{-9}
	n_1	4.6923	5.2095	-	2.3955	3.3200	2.5230	2.3547
	a_1	8.8839	8.7592	-	7.8504	3.1064	10.0511	14.7870
	a_2	−0.0115	−0.0114	-	−0.0101	−0.0039	−0.0130	−0.0195
	a_3	−1.8688	−4.4695	-	−1.7667	−1.4606	−1.5163	−1.7387
	a_4	0.0032	0.0070	-	0.0030	0.0024	0.0027	0.0031
Pearlite	k_3	27,988,177	76,767,479	16,492,913	31,355,038	43,316,272	56,142,707	93,514,410
	k_4	1.1450×10^{-11}	9.8057×10^{-5}	2.1918×10^{-19}	6.1325×10^{-12}	8.6385×10^{-17}	2.5088×10^{-17}	5.4094×10^{-12}
	k_5	51,728	−92,146	214,273	47,324	149,610	151,595	46,211
Bainite	k_6	−99,033	51,128	−101,946	−15,089	−77,442	−34,123	−16,704
	k_7	3.2308×10^8	2.4803×10^9	2.0943×10^9	2.5027×10^9	2.5966×10^6	2.2482×10^9	1.6819×10^9
	k_8	5.4488×10^7	1.1746×10^{20}	8.1265×10^7	8.0362×10^{14}	6.6340×10^{10}	2.7482×10^{12}	4.2311×10^{13}
	n_B	1.4923	2.30043	1.09856	1.58602	1.12141	1.51777	1.74637
	a_5	2.989028	1.000000	-	2.349733	2.794667	3.587317	2.881892
	a_6	−0.003041	0.00000	-	−0.002139	−0.002667	−0.003902	−0.002973

4. Discussion

The values of kinetic parameters given in Table 2 were verified by comparing the modeled curves of the isothermal transformation (IT) diagram of steel 42CrMo4, 36Cr6, Ck45, and 28NiCrMo74 with those obtained experimentally. In Figures 2–5, the dashed lines show the experimental IT diagram, while the mathematically determined times of start (incubation time) and times of finish of the isothermal austenite decomposition, $t_{0.01}$ and $t_{0.99}$, are shown by solid lines. Additionally, Figure 2 shows curves corresponding to bainite volume fraction of 25%, 50%, 75% and 90%.

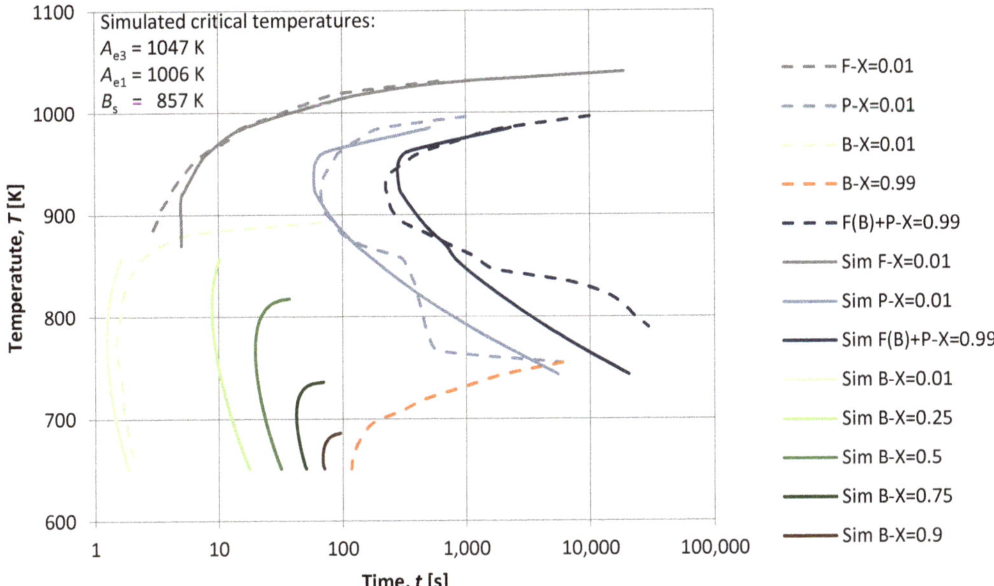

Figure 2. IT diagram of steel 42CrMo4.

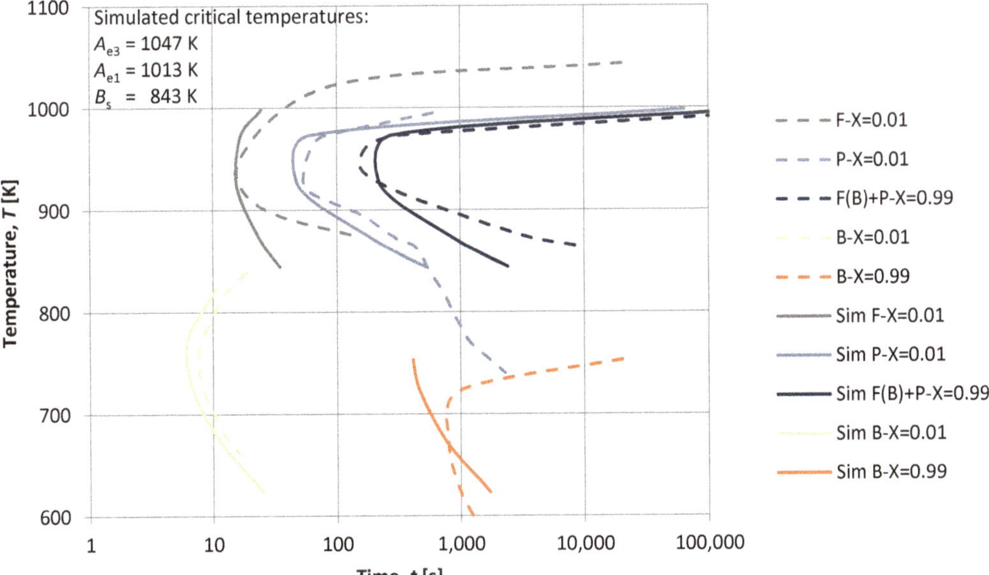

Figure 3. IT diagram of steel 36Cr6.

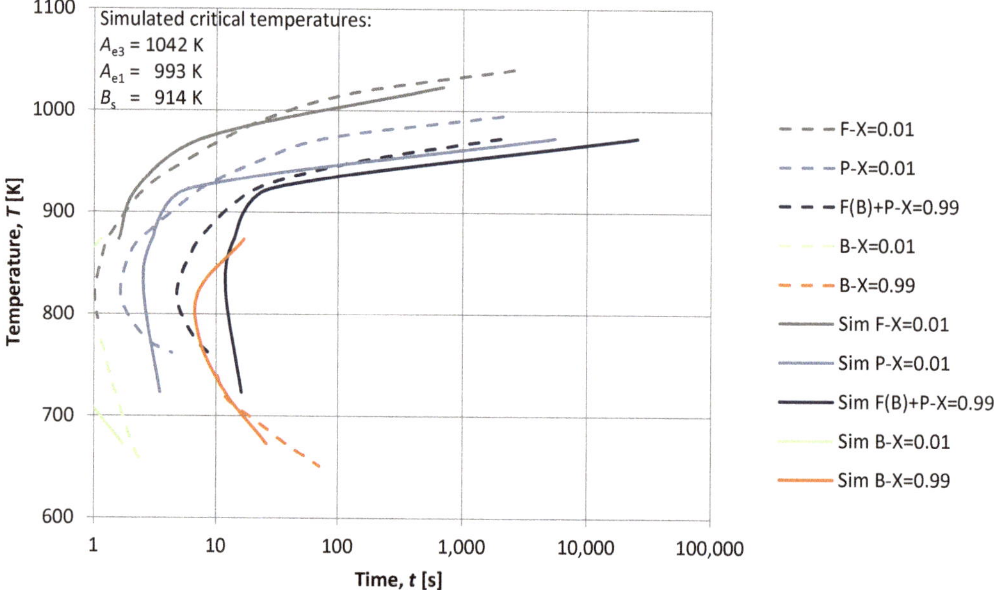

Figure 4. IT diagram of steel Ck45.

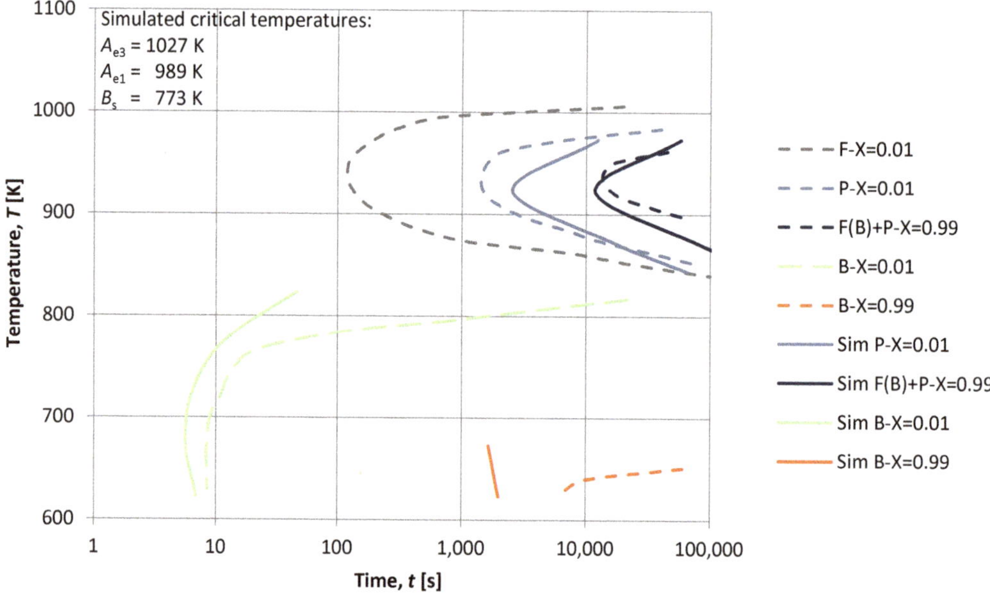

Figure 5. IT diagram of steel 28NiCrMo74.

The times of start and finish of isothermal decomposition of austenite were calculated based on Equations (30), (48) and (64), and the known values of the kinetic parameters.

Kinetic parameters of ferrite, bainite and pearlite transformation, k_F, k_B and k_P, were calculated by Equations (12), (45) and (53), respectively. Other physical quantities used in the developed mathematical model are shown in Table 3.

Table 3. Physical quantities used in modelling of austenite isothermal decomposition.

Quantity Value	Units	Description
$D_0 = 2.3 \times 10^{-5}$	$m^2\,s^{-1}$	Material constant
$Q_{dif} = 1.48 \times 10^5$	$J\,mol^{-1}$	Diffusion activation energy
$R = 8.314$	$J\,mol^{-1}\,K^{-1}$	Universal gas constant
$S = 170153$	m^{-1}	Surface of austenite grain suitable for nucleation
$c_\gamma = 1186.661\,\exp(-7.2834 \times 10^{-3}\,T)$	wt.% C	Concentration of austenite
$c_\alpha = 0.1592 - 1.3423 \times 10^{-4}\,T$	wt.% C	Concentration of ferrite
$c_{\gamma\alpha} = 9.6782 - 8.82 \times 10^{-3}\,T$	wt.% C	Concentration of austenite at the boundary with ferrite
$c_{\gamma Fe3C} = -0.5248 + 1.28 \times 10^{-3}\,T$	wt.% C	Concentration of austenite at the boundary with cementite

Figures 2–5 show that differences between times of transformations in experimentally and mathematically determined IT diagrams are not relevant. Therefore, it is seen that the kinetic parameters involved in an established mathematical model of ferrite, pearlite and bainite transformation can be successfully determined on the basis of Equations (68)–(78) with high accuracy. Developed model avoids the use of simple empirical expressions in predictions of isothermal decomposition of austenite.

Since the developed model is written in incremental form, it is suitable for predicting austenite decomposition during the continuous cooling of steel using Scheil's additivity rule. Additionally, it is very easy to extend this approach in the prediction of the kinetics of austenite decomposition for other types of steel.

5. Conclusions

In this paper, the equations for the estimation of microstructure constituents' volume fractions after the isothermal decomposition of austenite have been proposed. Isothermal decomposition of austenite implies quenching of steel from the austenite range to the temperature of isothermal transformation where all austenite decomposes at a constant temperature.

The inversion methods for the calculation of characteristic variables in the mathematical model of kinetics of austenite decomposition were developed.

The mathematical model was verified by the comparison of experimentally and mathematically determined IT diagrams of steel. It can be concluded that characteristic parameters included in the mathematical model of ferrite, pearlite and bainite transformation can be successfully evaluated by the proposed method.

Author Contributions: Conceptualization, B.S. and S.S.H.; investigation, D.I. and S.S.H.: methodology, B.S. and S.S.H.; validation, B.S., D.I., S.S.H. and K.H.; writing—original draft preparation, S.S.H. and B.S.; writing—review and editing, B.S., D.I., S.S.H. and K.H. All authors have read and agreed to the published version of the manuscript.

Funding: This research was funded in part by Croatian Science Foundation under the project IP-2020-02-5764 and in part by the University of Rijeka under the project number uniri tehnic-18-116.

Data Availability Statement: Data are available in the article.

Acknowledgments: This work has been supported in part by Croatian Science Foundation under the project IP-2020-02-5764 and in part by the University of Rijeka under the project number uniri tehnic-18-116.

Conflicts of Interest: The authors declare no conflict of interest.

References

1. Pan, J.; Gu, J.; Zhang, W. Industrial Applications of Computer Simulation of Heat Treatment and Chemical Heat Treatment. In *Handbook of Thermal Process Modeling of Steels*; Gür, C.H., Pan, J., Eds.; CRC Press: Boca Raton, FL, USA, 2009; pp. 673–701.
2. Reti, T.; Felde, I.; Guerrero, M.; Sarmiento, S. Using generalized time-temperature parameters for predicting the hardness change occurring during tempering. In Proceedings of the International Conference on New Challenges in Heat Treatment and Surface Engineering, Dubrovnik–Cavtat, Dubrovnik, Croatia, 9–12 June 2009; Smoljan, B., Matijević, B., Eds.; Listemann AG: Bendern, Switzerland, 2009; pp. 333–342.
3. Smokvina Hanza, S. Mathematical Modeling and Computer Simulation of Microstructure Transformations and Mechanical Properties During Steel Quenching. Ph.D. Thesis, University of Rijeka, Rijeka, Croatia, 2011. (In Croatian)
4. Smoljan, B.; Iljkić, D.; Totten, G. Mathematical modeling and simulation of hardness of quenched and tempered steel. *Met. Mater. Trans. B* **2015**, *46*, 2666–2673. [CrossRef]
5. Rose, A.; Wever, F. *Atlas zur Wärmebehandlung der Stähle I*; Verlag Stahleisen: Düsseldorf, Germany, 1954.
6. Hillert, M. Discussion of "a personal commentary on transformation of austenite at constant subcritical temperatures". *Metall. Mater. Trans. A* **2011**, *42*, 541–542. [CrossRef]
7. Zener, C. Kinetics of decomposition of austenite. *Trans. AIME* **1946**, *167*, 550–595.
8. Zener, C. Theory of growth of spherical precipitates from solid solution. *J. Appl. Phys.* **1949**, *20*, 950–953. [CrossRef]
9. Serajzadeh, S. A mathematical model for prediction of austenite phase transformation. *Mater. Lett.* **2004**, *58*, 1597–1601. [CrossRef]
10. Militzer, M.; Hawbolt, E.B.; Meadowcroft, T.R. Microstructural model for hot strip rolling of high-strength low-alloy steels. *Met. Mater. Trans. A* **2000**, *31*, 1247–1259. [CrossRef]
11. Avrami, M. Kinetics of phase change. I general theory. *J. Chem. Phys.* **1939**, *7*, 1103–1112. [CrossRef]
12. Simsir, C.; Gür, C.H. Simulation of quenching. In *Handbook of Thermal Process Modeling of Steels*; Gür, C.H., Pan, J., Eds.; CRC Press: Boca Raton, FL, USA, 2009; pp. 341–425.
13. Christian, J.W. *The Theory of Transformations in Metals and Alloys*, 1st ed.; Pergamon Press: Oxford, UK, 2002.
14. Suehiro, M. A mathematical model for predicting temperature of steel during cooling based on microstructural evolution. In Proceedings of the 11th Congress of IFHTSE and 4th ASM Heat Treatment and Surface Engineering Conference in Europe, Florence, Italy, 19–21 October 1998; Volume 1, pp. 11–20.
15. Offerman, S.; van Wilderen, L.; van Dijk, N.; Sietsma, J.; Rekveldt, M.; van der Zwaag, S. In-situ study of pearlite nucleation and growth during isothermal austenite decomposition in nearly eutectoid steel. *Acta Mater.* **2003**, *51*, 3927–3938. [CrossRef]
16. Whiting, M. A reappraisal of kinetic data for the growth of pearlite in high purity Fe-C eutectoid alloys. *Scr. Mater.* **2000**, *43*, 969–975. [CrossRef]
17. Davenport, E.S.; Bain, E.C. Transformation of austenite at constant subcritical temperatures. *Trans. Met. Soc. AIME* **1930**, *90*, 117–154.
18. Bhadeshia, H.K.D.H. *Bainite in Steels: Transformations, Microstructure and Properties*, 2nd ed.; IOM Communications: London, UK, 2001.
19. Fielding, L.C.D. The bainite controversy. *Mater. Sci. Technol.* **2013**, *29*, 383–399. [CrossRef]
20. Yang, Z.-G.; Fang, H.-S. An overview on bainite formation in steels. *Curr. Opin. Solid State Mater. Sci.* **2005**, *9*, 277–286. [CrossRef]
21. Hillert, M. The nature of bainite. *ISIJ Int.* **1995**, *35*, 1134–1140. [CrossRef]
22. Quidort, D.; Bréchet, Y.J.M. A model of isothermal and non isothermal transformation kinetics of bainite in 0.5% C steels. *ISIJ Int.* **2002**, *42*, 1010–1017. [CrossRef]
23. Rees, G.I.; Bhadeshia, H.K.D.H. Bainite transformation kinetics part 1 modified model. *Mater. Sci. Technol.* **1992**, *8*, 985–993. [CrossRef]
24. Pan, Y.-T. Measurement and Modelling of Diffusional Transformation of Austenite in C.-Mn Steels. Ph.D. Thesis, Institute of Materials Science and Engineering of the National Sun Yat-Sen University, Kaohsiung, Taiwan, 2001.
25. Rose, A.; Hougardy, H. *Atlas zur Wärmebehandlung der Stähle*; Verlag Stahleisen: Düsseldorf, Germany, 1972.
26. Lusk, M.T.; Lee, Y.-K. A global material model for simulating the transformation kinetics of low alloy steels. In Proceedings of the 7th International Seminar of IFHT: Heat Treatment and Surface Engineering of Light Alloys, Budapest, Hungary, 15–17 September 1999; Lendvai, J., Réti, T., Eds.; Hungarian Scientific Society of Mechanical Engineering: Budapest, Hungary, 1999; pp. 273–282.
27. Steven, W.; Haynes, A.G. The temperature of formation of martensite and bainite in low alloy steels. *J. Iron Steel Inst.* **1956**, *183*, 349–359.

Article

Mathematical Modeling of Induction Heating of Waveguide Path Assemblies during Induction Soldering

Vadim Tynchenko [1,2,3,*], Sergei Kurashkin [1,2], Valeriya Tynchenko [1,2], Vladimir Bukhtoyarov [1,2], Vladislav Kukartsev [1,2], Roman Sergienko [4], Viktor Kukartsev [1] and Kirill Bashmur [1]

[1] Department of Technological Machines and Equipment of Oil and Gas Complex, School of Petroleum and Natural Gas Engineering, Siberian Federal University, 660041 Krasnoyarsk, Russia; scorpion_ser@mail.ru (S.K.); 051301@mail.ru (V.T.); vladber@list.ru (V.B.); vlad_saa_2000@mail.ru (V.K.); vakukartsev@sfu-kras.ru (V.K.); bashmur@bk.ru (K.B.)

[2] Information-Control Systems Department, Institute of Computer Science and Telecommunications, Reshetnev Siberian State University of Science and Technology, 660037 Krasnoyarsk, Russia

[3] Marine Hydrophysical Institute, Russian Academy of Sciences, 299011 Sevastopol, Russia

[4] Gini Gmbh, 80339 Munich, Germany; roman@gini.net

* Correspondence: vadimond@mail.ru; Tel.: +7-95-0973-0264

Abstract: The waveguides used in spacecraft antenna feeders are often assembled using external couplers or flanges subject to further welding or soldering. Making permanent joints by means of induction heating has proven to be the best solution in this context. However, several physical phenomena observed in the heating zone complicate any effort to control the process of making a permanent joint by induction heating; these phenomena include flux evaporation and changes in the emissivity of the material. These processes make it difficult to measure the temperature of the heating zone by means of contactless temperature sensors. Meanwhile, contact sensors are not an option due to the high requirements regarding surface quality. Besides, such sensors take a large amount of time and human involvement to install. Thus, it is a relevant undertaking to develop mathematical models for each waveguide assembly component as well as for the entire waveguide assembly. The proposed mathematical models have been tested by experiments in kind, which have shown a great degree of consistency between model-derived estimates and experimental data. The paper also shows how to use the proposed models to test and calibrate the process of making an aluminum-alloy rectangular tube flange waveguide by induction soldering. The Russian software, SimInTech, was used in this research as the modeling environment. The approach proposed herein can significantly lower the labor and material costs of calibrating and testing the process of the induction soldering of waveguides, whether the goal is to adjust the existing process or to implement a new configuration that uses different dimensions or materials.

Keywords: induction heating; mathematical modeling; process; control; automation; optimization; waveguide

Citation: Tynchenko, V.; Kurashkin, S.; Tynchenko, V.; Bukhtoyarov, V.; Kukartsev, V.; Sergienko, R.; Kukartsev, V.; Bashmur, K. Mathematical Modeling of Induction Heating of Waveguide Path Assemblies during Induction Soldering. *Metals* **2021**, *11*, 697. https://doi.org/10.3390/met11050697

Academic Editors: Francesco De Bona and António Bastos Pereira

Received: 1 April 2021
Accepted: 23 April 2021
Published: 24 April 2021

Publisher's Note: MDPI stays neutral with regard to jurisdictional claims in published maps and institutional affiliations.

1. Introduction

Modeling the process of induction soldering is one of the easier and more effective ways to improve the quality of process control, which helps to enhance the ultimate product. The authors in [1–3], describe how such modeling can be used for quality improvement in the photovoltaic industry. The authors of [1] present a multiphysical model of induction soldering for making modular solar panel systems. The model presented in the paper can be used to optimize the process of induction soldering by adjusting the geometry of the head to heat the solder more efficiently and evenly (the "head" hereafter refers to the work-head of the induction soldering unit). The model also makes adjustments for deviations resulting from the specifics of materials in use.

The authors in [2,3] use modeling to analyze the effects of cell cracking on the performance of solar cells. Their simulations show that cracks do not necessarily compromise the performance of photovoltaic modules. The authors in [4] present an induction heating model implemented in Cedrat Flux 10.3, a commercial package. Experimental tests prove the model to be accurate. The simulation results are consistent with the experimental temperature profile of the specified surface points. The model has the advantage of being able to predict such parameters as current density and magnetic flux field inside a workpiece—these parameters are difficult to measure directly.

The authors in [5] show that reduced-order models constitute a fairly effective and promising tool for controlling the induction soldering process by means of indirect measurements. A fourth-order system is obtained by proper orthogonal decomposition. The model does enable the process to attain the desired temperature; at the same time, it is sufficiently simple from the computational standpoint to run on a relatively cheap microcontroller.

As part of the research presented in [6], the authors developed a thermal model of the infrared soldering process. The model is able to predict thermal effects from the parameters of convection in an infrared furnace to the detailed thermal response including the solid–liquid transition of the solder. The authors in [7–9] present a neural-network model of induction soldering process control. The model greatly improves the quality of controlling the process of induction soldering for spacecraft waveguides. The authors in [10] describe the use of a neural-network model developed to control the process of oil-and-gas equipment repair and maintenance.

The authors of [11,12] present a mathematical model of the process of spacecraft waveguide induction soldering. These papers clearly demonstrate the high quality of the proposed models. The proposed models have also been proven to be quite efficient for this application.

The authors in [13] present a mathematical model, which is a neural-fuzzy controller for the process of waveguide induction soldering. Experimental tests in kind prove the model to be accurate.

The simulation modeling of the soldering process is the topic of many works. Satheesh et al. [14] carry out the numerical estimation of soldering process in terms of strain fields and localized transient temperature. Works [15–17] are devoted to the issues of modeling and analysis of induced stresses, as well as the effect of the soldered joint geometry on such parameter. The authors of [18] propose a mathematical model of the process of induction brazing of products made of nonmagnetic metals, considering the existing nonlinearities of the process under consideration and using the orthogonal expansion to reduce the complexity of the numerical study of the proposed model. Given how commonly simulation modeling is used to optimize induction heating-based processes in a variety of mechanical engineering applications, it is clear that the approach is highly efficient when it comes to modeling such processes [19–21]. The author in [22] uses COMSOL Multiphysics simulation system to investigate how the head parameters affect the efficiency of heating parts and the maximum heating zone. The electromagnetic field parameters are modeled there at 22 to 100 kHz. The simulation results show that heating is fastest at lower frequencies; as such, electromagnetic fields penetrate deeper into the material of the parts.

However, in the works available in the public domain at the moment, the issues of modeling the induction heating process of special structures that are widely used in communication systems are not considered. The present work is intended to close the existing knowledge gap and to present a mathematical apparatus that provides high-quality modeling of the process of waveguide path soldering using induction heating.

The first steps of simulating the induction soldering process can be conveniently done in ready-made modeling software such as Simulink (a MATLAB extension) [23], which can be used for modeling both static [24] and dynamic systems [25]. Further, it is possible to verify the created models both in the COMSOL Multiphysics environment [26–29] and in the well-known ANSYS system [30–33]. As an alternative to the Simulink package, the present work uses the Russian software SimInTech [34–36]. SimInTech allows not only to implement models of complex dynamic systems [37], but also to test systems for automated control of complex objects and technological processes [38].

In this work, the resulting mathematical models were used in the development of an adaptive (smart) process control system for the induction soldering of waveguide assemblies in order to obtain an even heating of soldered elements in order to make a high-quality permanent joint. The applicability and usability of the proposed algorithm is tested both by computational and in-kind experiments.

2. Materials and Methods

2.1. Mathematics Behind Models

In order to model the induction heating process of waveguides, it was necessary to develop mathematical models for individual waveguide assembly components and for entire waveguide assemblies. A waveguide assembly consists of a waveguide rectangular tube and a flange or a coupler.

As mathematical models for heating assembly elements, which are used to work out the technological process of the induction soldering of thin-walled aluminum waveguide paths, we write the expression of the temperature field with a continuously operating stationary source (1):

$$T(x,t) = \int_0^t \frac{Q}{c\rho F \sqrt{4\pi at}} e^{\left(-\frac{x^2}{4at} - bt\right)} dt \tag{1}$$

$$b = \frac{\alpha p}{c\rho F} \tag{2}$$

where Q is the amount of heat [J], F is the cross-section of a waveguide element [m^2], x is the distance from the heat source [m], cp is the volumetric heat capacity [J/m^3], t is time [s], b is the coefficient of convective heat transfer from the rod surface to the environment (see Equation (2)), a is the thermal conductivity coefficient and p is the cross-sectional perimeter.

The cross-section of an element of the waveguide path can be represented in the following form (3):

$$F = AB - A'B' \tag{3}$$

where A and B are the length and width of the product, respectively, and A' and B' are the length and width of the inner hole in the product, respectively.

For the tube, the assumption is made that the cross-section is heated evenly. Accordingly, an expression can be formulated for calculating the temperature field from the action of an instantaneous heating source in a flat rod, considering the input of geometric constraints (reflections), which is applicable for bodies bounded by mutually perpendicular planes (a parallel tubed, rectangular plate), for a rod of finite length and for an infinite wedge with an opening angle π/n, where n is an integer [39].

The essence of the reflection method is reduced to the expansion of the bounded body along the corresponding coordinate to infinity and the selection of additional sources in the extended area so that the boundary conditions on the surface of the bounded body are satisfied [39].

Based on the fact that the waveguide tube is limited on two sides, one side situated in the immediate vicinity of the induction heating source is as follows (4):

$$T(x,t) = \int_{i=-\infty}^{\infty} \int_{j=-1,1} \frac{Q}{c\rho F \sqrt{4\pi at}} e^{\left(-\frac{(x-jl-2iL)^2}{4at} - bt\right)} \tag{4}$$

where x is the distance from the left side [m], t is time [s], a is the thermal conductivity coefficient, L is the rod length [m], l is the end-to-heat source distance and j is the number of reflections taken into account in the calculation, selected in such a way that, for $j + 1$ for any x and t, $T(x,t) \leq \varepsilon$ is valid as $\varepsilon \to 0$.

Equation (4) rapidly converges, and only a few terms are considered for real bodies. As the temperature is equalized, the number of retained members of the series increases [39]. An illustration of the computational domain is presented in Figure 1 describing the distribution of heat sources in the heated element of the waveguide path (tube/flange/coupling).

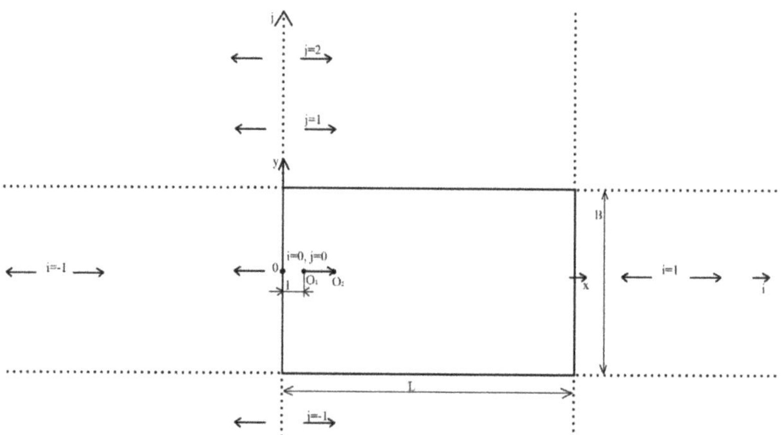

Figure 1. Heat source distribution scheme.

The diagram illustrates the movement of the energy source and point O_1 to point O_2 along the x-axis in time t. In this case, this source is reflected relative to the planes $x = 0$, $x = L$ and $y = B/2$ and $y = -B/2$.

Equation (4) makes it possible to represent the new mathematical models describing the heating of elements and assemblies of waveguide paths, which differ from the known ones considering the geometry of the products and allow the development of the parameters of the induction soldering process.

2.2. Waveguide Assembly: Tube Model

For a waveguide tube, the following assumptions must be made:

1. The tube is a sufficiently long body of a homogeneous material;
2. The cross-section of the waveguide tube along the entire length is constant;
3. The tube is similar to the rod in terms of heat transfer and thermal conductivity.

The conclusion here is that a mathematical model of a planar heat source in a rod holds true for a planar heat source in the rectangular tube of a waveguide assembly. This effectively introduces a geometric constraint on one side of the rod, whereas we denote and take into account the finiteness of the tube where the flange overhangs it when a respective joint is made. Another assumption is that the waveguide tube heats evenly over its cross-section as a result of being thin (<2 mm) and that the head is designed in a way to enable the even heating of the tube along its perimeter.

A realistic image of a standard waveguide assembly tube is shown in Figure 2. With the introduction of restrictions and the projection of the tube (Figure 3), the calculation in Equation (5) for the process of heating the waveguide tube, considering the geometric dimensions of the product, can therefore be written as follows:

$$T(x,t) = \sum_{j=-1,1} \frac{Q}{Fc\rho\sqrt{4at}} e^{\left(-\frac{(x+jl)^2}{4at} - bt\right)} \tag{5}$$

Figure 2. Photorealistic image of a waveguide assembly tube.

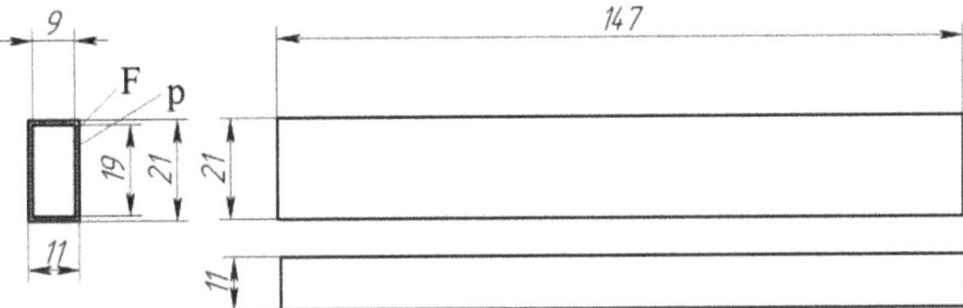

Figure 3. Projections of a waveguide assembly tube with its dimensions. F is the cross-sectional area of the tube; p is the cross-sectional perimeter.

To test the effectiveness of the proposed approach, the curves produced by the waveguide assembly tube heating model for various power levels of the induction heater are obtained (Figure 4). Thus, this calculation procedure is sufficiently consistent with the process of induction heating for waveguide assembly tubes.

2.3. Waveguide Assembly: Flange Model

Flange (Figure 5) heating is simulated in a similar manner to the methodology described above. A flange is a relatively small body, which means heat is withdrawn from it at a relatively low rate. This means that, at certain heating rates, the temperature will be uniform along the sectional axis near the soldering zone, which in turn means that no adjustment for temperature distribution in that plane is necessary. However, a flange is quite thin along its other axis, and therefore its heat distribution along that axis cannot be even. Unlike a tube, a flange is finite on either side, which means that the heat reflection off the finite boundaries of the body needs to be adjusted for.

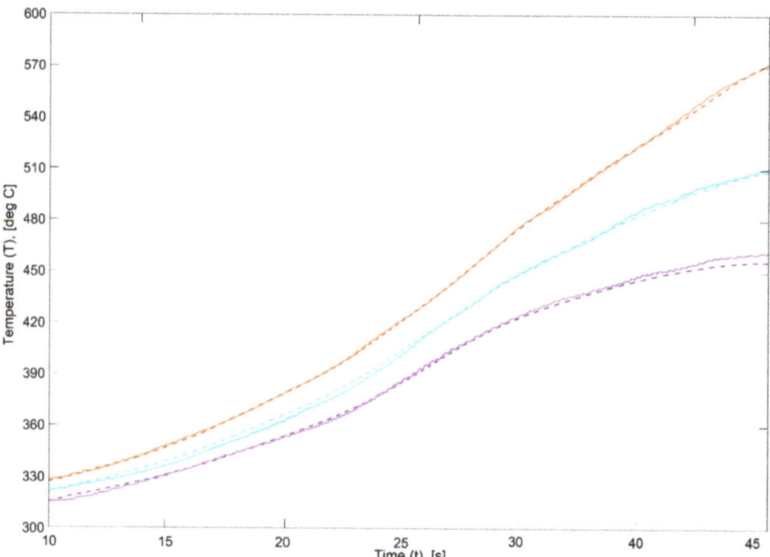

Figure 4. Curves produced by the waveguide assembly tube heating model, where dotted lines are model-derived data, solid lines are experimental data, orange curve is for 11 kW heating, blue curve is for 5 kW heating, and lilac curve is for 3 kW heating.

Figure 5. Photorealistic image of a flange.

Thus, with the introduction of constraints and the projection of the flange/coupling (Figure 6), in this study, a new calculation in Equation (6) for the heating process of the flange/coupling of the waveguide assembly is proposed:

$$T(x,t) = \sum_{i=0}^{j} \frac{2Q}{Fc\rho\sqrt{4at}} e^{\left(-\frac{(x+2iL)^2}{4at} - bt\right)} \tag{6}$$

where x is the distance from the left side [m] and L is the flange/coupler length [m].

The curves produced by the waveguide assembly flange heating model for various power levels of the induction heater are obtained (Figure 7). Thus, this calculation procedure is sufficiently consistent with the process of induction heating for waveguide assembly flanges.

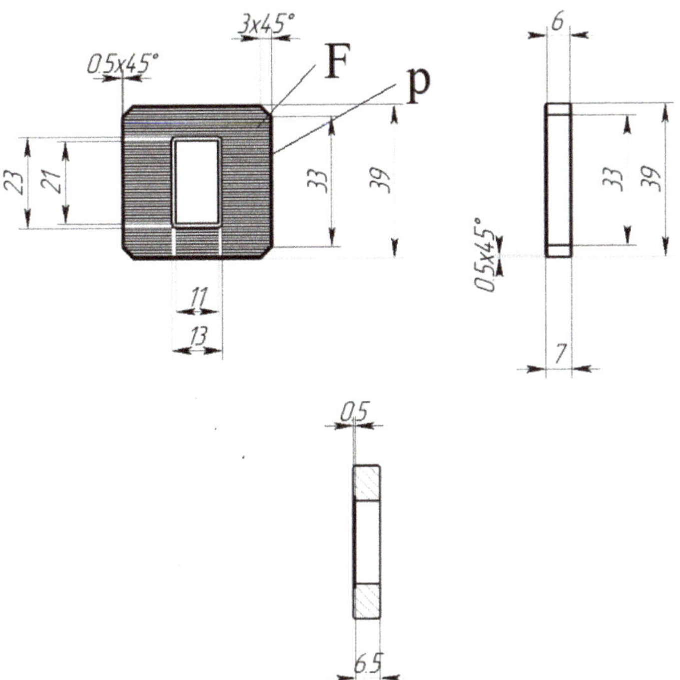

Figure 6. Projections of a waveguide assembly flange with its dimensions. F is the cross-sectional area of the flange; p is the cross-sectional perimeter.

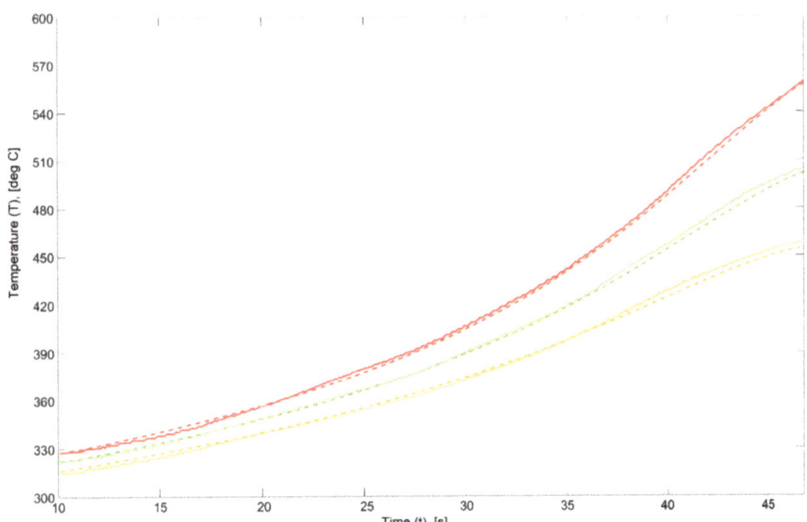

Figure 7. Curves produced by the waveguide assembly flange heating model, where dotted lines are model-derived data, solid lines are experimental data, red curve is for 11 kW heating, green curve is for 5 kW heating, and yellow curve is for 3 kW heating.

2.4. Waveguide Assembly Model

When modeling the distribution of heat energy between assembly components (Figure 8), the assumption is that there is a certain law of energy distribution that is bound to both the dimensions and the configuration of the head.

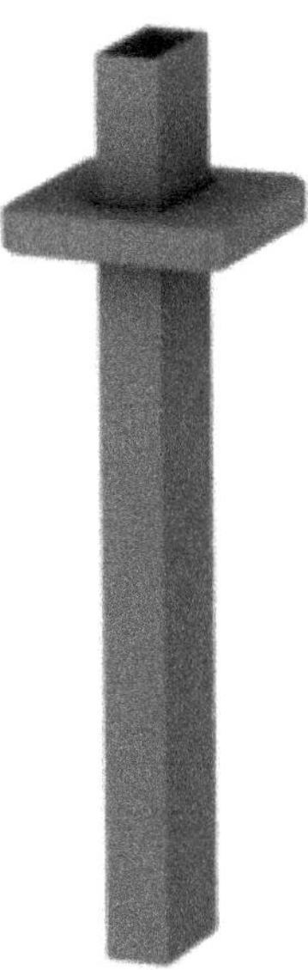

Figure 8. Standard tube–flange waveguide assembly.

The induction soldering of waveguides requires a head with a slanted opening to localize peak heat near the soldering zone. Given this specific feature of the technology, it can be assumed for modeling that all the head-transferred energy is released in the soldering zone. Thus, let us assume for convenience that all of the generator's energy will be transferred to the assembly being soldered. Therefore, the law of heat distribution between the waveguide assembly components will be as follows:

$$q(t) = q(t)M(x) + q(x)(1 - M(x)) \tag{7}$$

where $q(t)$ is a permanent heat source, $M(x)$ is the coefficient of heat distribution between the assembly components, $M(x)$ belongs to $[0, \ldots, 1]$ and x belongs to $[n, \ldots, m]$, where n and m are, respectively, the upper boundary and the lower boundary, and x is the flange/coupler to head distance.

For this research, the distribution function (7) was derived empirically. A design diagram (Figure 9) could explain the essence of the energy distribution during the heating of the waveguide array.

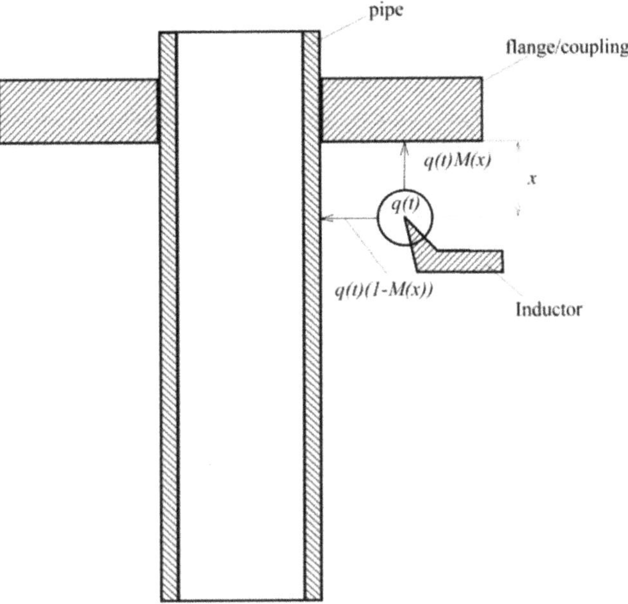

Figure 9. Diagram of the heat distribution of a continuously operating source of the waveguide assembly.

3. Experimental Study and Discussion

3.1. Verification of Proposed Mathematical Models

For the verification of the compliance of the developed mathematical models with the real technological process of the induction heating of waveguide paths, a series of experiments was conducted using an experimental installation for the induction soldering of waveguide paths for spacecraft, which includes a high-frequency generator (66 kHz), a voltage source (up to 10 V, with an output current of up to 150 A and a power of up to 15 kW), a matching device, a flat inductor with a working window of a 26 mm × 15 mm rectangular section and a manipulator–positioner.

Non-contact temperature sensors AST A250 are used to measure the temperature in the system. These pyrometers have the following characteristics: spectral range is 1.6 μm, accuracy is ±0.3% of the measured value + 1°C, distance to spot size ratio is 200:1 (350 °C–1800 °C). The temperature is measured from the flat surface of the flange in the area of the proposed connection with the pipe. On the pipe, measurements were carried out at a distance of 1 cm from its side in the area of maximum energy application formed by the magnetic field of the inductor. The layout of pyrometric sensors is presented in previous work [40].

The experiments were conducted for a tube and a flange for a waveguide path with a standard size of 22 mm × 11 mm by heating them and fixing the temperature values by means of contactless pyrometry. Summary graphs (Figure 10) could be used to compare the model-generated curves and real-world process curves of induction heating; the graphs are specific to the assembly components of spacecraft waveguides.

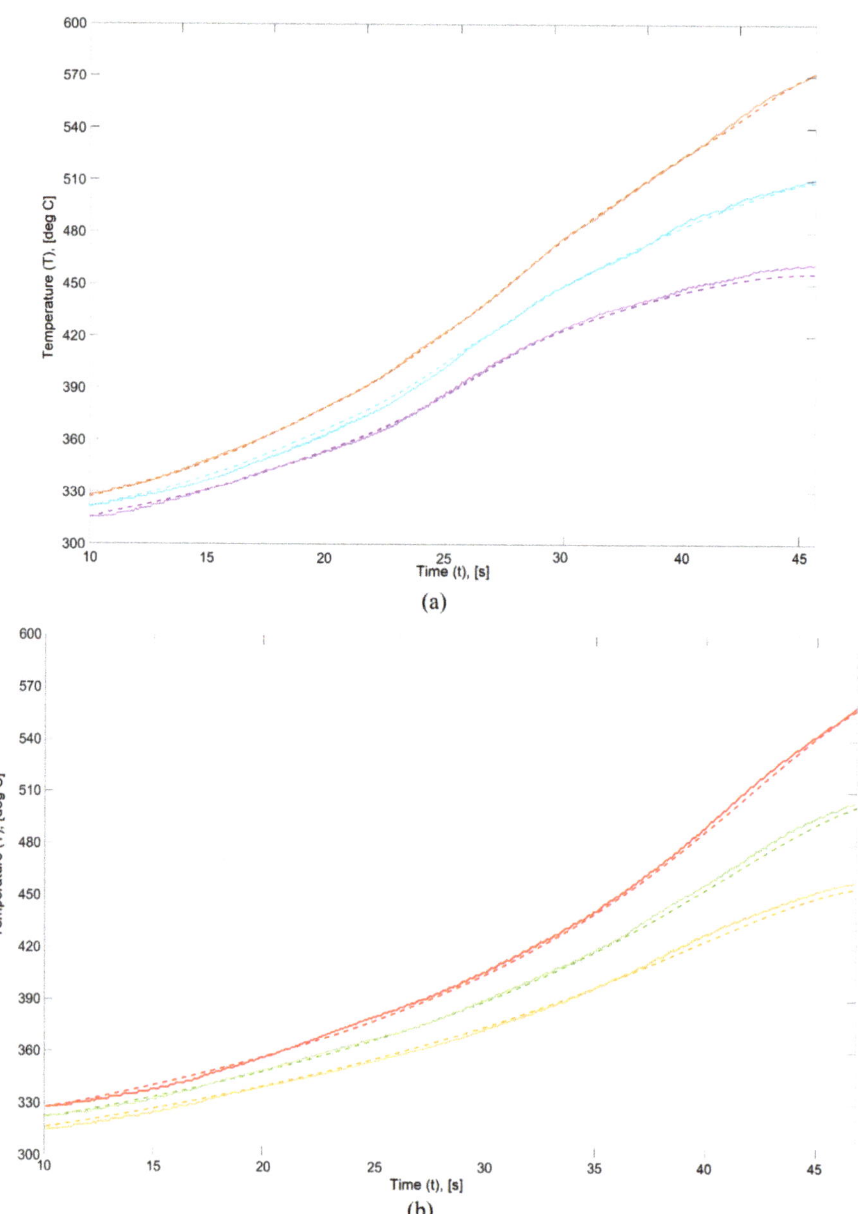

Figure 10. Waveguide component heating curves: (**a**) 22 mm × 11 mm tubes; (**b**) 22 mm × 11 mm flanges, where dotted lines are model-derived data, solid lines are experimental data, orange and red curves are for 11 kW heating, blue and green curves are for 5 kW heating, and lilac and yellow curves are for 3 kW heating.

From the graphs shown in Figure 10, it can be seen that the mathematical models proposed in this work repeat the real technological process of heating the elements of the waveguide path with a high degree of reliability. In this case, the waveguide tube experiences more intense heating in comparison with the flange, which is due to the small thickness of the tube.

As can be seen in Figure 10 as well as in the standard error values given in Table 1, the induction heating models developed for thin-walled aluminum waveguide assemblies were proven to be accurate in simulating this process. In-kind and simulation experiments showed that the models could indeed be used to test and calibrate the process parameters for the induction soldering of thin-walled aluminum spacecraft waveguides.

Table 1. Standard deviation: simulation vs. real-world processes.

Waveguide Component	Heating Power P, kW		
	3	5	10
Tube	1.4	1.6	1.5
Flange	1.5	1.8	1.7

3.2. Implementation of the Proposed Models to Calibrate the Induction Soldering Control Algorithms

To test and calibrate the process of making a permanent joint, it is necessary to implement a model in the simulation system of choice, which in this case was SimInTech [34,35].

SimInTech (Simulation in Technic) [36] is an environment for the dynamic simulation of technical systems; it is designed to validate process control systems for complex facilities. SimInTech simulates processes found in various industries and can also model control systems to help improve the design of such systems by testing every decision at any stage of the project [37]. SimInTech is designed to investigate and analyze in detail nonstationary processes in nuclear and thermal power plants, automatic control systems, tracking drives and robots or any other systems whose dynamics can be described by a system of differential algebraic equations and/or implemented by structural modeling methods. SimInTech is mainly intended for making models, designing control algorithms and debugging them on the object model, as well as for generating source codes in C for programmable controllers [38].

The model the authors implemented in SimInTech was an automated control system that comprised a control action, actuators, the controlled object and feedback (Figure 11).

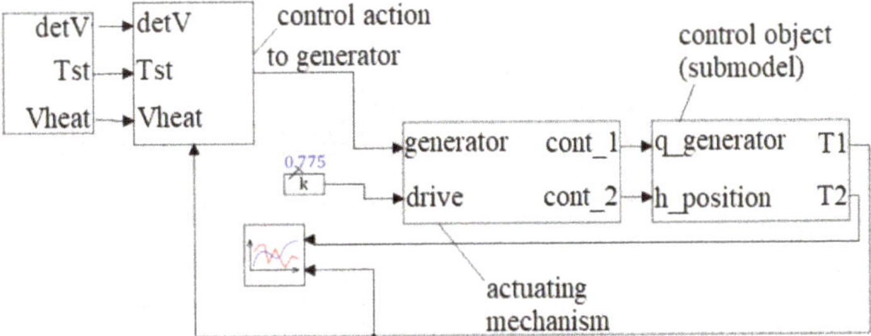

Figure 11. Overview of the SimInTech model, where detV is the deviation in heating rates, Tst is the stabilization temperature, Vheat is the workpiece heating rate, generator is the actuator input, q_generator is the control action, T1 is the waveguide tube temperature and T2 is the flange/coupler temperature.

The induction soldering of thin-walled waveguides was implemented as a proportional-integral-differentiating (PID) controller (Figure 12), which uses the set of actuators (Figure 13). The amplification factor is marked with a blue number in the Figure 13 and at the given moment is equal to 1. The upper circuit is the transfer function of a standard generator; the lower circuit is the transfer function of a standard feedback-enabled motor. The deconstructed controlled object in SimInTech shown in Figure 14 is of particular interest.

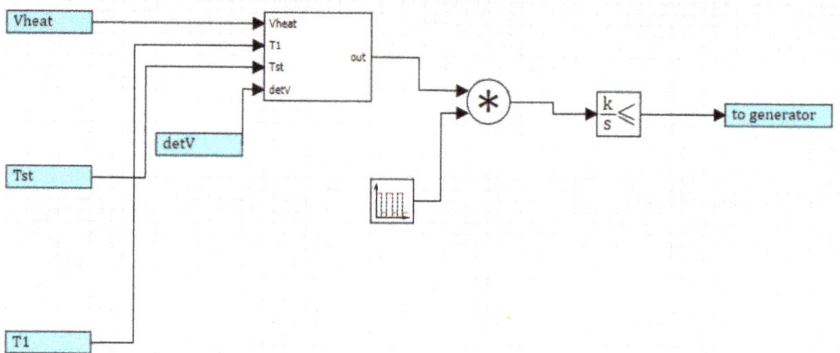

Figure 12. Deconstruction of the control action block in SimInTech, where detV is the deviation in heating rates, Tst is the stabilization temperature, Vheat is the workpiece heating rate, generator is the actuator input, q_generator is the control action, T1 is the waveguide tube temperature, * is the multiplier, k/s is the integrating link and to_generator is the control action output.

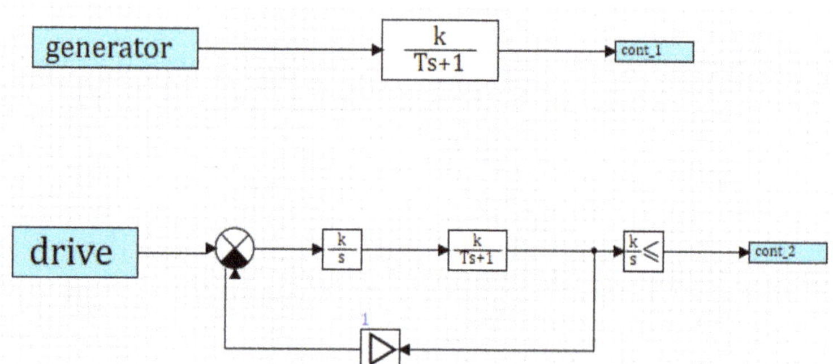

Figure 13. Deconstruction of actuators in SimInTech, where generator is the actuator input, drive is the input of the positioner, cont_1 is the control action for heating and cont_2 is the control action for workpiece positioning.

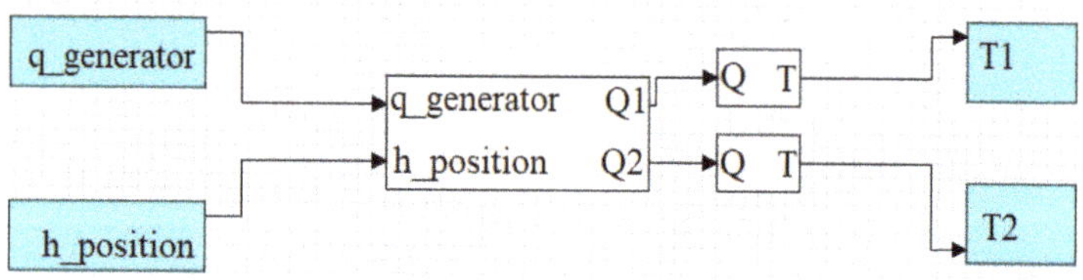

Figure 14. Deconstruction of the controlled object in SimInTech.

This submodel centers around a block that simulates the distribution of heat between the waveguide assembly components. It has the following inputs: q_generator and h_position. The block is followed by blocks that implement the waveguide assembly components: the upper one is used for a tube and the lower one for a flange. The outputs are current temperatures of the corresponding waveguide assembly components.

Curves of the control process for soldering a 22 × 11 tube–flange assembly plotted by SimInTech is shown in Figure 15. The head to assembly distance was set to 4 mm. The assembly heating rate was altered, as the process was controlled by the flange temperature: 18 °C/s (Figure 15a), 14 °C/s (Figure 15b) and 10 °C/s (Figure 15c).

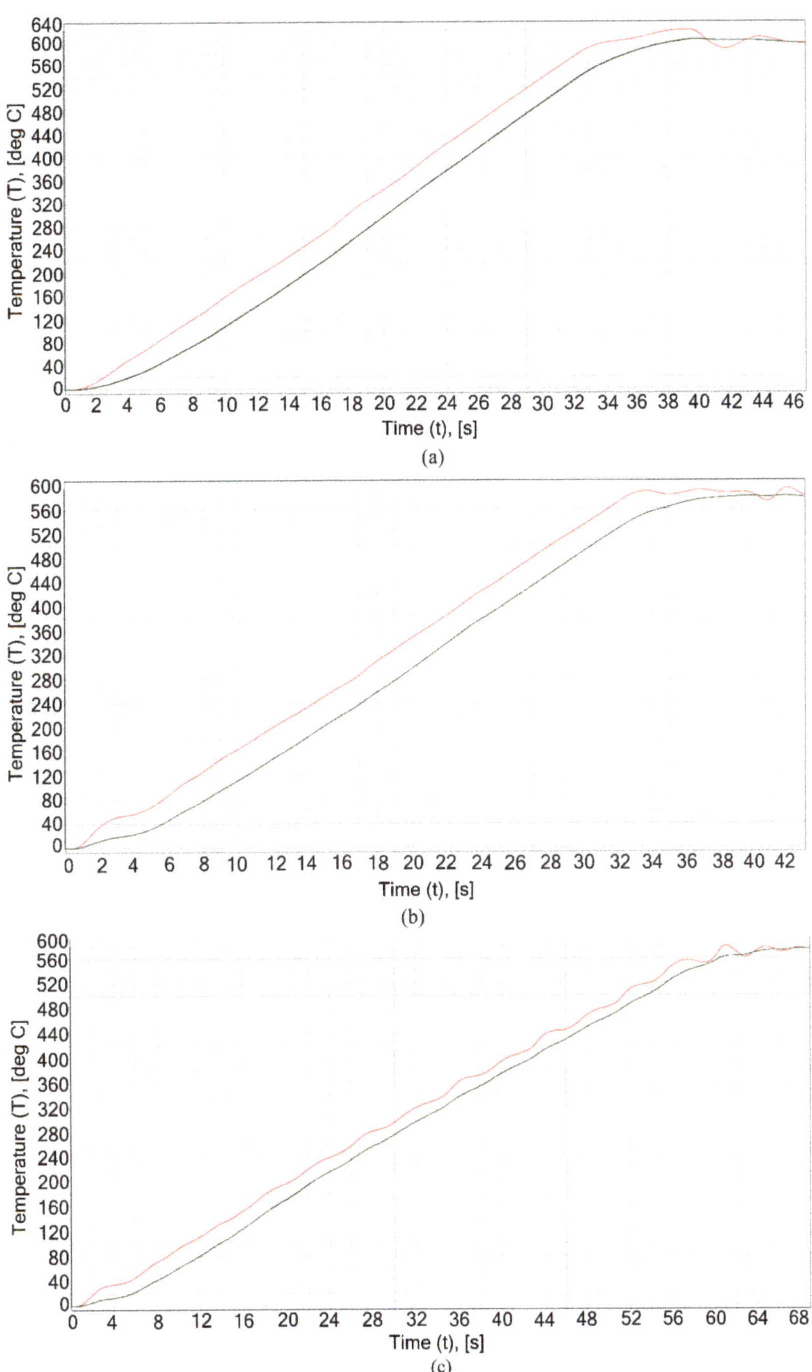

Figure 15. Waveguide induction soldering process control curves, with a distance of 4 mm to the inductor window: (**a**) 18 °C/s, (**b**) 14 °C/s, (**c**) 10 °C/s. The red curve is for the waveguide tube temperature (T^{Tb}), and the green curve is for the flange temperature (T^{Fl}).

The simulation experiment showed that the temperatures of the soldered components had the least scatter at 10 °C/s. Simulations at a lower heating rate were considered unadvisable due to the peculiarities of the soldering technology, as flux would only be active for a limited time.

To eliminate the temperature scatter, soldering was again simulated at 10 °C/s but, this time, with a 3 mm distance between the waveguide assembly and the inductor window; this redistributed temperatures from the waveguide tube to the flange. Figure 16 shows the simulation results that prove the process to be of high quality.

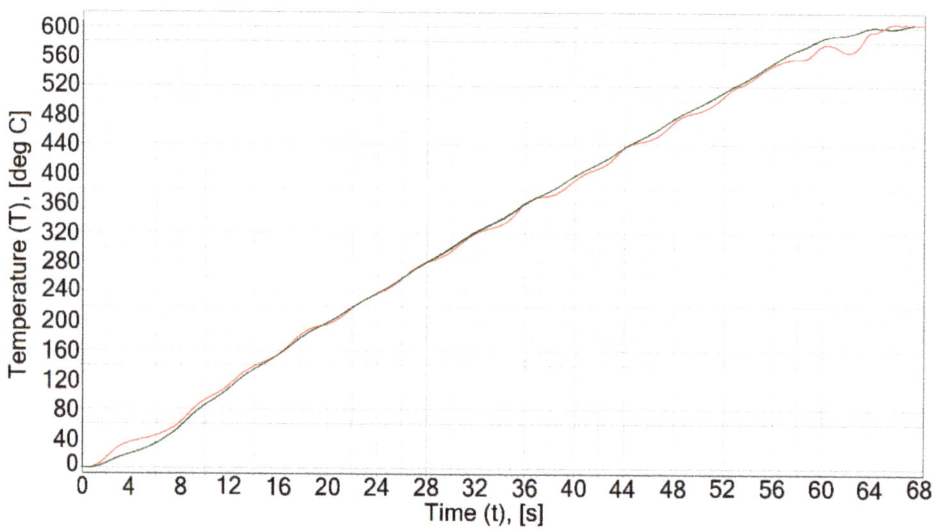

Figure 16. Waveguide inductions' soldering process control curves, with a distance of 3 mm to the inductor window, 10 °C/s: the red curve is for the waveguide tube temperature (T^{Tb}), and the green curve is for the flange temperature (T^{Fl}).

In order to exclude the spread of temperatures of the elements to be soldered, the soldering process was simulated at a rate of 10 °C/s, but the distance from the waveguide assembly to the inductor window was reduced to 3 mm for the temperature redistribution from the waveguide tube on the flange. The simulation results (Figure 16) show a good fit of the model to the real process.

Table 2 summarizes the numerical values of the operation quality of the automatic control system of induction soldering in the process of optimizing the control parameters: the heating rate of the waveguide assembly and the distance between the flange/coupling and the inductor window. The standard deviation is evaluated by the next Equation (8):

$$SD = \sqrt{\frac{1}{n}\sum_{i=1}^{n}\left(T_i^{Diff} - \overline{T}^{Diff}\right)^2} \qquad (8)$$

where $T_i^{Diff} = \left|T_i^{Tb} - T_i^{Fl}\right|$ is the absolute waveguide elements temperature difference, $\overline{T}^{Diff} = \frac{1}{n}\sum_{i=1}^{n}\left|T_i^{Tb} - T_i^{Fl}\right|$ is the mean value of waveguide elements temperature difference, n is the number of temperature control points, T_i^{Tb} is the tube temperature in i-th point, T_i^{FL} is the flange temperature in i-th point.

Table 2. Overregulation (OR) and standard deviation (SD): flange/coupling temperature vs. tube temperature (in °C).

Distance from the Waveguide Assembly to the Inductor Window, mm	Heating Rate, °C/s					
	18		14		10	
	OR	SD	OR	SD	OR	SD
3	-	-	-	-	0.8	4.3
4	18.2	39.7	10.8	38.4	4.8	21.1

A multitude of experiments can be run to test soldering parameters on various waveguide assembly configurations by using the mathematical models developed here for testing and calibrating the induction soldering process for thin-walled aluminum waveguides. The use of the models can reduce R&D costs as they enable pre-calibrated experimentation in kind, which effectively cuts the costs of consumables.

4. Conclusions

The goal of this work was to develop mathematical models of induction soldering for waveguide assembly components, which could help in testing and calibrating the induction soldering process for the thin-walled aluminum waveguides found in spacecraft. To verify the developed models, the research team carried out simulations and tests in kind, which showed the models were indeed accurate in predicting the actual induction heating processes observed in the induction soldering of thin-walled aluminum waveguides found in spacecraft.

The standard deviation of the temperature difference between simulation and real-world processes did not exceed 2 °C, which is satisfactory in terms of induction soldering technology.

In addition, the authors studied the application of the developed mathematical apparatus for calibrating the control of the induction soldering process. The results showed that the found effective values of the heating rate and the distance from the wave-guide assembly to the inductor window allowed the control of the technological process with low overregulation (0.8 °C) and for a low standard deviation between the temperatures of the soldered elements to be maintained (4.3 °C).

In future, we aim to extend this research to achieve the following:

- To develop a set of adaptive induction soldering process control methods that employ state-of-the-art data mining algorithms;
- To implement a prototype spacecraft waveguide induction soldering control system based on the developed mathematical models and algorithms;
- To test the applicability, usability and effectiveness of such a prototype system with computational and in-kind experiments;
- To design a process diagram to integrate the proposed software prototype into an existing experimental system for the induction soldering of spacecraft waveguides.

Integrating the developed models in the existing process control hardware and software for spacecraft waveguide induction soldering will help to cut costs due to the use of the models to test and calibrate different induction soldering parameters on a variety of waveguide configurations and sizes.

Author Contributions: Conceptualization, V.T. (Vadim Tynchenko); Data curation, V.B. and V.K. (Viktor Kukartsev); Formal analysis, S.K., V.T. (Valeriya Tynchenko), V.B., V.K. (Vladislav Kukartsev) and R.S.; Investigation, V.T. (Vadim Tynchenko), S.K., V.B. and V.K. (Viktor Kukartsev); Methodology, V.K. (Vladislav Kukartsev) and R.S.; Project administration, V.T. (Vadim Tynchenko); Resources, V.K. (Vladislav Kukartsev) and K.B.; Software, V.T. (Valeriya Tynchenko), V.B. and V.K. (Vladislav Kukartsev); Supervision, V.T. (Vadim Tynchenko); Validation, V.T. (Vadim Tynchenko) and V.T. (Valeriya Tynchenko); Visualization, V.T. (Vadim Tynchenko), S.K. and K.B.; Writing—original draft, V.T. (Vadim Tynchenko), S.K., V.T. (Valeriya Tynchenko), V.B., V.K. (Vladislav Kukartsev), R.S., V.K. (Viktor Kukartsev) and K.B.; Writing—review and editing, V.T. (Vadim Tynchenko), S.K., V.T. (Valeriya Tynchenko), V.B., V.K. (Vladislav Kukartsev), R.S., V.K. (Viktor Kukartsev) and K.B. All authors have read and agreed to the published version of the manuscript.

Funding: This research received no external funding.

Institutional Review Board Statement: Not applicable.

Informed Consent Statement: Not applicable.

Conflicts of Interest: The authors declare no conflict of interest.

References

1. Zeller, U.; Lohmeier, M.; Pander, M.; Lausch, D. Multiphysics simulation of induction soldering process. In Proceedings of the 2018 IEEE 7th World Conference on Photovoltaic Energy Conversion (WCPEC) (A Joint Conference of 45th IEEE PVSC, 28th PVSEC & 34th EU PVSEC), Waikoloa, HI, USA, 10–15 June 2018; pp. 654–659. [CrossRef]
2. Papargyri, L.; Theristis, M.; Kubicek, B.; Krametz, T.; Mayr, C.; Papanastasiou, P.; Georghiou, G.E. Modelling and experimental investigations of microcracks in crystalline silicon photovoltaics: A review. *Renew. Energy* **2020**, *145*, 2387–2408. [CrossRef]
3. Zhu, T.Z.; Feng, P.; Li, X.; Li, F.; Rong, Y. The study of the effect of magnetic flux concentrator to the induction heating system using coupled electromagnetic-thermal simulation model. In Proceedings of the 2013 International Conference on Mechanical and Automation Engineering, Jiujang, China, 21–23 July 2013; pp. 123–127. [CrossRef]
4. Pánek, D.; Orosz, T.; Kropík, P.; Karban, P.; Doležel, I. Reduced-order model based temperature control of induction brazing process. In Proceedings of the 2019 Electric Power Quality and Supply Reliability Conference (PQ) & 2019 Symposium on Electrical Engineering and Mechatronics (SEEM), Kärdla, Estonia, 12–15 June 2019; pp. 906–911. [CrossRef]
5. Eftychiou, M.A.; Bergman, T.L.; Masada, G.Y. A detailed thermal model of the infrared reflow soldering process. *ASME J. Electron. Packag.* **1993**, *115*, 55–62. [CrossRef]
6. Milov, A.V.; Tynchenko, V.S.; Murygin, A.V. Neural Network Modeling to Control Process of Induction Soldering. In Proceedings of the 2019 International Conference on Industrial Engineering, Applications and Manufacturing (ICIEAM), Sochi, Russia, 25–29 March 2019; pp. 1–5. [CrossRef]
7. Milov, A.V.; Tynchenko, V.S.; Petrenko, V.E. Algorithmic and software to identify errors in measuring equipment during the formation of permanent joints. In Proceedings of the 2018 International Multi-Conference on Industrial Engineering and Modern Technologies (FarEastCon), Vladivostok, Russia, 3–4 October 2018; pp. 1–5. [CrossRef]
8. Milov, A.V.; Tynchenko, V.S.; Kukartsev, V.V.; Tynchenko, V.V.; Bukhtoyarov, V.V. Use of artificial neural networks to correct non-standard errors of measuring instruments when creating integral joints. *J. Phys. Conf. Ser.* **2018**, *1118*, 012037. [CrossRef]
9. Bukhtoyarov, V.V.; Tynchenko, V.S.; Petrovskiy, E.A.; Tynchenko, S.V.; Milov, A.V. Intelligently informed control over the process variables of oil and gas equipment maintenance. *Int. Rev. Autom. Control* **2019**, *12*, 59–66. [CrossRef]
10. Bocharova, O.A.; Tynchenko, V.S.; Bocharov, A.N.; Oreshenko, T.G.; Murygin, A.V.; Panfilov, I.A. Induction heating simulation of the waveguide assembly elements. *J. Phys. Conf. Ser.* **2019**, *1353*, 012040. [CrossRef]
11. Murygin, A.V.; Tynchenko, V.S.; Laptenok, V.D.; Emilova, O.A.; Seregin, Y.N. Modeling of thermal processes in waveguide tracts induction soldering. *IOP Conf. Ser. Mater. Sci. Eng.* **2017**, *173*, 012026. [CrossRef]
12. Milov, A.; Tynchenko, V.; Petrenko, V. Intellectual Control of Induction Soldering Process using Neuro-fuzzy Controller. In Proceedings of the 2019 International Russian Automation Conference (RusAutoCon), Sochi, Russia, 8–14 September 2019; pp. 1–6. [CrossRef]
13. Pánek, D.; Karban, P.; Doležel, I. Calibration of numerical model of magnetic induction brazing. *IEEE Trans. Magn.* **2019**, *55*, 1–4. [CrossRef]
14. Satheesh, A.; Kattisseri, M.; Vijayan, V. Numerical estimation of localized transient temperature and strain fields in soldering process. In Proceedings of the 2018 7th Electronic System-Integration Technology Conference (ESTC), Dresden, Germany, 8–21 September 2018; pp. 1–5. [CrossRef]
15. Tan, J.S.; Khor, C.Y.; Ishak, M.I.; Rosli, M.U.; Jamalludin, M.R.; Nawi, M.A.M.; Syafiq, A.M.; Aziz, M.A. Effect of Solder Joint Width to the Mechanical Aspect in Thermal Stress Analysis. *IOP Conf. Ser. Mater. Sci. Eng.* **2019**, *551*, 012105. [CrossRef]
16. Dudek, R.; Döring, R.; Rzepka, S.; Herberholz, T.; Feil, D.; Seiler, B.; Christian, S.; Fritzsche, S. Stress Analyses in HPC-Soldered Assemblies by Optical Measurement and FEA. In Proceedings of the 2018 7th Electronic System-Integration Technology Conference (ESTC), Dresden, Germany, 18–21 September 2018; pp. 1–6. [CrossRef]

17. Ribes-Pleguezuelo, P.; Zhang, S.; Beckert, E.; Eberhardt, R.; Wyrowski, F.; Tünnermann, A. Method to simulate and analyse induced stresses for laser crystal packaging technologies. *Opt. Express* **2017**, *25*, 5927–5940. [CrossRef]
18. Karban, P.; Pánek, D.; Doležel, I. Model of induction brazing of nonmagnetic metals using model order reduction approach. *Compel Int. J. Comput. Math. Electr. Electron. Eng.* **2018**, *37*, 1515–1524. [CrossRef]
19. Seehase, D.; Kohlen, C.; Neiser, A.; Novikov, A.; Nowottnick, M. Selective soldering on printed circuit boards with endogenous induction heat at appropriate susceptors. *Period. Polytech. Electr. Eng. Comput. Sci.* **2018**, *62*, 172–180. [CrossRef]
20. Wang, S.; Xu, H.; Yao, Y. Annealing optimization for tin–lead eutectic solder by constitutive experiment and simulation. *J. Mater. Res.* **2017**, *32*, 3089–3099. [CrossRef]
21. Sarhadi, A.; Bjørk, R.; Pryds, N. Optimization of the mechanical and electrical performance of a thermoelectric module. *J. Electron. Mater.* **2015**, *44*, 4465–4472. [CrossRef]
22. Lanin, V.L. Sizing up the efficiency of induction heating systems for soldering electronic modules. *Surf. Eng. Appl. Electrochem.* **2018**, *54*, 401–406. [CrossRef]
23. Yakimenko, O.A. *Engineering Computations and Modeling in MATLAB®/Simulink®*; American Institute of Aeronautics and Astronautics, Inc.: Reston, VA, USA, 2019.
24. Chaturvedi, D.K. *Modeling and Simulation of Systems Using MATLAB and SIMULINK*; CRC Press: Boca Raton, FL, USA, 2017.
25. Klee, H.; Allen, R. *Simulation of Dynamic Systems with MATLAB and Simulink*; CRC Press: Boca Raton, FL, USA, 2016.
26. Zhou, S.; Rabczuk, T.; Zhuang, X. Phase field modeling of quasi-static and dynamic crack propagation: COMSOL implementation and case studies. *Adv. Eng. Softw.* **2018**, *122*, 31–49. [CrossRef]
27. Panagakos, G.; Ajayi, O.; Yang, T.; Brunello, G.; Mason, J.H.; Abernathy, H.W.; Miller, D.C. *Modeling Multi-Physics Problems for Energy Applications with Comsol Multi-Physics*; No. NETL-PUB-21189; NETL: Pittsburgh, PA, USA, 2017.
28. Wei, Z.; Weavers, L.K. Combining COMSOL modeling with acoustic pressure maps to design sono-reactors. *Ultrason. Sonochem.* **2016**, *31*, 490–498. [CrossRef]
29. Sezgin, B.; Caglayan, D.G.; Devrim, Y.; Steenberg, T.; Eroglu, I. Modeling and sensitivity analysis of high temperature PEM fuel cells by using Comsol Multiphysics. *Int. J. Hydrog. Energy* **2016**, *41*, 10001–10009. [CrossRef]
30. Stolarski, T.; Nakasone, Y.; Yoshimoto, S. *Engineering Analysis with ANSYS Software*; Butterworth-Heinemann: Oxford, UK, 2018.
31. Lee, H.H. *Finite Element Simulations with ANSYS Workbench 18*; SDC Publications: Mission, KS, USA, 2018.
32. Alawadhi, E.M. *Finite Element Simulations Using ANSYS*; CRC Press: Boca Raton, FL, USA, 2015.
33. Menter, F.R.; Lechner, R.; Matyushenko, A. *Best Practice: Generalized k-ω Two-Equation Turbulence Model in ANSYS CFD (GEKO)*; Technical Report; ANSYS: Nurnberg, Germany, 2019.
34. Smagin, D.I.; Starostin, K.I.; Saveliev, R.S.; Kobrinets, T.A.; Satin, A.A.E. Application of this software Simintech for mathematical modeling of various onboard systems of aircraft. *Comput. Nanotechnol.* **2018**, *3*, 9–15.
35. Smagin, D.I.; Starostin, K.I.; Saveliev, R.S.; Kobrinets, T.A.; Satin, A.A.E.; Suvorov, A.V.E.; Medvedev, P.I. Simulation of air conditioning system perspective of a passenger plane in the software package SiminTech. *Comput. Nanotechnol.* **2018**, *3*, 24–31.
36. Smagin, D.I.; Starostin, K.I.; Saveliev, R.S.; Satin, A.A.E.; Pritulkin, A.A. Modeling of failures of the electricity system (SES) ac long-haul passenger plane in the software package SimInTech. *Comput. Nanotechnol.* **2019**, *2*, 63–70. [CrossRef]
37. Smagin, D.I.; Starostin, K.I.; Saveliev, R.S.; Kobrinets, T.A.; Satin, A.A.E. The technique of creating a dynamic mathematical model of the neutral gas system for a promising aircraft in the program complex SimInTech. *Comput. Nanotechnol.* **2018**, *2*, 21–27.
38. Abalov, A.A.; Nosachev, S.V.; Zharov, V.P.; Minko, V.A. Using the SimInTech dynamic modeling environment to build and check the operation of automation systems. In *MATEC Web of Conferences*; EDP Sciences: Les Ulis, France, 2018; Volume 226. [CrossRef]
39. Karkhin, V.A. *Fundamentals of Heat Transfer in Welding and Brazing*; Textbook for Universities; SPb SPU Press: Saint Petersburg, Russia, 2011.
40. Murygin, A.V.; Laptenok, V.D.; Tynchenko, V.S.; Emilova, O.A.; Seregin, Y.N. Development of an automated information system for controlling the induction soldering of aluminum alloys waveguide paths. In Proceedings of the 2018 3rd Russian-Pacific Conference on Computer Technology and Applications (RPC), Vladivostok, Russia, 18–25 August 2018; pp. 1–5. [CrossRef]

Article

Warpage Analysis and Control of Thin-Walled Structures Manufactured by Laser Powder Bed Fusion

Xufei Lu [1], Michele Chiumenti [1], Miguel Cervera [1,*], Hua Tan [2,*], Xin Lin [2] and Song Wang [2]

[1] International Center for Numerical Methods in Engineering, Technical University of Catalonia, 08034 Barcelona, Spain; xufei.lu@upc.edu (X.L.); michele.chiumenti@upc.edu (M.C.)

[2] State Key Laboratory of Solidification Processing, Northwestern Polytechnical University, Xi'an 710072, China; xlin@nwpu.edu.cn (X.L.); m15702930675@163.com (S.W.)

[*] Correspondence: miguel.cervera@upc.edu (M.C.); tanhua@nwpu.edu.cn (H.T.); Tel.: +34-934-016-492 (M.C.); +86-029-8849-4001 (H.T.)

Abstract: Thin-walled structures are of great interest because of their use as lightweight components in aeronautical and aerospace engineering. The fabrication of these components by additive manufacturing (AM) often produces undesired warpage because of the thermal stresses induced by the manufacturing process and the components' reduced structural stiffness. The objective of this study is to analyze the distortion of several thin-walled components fabricated by Laser Powder Bed Fusion (LPBF). Experiments are performed to investigate the sensitivity of the warpage of thin-walled structures fabricated by LPBF to different design parameters such as the wall thickness and the component height in several open and closed shapes. A 3D-scanner is used to measure the residual distortions in terms of the out-of-plane displacement. Moreover, an in-house finite element software is firstly calibrated and then used to enhance the original design in order to minimize the warpage induced by the LPBF printing process. The outcome of this shows that open geometries are more prone to warping than closed ones, as well as how vertical stiffeners can mitigate component warpage by increasing stiffness.

Keywords: additive manufacturing; laser powder bed fusion; thin-walled structures; warpage; finite element analysis

Citation: Lu, X.; Chiumenti, M.; Cervera, M.; Tan, H.; Lin, X.; Wang, S. Warpage Analysis and Control of Thin-Walled Structures Manufactured by Laser Powder Bed Fusion. *Metals* **2021**, *11*, 686. https://doi.org/10.3390/met11050686

Academic Editor: Mohammad Jahazi

Received: 23 March 2021
Accepted: 20 April 2021
Published: 22 April 2021

Publisher's Note: MDPI stays neutral with regard to jurisdictional claims in published maps and institutional affiliations.

1. Introduction

Additive manufacturing (AM) is an industrial process increasingly integrated in the production chain. Nowadays, the aerospace and aeronautical industries are "printing" several high-performance metallic components [1,2]. Laser Powder Bed Fusion (LPBF), also known as Selective Laser Melting (SLM), is the most widely adopted AM technology to fabricate complex structures through a layer-by-layer metal deposition sequence, as it provides remarkable geometrical accuracy and satisfactory mechanical properties from a wide range of metallic powders [3,4]. The LPBF process is carried out in a closed chamber with controlled atmosphere (e.g., argon) to prevent oxidation. The 3D-printing process consists of the following steps: (i) a new powder layer (30~60 μm) is spread, (ii) the high-energy laser beam selectively melts the power bed according to a user defined scanning sequence, and, (iii) the base-plate is lowered to allow for the deposition of a new powder layer. This sequence is repeated until the building process is completed. At the end of the AM process, the loose powder is removed, allowing for the final cooling stage at ambient temperature.

In AM, the material undergoes repeated heating and cooling cycles, high temperature gradients are generated and, consequently, large thermal deformations are induced. The temperature gradient between layers produces residual stresses and plastic deformations because the material cannot freely expand and contract. As a result, the accumulated stresses generate warpage and, eventually, cracking in thin-walled structures, compromising both the geometrical accuracy and the structural integrity of the AM component [5–8].

Part warpage and the accumulation of residual stresses are closely related to the actual parameters of the printing process, the scanning strategy and the geometrical features of the built. Compared to the blown powder or wire feeding AM techniques, in LPBF processes the size of the melting pool is smaller, the scanning speed is higher and, consequently, the thermal gradients are also higher (up to 10^7 °C/m), particularly when printing thin-walled structures [9–13].

Several experimental and numerical analyses have been performed to mitigate both residual stress and distortion in AM [14–21]. Yakout et al. [15] studied the effect of the thermal expansion coefficient and the thermal diffusivity in the Thermo-Mechanically Affected Zone (TMAZ) of SLM samples by experimental measurements and finite element (FE) analysis. Lu et al. [22] proposed a substrate design strategy to mitigate the residual stresses in AM processes. Ramos et al. [23] optimized the building strategy to reduce both deformations and residual stresses by controlling heat concentration during the LPBF process. LevkulichL et al. [24] performed sensitivity analysis of the process parameters in SLM to study their effects on the stress evolution of Ti-6Al-4V built. Their results showed that, by increasing the laser power and reducing the scanning speed, it is possible to mitigate the residual stresses of metal deposition. Contrarily, Li et al. [25] found that similar process parameters induce higher residual stresses at the single-track end of SLM Ti-6Al-4V. Cao et al. [6] investigated the effect of substrate preheating. They found that by increasing the preheating temperature it is possible to minimize part distortion and residual stresses. Recently, some researches introduced geometric compensation as an effective strategy to counterbalance the part distortions induced by the thermal field in AM [26–28].

Several works [29–31] have studied the warpage of single-walled parts by LPBF. Among them, Li et al. [29] showed the influence of the scanning sequence and, particularly, the scanning length on the warpage of components. Chen et al. [30] analyzed the maximum warpage of Ti6Al4V adopting different wall thicknesses. Ahmed et al. [31] investigated the dimensional accuracy of AlSi10Mg parts. Contrariwise, the warpage mechanism of thin-walled components of greater geometrical complexity has scarcely been investigated. Furthermore, no strategy for warpage control has been developed for them. This is the main objective of the present investigation.

In this work, a variety of thin-walled components are printed by LPBF, varying their wall-thickness and building-height. Different open and closed shapes are compared and a 3D-scanner is used to measure their actual warpage. These results are used to calibrate a thermo-mechanical FE model implemented in an in-house 3D-printing FE software used for the numerical simulation of the AM process. Thereby, this numerical tool is used to define a structural optimization strategy to mitigate the warpage of thin-walled parts printed by LPBF.

2. Experimental Campaign

The experimental campaign is carried out using a Concept Laser M2 powder bed machine (Concept Laser, Lichtenfels, Germany). This equipment uses a Yb-fibre laser with a maximum input power of 400 W (D4 Sigma Gaussian beam). The building process takes place in a closed chamber ($250 \times 250 \times 280$ mm^3) with an annealed Ti-6Al-4V base-plate ($250 \times 250 \times 25$ mm^3) and a controlled atmosphere of pure argon (the oxygen content is restricted to 100 ppm).

In this research, the gas atomized Ti-6Al-4V powder used for building has a spherical shape, with a size of 15~53 μm. The chemical composition of this powder is given in Table 1. Before the printing process, the powder is dried in a vacuum oven (Fengxun, Shanghai, China) (vacuum degree less than 80 kPa) at 125 °C during 2.5 h.

Table 1. Composition of Ti-6Al-4V alloy powder (wt.%).

Al	V	O	H	N	C	Fe	Si	Ti
6.28	3.90	0.098	0.002	0.020	0.008	0.022	0.026	Balance

The 3D-printing is characterized by a 90° rotating scan pattern. Table 2 reports the processing parameters adopted for the manufacturing process. In addition, no preheating is used in this work.

Table 2. Processing parameters used in the PLBF process.

Laser Power (W)	Layer Thickness (μm)	Scan Speed (mm/s)	Hatch Spacing (μm)	Laser BeamDiameter (μm)
200	30	1000	100	100

Figure 1 shows the thin-walled structures printed on the LPBF base-plate after removing the loose powder. These include:

- Single-wall structures of different thicknesses and heights;
- Cylindrical structures of different diameters and heights;
- Square-section structures of several thicknesses;
- Open-section structures (e.g., semi-cylinder, L-shape, etc.).

Figure 1. Typical thin-walled structures printed by LPBF.

Figure 2 shows the *Breuckmann SmartSCAN3D* scanner (Aniwaa, Central Singapore, Singapore) with the measurement accuracy of 0.015 mm, used to measure the actual distortion of all the components after the 3D-printing process. The relative component warpage is calculated by the Geomagic Qualify software (3D Systems, Rock Hill, Washington, DC, USA) by comparing the experimental 3D-scan images with the nominal CAD geometries. It should be mentioned that all the measurements are done before the removal of the substrate (cutting phase).

Figure 2. *Breuckmann SmartSCAN3D* scanner for the warpage measurements.

Scale Effect on the Part Warpage

In this section, the sensitivity to the dimensions (e.g., wall-thickness, building height and component size) of the thin-walled components is examined. Therefore, the warpage is measured by 3D-scanning to obtain its relation with the actual dimensions of the 3D-printed structures.

The first assessment study considers a single-wall geometry with a base length of 50 mm and a total height of 50 mm. Three thicknesses of 1 mm, 2 mm and 5 mm, respectively, are investigated.

Figure 3 compares the final warpage distributions as measured for the three wall thicknesses used. It can be seen that the thinnest wall presents the more pronounced warpage, up to 0.41 mm in the middle of the final cambered surface. As the wall-thickness increases, the warpage rapidly decreases; the maximum out-of-plane displacement for the 5 mm thickness wall is only of about 0.03 mm. Hence, as the thickness is increased, the wall warpage is restrained by the enhanced structural stiffness of the component. Simultaneously, higher residual stresses are induced by the printing process.

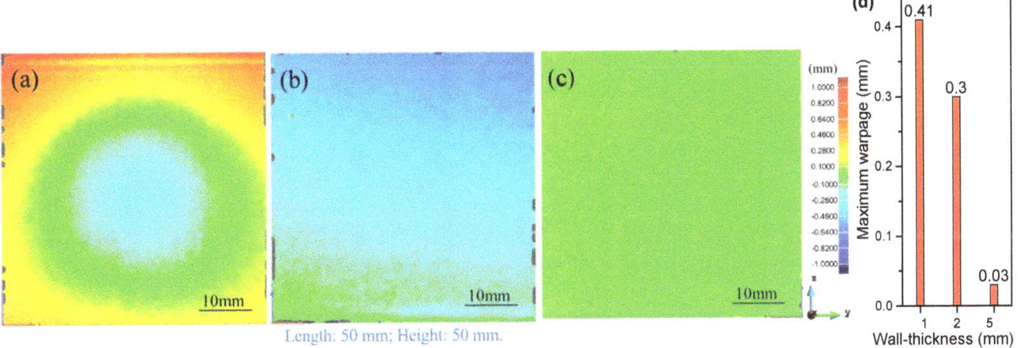

Figure 3. Residual warpage (out-of-plane displacement) for the single-wall structures for different wall-thicknesses: (**a**) 1 mm; (**b**) 2 mm; (**c**) 5 mm, respectively. (**d**) Maximum warpage recorded in the middle section.

A second study analyses three different wall heights (37 mm, 43 mm and 50 mm, respectively) for the same single-walled geometry and a fixed thickness of 1 mm. Figure 4 compares the residual warpages of the three thin-walled structures. It is possible to observe that increasing the component height results in larger part warpage. This is not only because of the reduced wall thickness, but also because the higher the component, the less sensitive it is to the clamping conditions at the base and, thus, the more free it is to deform.

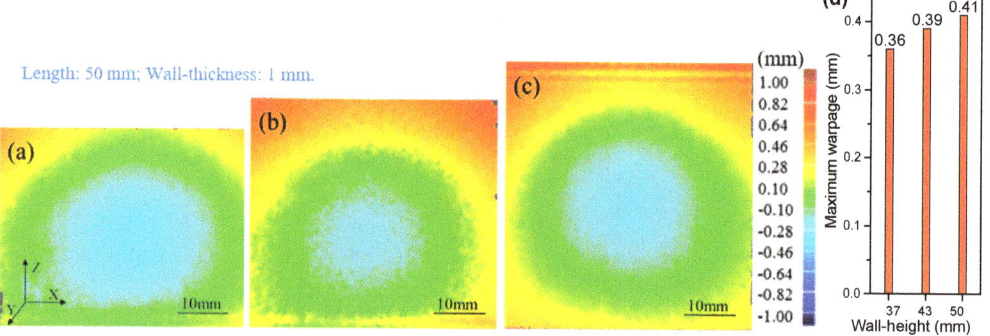

Figure 4. Residual warpage contour-fills of the thin-walled structures with different heights: (**a**) 37 mm; (**b**) 43 mm; (**c**) 50 mm, respectively. (**d**) Maximum warpages recorded in the middle section.

Next, a closed thin-wall cylindrical structure is studied. The diameter is 50 mm and the wall thickness is to 1 mm. The cylinder heights are of 30 mm, 50 mm and 70 mm, respectively.

Figure 5 shows the contour-fills of the warpage for the three cylindrical structures with different heights. It can be found that increasing the height leads to larger warpages as observed for the single-wall. However, the maximum warpage is remarkably lower than in both the single-wall and the square thin-wall structures. In this case, the (cylindrical) shape noticeably improves the structural stiffness of the printed structure for the same reduced wall thickness and component height.

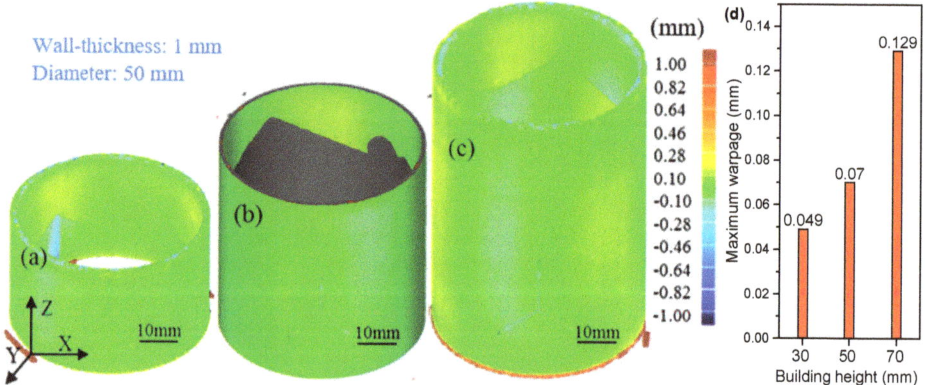

Figure 5. Contour-fills of the final warpage of the cylindrical thin-wall structures with different building heights: (**a**) 30 mm; (**b**) 50 mm; (**c**) 70mm, respectively. (**d**) Maximum warpages recorded in the middle section.

A further assessment study is performed assuming a thin-wall cylindrical geometry with a thickness of 1 mm and a height of 50 mm. Different cylinder diameters of 50 mm, 70 mm and 90 mm are studied. Figure 6 depicts the contour-fills of the final warpage of the three cylinders. As expected, by increasing the diameter, the warpage is also more pronounced. Eventually, if very large diameters are considered, the results of the open single-wall should be replicated.

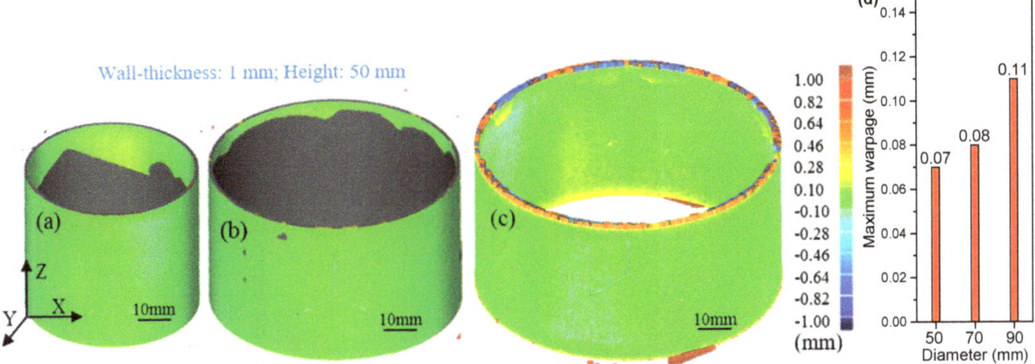

Figure 6. Contour-fills of the final warpage of the cylindrical thin-wall structures with different diameters of: (**a**) 50 mm; (**b**) 70 mm; (**c**) 90 mm; (**d**) Maximum warpages recorded in the middle section.

Finally, the last assessment studies a closed square thin-wall section. The same base length and height of 50 mm are used, while wall thicknesses of 1 mm, 2 mm and 5 mm are investigated. Figure 7 shows the contour-fills of the residual warpage for the three AM structures. Once again, it can be observed that the thicker the wall-thickness, the smaller the observed warpage. Therefore, a pronounced bulging is shown by the component with the thinnest wall. Nevertheless, the overall structural stiffness of the closed section allows for a reduction of the maximum out-of-plane displacements: 0.37 mm compared to 0.41 mm as measured for the open single-wall. This is because the deformation of each one of the walls conforming the square thin-wall section is constrained by the substrate and also laterally. Notably, the cylinder of similar dimensions shows a much better performance than the square thin-wall section of similar dimensions, Figures 5b and 7a, respectively. This is explained as follows. In LPBF, high tensile stresses in the hoop/wall direction exist at the top for two components. Large tensile stresses also happen along the corners of the square part due to the high cooling rates at such locations; this builds up compressive stresses in the central region of each square wall so that any small disturbance can favour the bulging perpendicular to the wall. This is not the case for the cylinder.

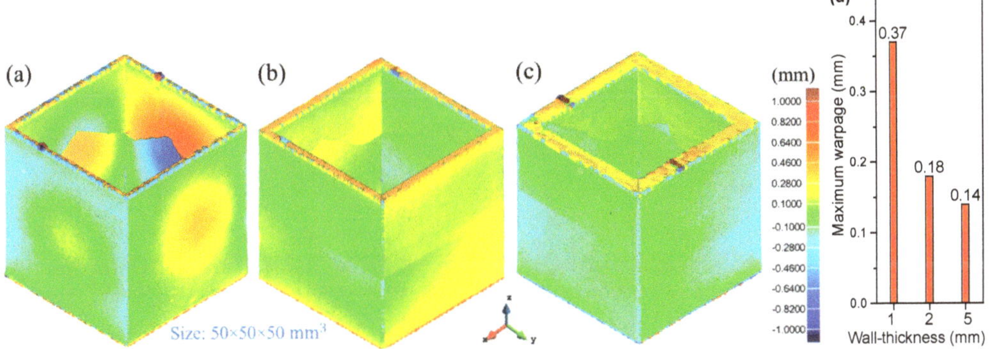

Figure 7. Contour-fills of the final warpage of the square thin-wall section with different wall-thicknesses: (**a**) 1 mm; (**b**) 2 mm; (**c**) 5 mm, respectively. (**d**) Maximum warpages recorded in the middle section.

3. Numerical Modeling

In LPBF the following two operations are repeated for each layer (see Figure 8):

1. *Recoating*: the baseplate is lowered to accommodate a new powder layer;
2. *Laser-scanning*: the laser heat source generates a molten pool which follows a user-defined trajectory to selectively melt the powder bed consolidating the new layer comprising the AM build.

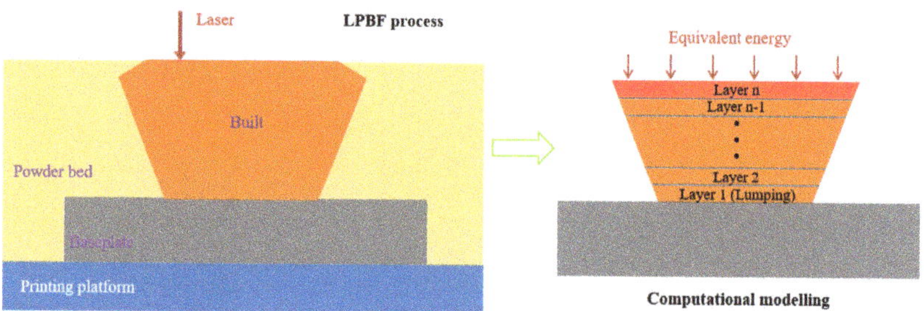

Figure 8. The schematic of the used layer-by-layer model for LPBF.

The most suitable numerical strategy to approach the simulation of the LPBF process is by the inherent strain method [32]. This method allows for a fast prediction of the distortions and residual stresses of the fabricated component. Therefore, the detailed simulation of the laser movement along the scanning trajectory is replaced by a layer-by-layer (or even multi-layer) analysis. This assumption is generally accepted when simulating LPBF processes because the scanning speed is very high (up to 1 m/s) compared to other AM technologies and the TMAZ is very localized, and it affects only the last few (2–3) layers. As a consequence, thermo-plastic (inherent) strains are only generated within the TMAZ.

This given, the original CAD geometry is firstly sliced according to the thickness of the powder-bed and then the FE discretization is generated accordingly. The birth-death FE activation technique is adopted [19,33] to progressively add the elements belonging to each new layer to the computational domain, Ω. In the initial FE discretization of the nominal CAD geometry, all elements are inactive. Inactive elements are not computed or assembled into the global system of equations. The numerical simulation of the building process consists of a sequence of layer activations. Hence, at each time-step, the elements belonging to the next layer are switched on (activated), becoming part of the current computational domain. For each time-step, the thermo-mechanical analysis is performed for all the active elements, until the whole layer-by-layer deposition sequence is completed.

3.1. Mechanical Problem

The stress analysis enforces the conservation of the balance of momentum and the continuity equations within the current active domain:

$$\bigtriangledown \mathbf{s} + \bigtriangledown p + \mathbf{b} = 0 \tag{1}$$

$$(\bigtriangledown \mathbf{u} - e^{T}) - p/K = 0 \tag{2}$$

where $\mathbf{b}$ are the body forces per unit of volume and $K(T)$ is the temperature-dependent bulk modulus and p and $\mathbf{s}$, are the hydrostatic (pressure) and the deviatoric components of the Cauchy stress tensor, σ. Thus:

$$\sigma = p\mathbf{I} + \mathbf{s} \tag{3}$$

The strain tensor, $\varepsilon(\mathbf{u}) = \bigtriangledown^{s}\mathbf{u}$, can also be split in its volumetric and deviatoric parts, $e_{vol} = \bigtriangledown \cdot \mathbf{u}$ and $\mathbf{e}$, respectively, so that:

$$\varepsilon = e_{vol}\mathbf{I}/3 + \mathbf{e} \tag{4}$$

The deviatoric stress tensor is computed as:

$$\mathbf{s}(\mathbf{u}) = 2G(\mathbf{e}(\mathbf{u}) - \mathbf{e}^{inh})$$

(5)

where $G(T)$ is the temperature-dependent shear modulus and $\mathbf{e}^{inh}$ is the inherent strain tensor, as:

$$\mathbf{e}^{inh} = e^T\mathbf{I} + \mathbf{e}^{vp}$$

(6)

which typically includes both the thermal deformation, $e^T(T)$, and the visco-plastic strains, $\mathbf{e}^{vp}$.

Commonly, the inherent strain method does not contemplate the solution of the thermal problem. Therefore, the inherent strain tensor is assumed as a (constant) user-defined material parameter depending on the material and process characterization. Its value can be obtained either experimentally or numerically, through the detailed simulation of the molten pool solidification.

In this work, a staggered thermo-mechanical solution is adopted to solve the AM problem [34,35]. Therefore, both the thermal and the mechanical analyses are carried out in a layer-by-layer manner. On the one hand, the thermal deformations can be computed from the actual (non-uniform) temperature field (inherent shrinkage method). On the other hand, the (inherent) visco-plastic deformations are defined as:

$$\mathbf{e}^{vp} = \mathrm{diag}(e_l, e_t, e_z) = \mathbf{I}(e_l, e_t, e_z)^{\mathrm{T}}$$

(7)

where e_l, e_t and e_z are the longitudinal (aligned with the scanning direction), transversal and vertical (building direction) visco-plastic components, respectively [33–35]. In LPBF, the thickness of the deposited layer is very small, so e_z can be neglected. Additionally, thin-walled parts can be assumed to be in a quasi-plane-stress condition, so that the only inherent strain component to be considered is the one along the mid-line of the cross-section, regardless of the scanning pattern.

3.2. Thermal Problem

H being the hatch spacing, the total travelled scanning length for each layer is:

$$l_{scan} = A/H$$

(8)

where A is the area of the current layer. The volume of the Heat Affected Zone (HAZ) is given by:

$$V_{HAZ} = l_{scan} \times d \times s = (d/H) \times (s/t) \times V$$

(9)

where d is the laser beam diameter and s is the penetration (depth) of the melt-pool, while $V = A \cdot t$ is the volume of the layer, being t its thickness. In this work, $d = H$ (no overlapping) and $s = t$ (the melt-pool depth is the layer thickness). Thus, the volume of the HAZ corresponds to the volume of the layer being activated.

Hence, the heat source density accounting for the total energy input during the scanning sequence of the whole layer is computed as:

$$q = \eta P/V$$

(10)

where η is the laser absorptivity and P is the laser power. This heat source is delivered during the whole scanning phase, which is characterized by a time interval, Δt_{scan} computed as:

$$\Delta t_{scan} = l_{scan}/V_{scan} = A/(H \times V_{scan})$$

(11)

where V_{scan} is the scanning velocity. Observe that the LPBF process also includes the recoating operation, that is a cooling phase of duration Δt_{cool}.

Often, the thermo-mechanical analysis is accelerated by lumping several layers into a single time-step. Therefore, the equivalent time-step of the analysis results in:

$$\Delta t_{eq} = n \cdot (\Delta t_{scan} + \Delta t_{cool}) \tag{12}$$

where n is the number of lumped layers (in this work n = 10) and the power input is uniformly spread in the lumped volume, V_{lump}, so that:

$$q_{eq} = \eta P / V_{lump} \tag{13}$$

3.3. Geometrical Models and FE Meshes

The coupled thermo-mechanical analysis is performed using the in-house FE software, *COMET*, developed at the International Centre for Numerical Methods in Engineering (CIMNE) [36,37]. The definition of the CAD geometries, and the corresponding the FE mesh generation as well as the pre/post-processing operations are carried out using the in-house pre/post-processor GiD [38]. The hardware used to run the simulation incorporates an Intel(R) Core (TM) i7-9700, 3.0 GHz processor and 16.0 GB of RAM.

Figure 9 shows the FE meshes of three different thin-walled geometries including both open and closed structures with a fixed thickness of 1 mm. The dimensions of these structures and the corresponding numbers of the hexahedral elements and nodes are listed in Table 3.

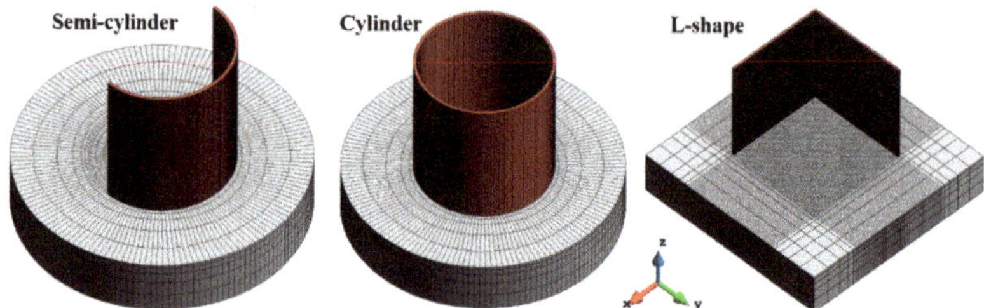

Figure 9. 3D FE mesh models of thin-walled structures.

Table 3. Dimensions of the proposed thin-walled structures and the numbers of FE elements and nodes.

Thin-Walled Parts	Dimensions	Number of FE Elements	Number of Nodes
Semi-cylindrical part	Φ50 mm × H50 mm	21,606	37,618
Cylindrical part	Φ50 mm × H50 mm	34,632	63,336
L-shaped part	L50 mm × W50 mm × H50 mm	34,366	55,392

The average mesh size in all cases is of about 1×1 mm^2 within the horizontal XY plane, and 0.3 mm in the building direction (a lumping layer-height). All these thin-walled work-pieces have the same height of 50 mm. The LPBF process is simulated with a sequence of 167 time-steps through a layer-by-layer activation strategy. Each layer used for the simulation has a thickness of 0.3 mm, thus including 10 physical layers (30 μm of loose powder is spread at each recoating step).

The temperature-dependent thermal and mechanical properties used to feed the constitutive model of Ti-6Al-4V alloy have been calibrated in a previous work [5], as shown in Table 4.

The thermal boundary conditions adopted in all the proposed analyses are defined through a heat convection coefficient of 12.7 W/(°C·m^2) and an emissivity of 0.35 [39]. The ambient temperature is set to 26°C during the whole LPBF process. The efficiency of the LPBF process, when Ti-6Al-4V loose powder is used, is defined in terms of the heat power absorption coefficient fixed to $\eta = 0.4$ [39].

Table 4. Temperature-dependent material properties of Ti-6Al-4V alloy [5].

Temperature (°C)	Density (kg/m³)	Specific Heat (J/(kg·°C))	Thermal Conductivity (W/(m·°C))	Poisson's Ratio	Young's Modulus (GPa)	Thermal Dilatancy (μm/m/°C)	Yield Stress (MPa)
20	4420	546	7	0.345	110	8.78	850
205	4395	584	8.75	0.35	100	10	630
500	4350	651	12.6	0.37	76	11.2	470
995	4282	753	22.7	0.43	15	12.3	13
1100	4267	641	19.3	0.43	5	12.4	5
1200	4252	660	21	0.43	4	12.42	1
1600	4198	732	25.8	0.43	1	12.5	0.5
1650	3886	831	35	0.43	0.1	12.5	0.1
2000	3818	831	35	0.43	0.01	12.5	0.01

All the thin-walled structures are printed at the same time (see Figure 1). Nevertheless, the simulation strategy assumes that each component is printed separately. This is because the heat conductivity of the loose powder is extremely low, so that most of the heat is evacuated by radiation through the upper surface of the layer being printed as well as by heat conduction through the baseplate. Therefore, the in-plane (XY) heat flux through the loose powder is negligible compared to the heat flow in the vertical (building) direction.

4. Calibration and Results

4.1. Calibration of the Numerical Model

In order to calibrate the numerical model used for LPBF, the warpage profiles of the thin-wall semi-cylindrical, cylindrical and L-shaped structures are computationally predicted and compared with the experimental 3D-scan results.

Figure 10a,b shows the out-of-plane displacements of the semi-cylindrical and cylindrical parts, respectively. The displacements at several control points are specifically used as a reference to assess the accuracy of the simulation. Similarly, Figure 11 compares the experimental and computed displacements of the L-shaped part. Once calibrated, the numerical model is able to faithfully reproduce the experimental results obtained by 3D scanning.

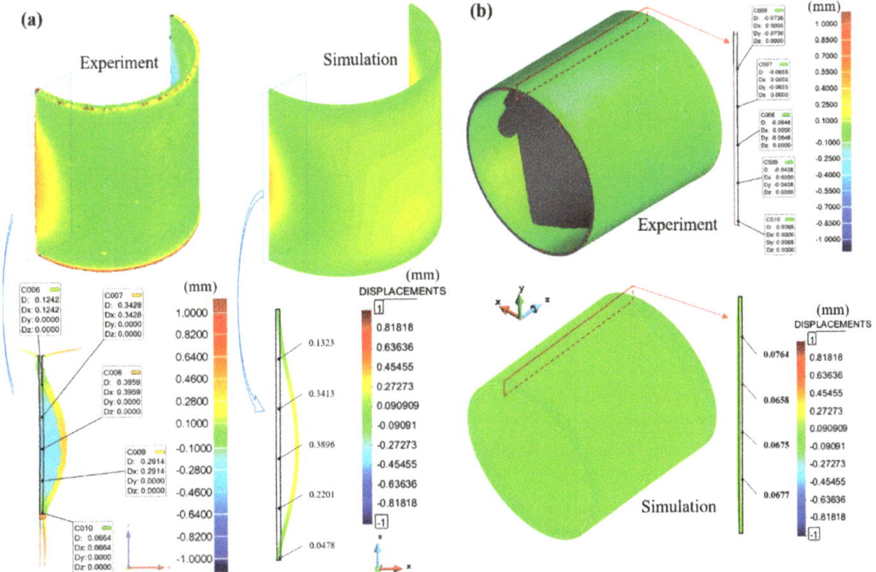

Figure 10. Comparison of the out-of-plane displacement fields obtained through 3D scanning and numerical simulation: (**a**) semi-cylinder; (**b**) cylinder.

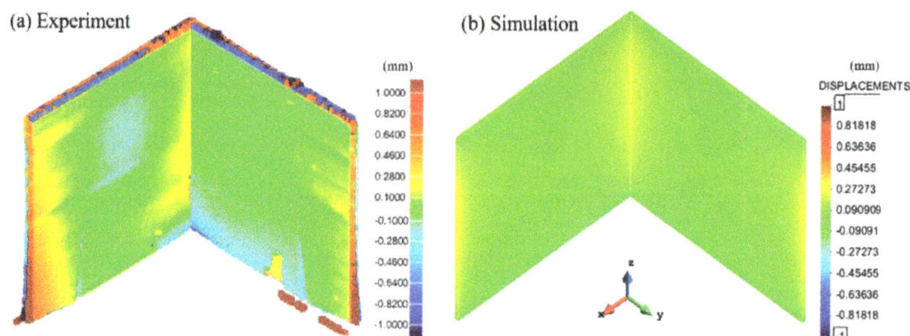

Figure 11. Comparison of the out-of-plane displacement fields obtained through (**a**) 3D scanning and (**b**) numerical simulation for the L-shaped structure.

It can be seen in Figure 10a that a visible bulge (of approximately 0.4 mm) occurs at opposite sides of the semi-cylindrical part, while the cylindrical geometry just shows very small radial displacements (less than 0.08 mm), these being more pronounced in the mid-section. The open L-shaped thin-wall structure also shows large deformations at the edges of each wall, as shown in Figure 11 and successfully predicted by the numerical model.

The slight discrepancies between the experimental and computed results are attributed to the mutual thermal influence among all the components as they are printed together (see Figure 1); this influence is disregarded in the FE simulation.

4.2. Results and Discussion on the Warpage Mechanism of the Thin-Walled Structures

Understanding the mechanical response of the 3D-printed geometries is necessary to establish a procedure to alleviate the warpage phenomenon in LPBF manufacture. As an example, the above results clearly illustrate that open thin-wall structures are more prone to deformation than closed ones.

Focusing on the open thin-walled semi-cylindrical geometry (Figure 10a), it is observed that the formation of the bulging progresses as the LPBF printing process moves away from the substrate as shown in Figure 12a. The bulging gets more pronounced when printing the upper half of the component because the mechanical constraining from the substrate restrains the development of the distortion in the bottom half. Figure 12b,c shows the evolution of the longitudinal and vertical stresses, σ_{xx} and σ_{zz}, respectively, at two stages of the manufacturing. It can be noted in Figure 12b that the internal surface sustains compressive stresses while tensile stresses are induced in the external surface. As a consequence, the wall deforms outwards. The vertical tensile stresses induced by the high cooling rate and the small bending stiffness of these structures also promote the bulging as shown in Figure 12c.

Contrarily, the closed thin-walled cylindrical part presents a much higher structural stiffness, and thus the AM process only produces a small deformation in the radial direction.

Dunbar et al. [40] studied the warpage of two thin-walled cylinders built by SLM. They also found that the warpage is relatively small. Throughout the printing process, the maximum warpage occurs several layers below the most recently deposited layer. This confers a final concave shape to the component.

Next, the warpage mechanism of the thin-wall L-shape structure is analyzed. Figure 13 shows the evolution of the longitudinal stresses (σ_{xx}) and the vertical stresses (σ_{zz}). During the building process, large tensile stresses develop at the top and two side edges while compressive stresses are produced in the central area of the built. As a result, warpage is visible along the lateral edges (see Figure 11). Note that a similar stress distribution and bulging phenomenon also take place when fabricating the single-wall and the square thin-walled section (see Figures 4 and 5).

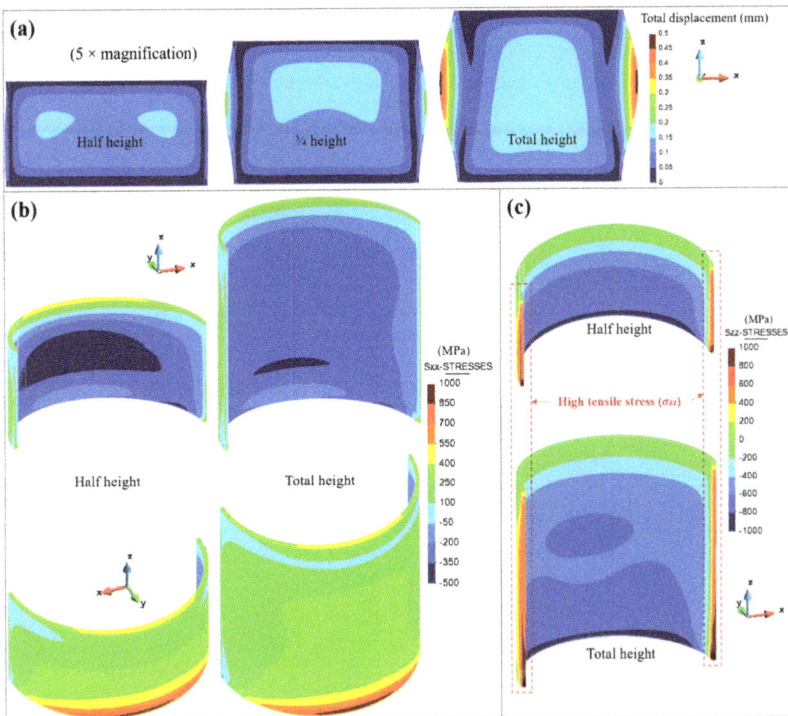

Figure 12. Semi-cylindrical part: the evolutions of (**a**) total displacements and (**b**) longitudinal stresses (σ_{xx}) as well as (**c**) vertical stresses (σ_{zz}).

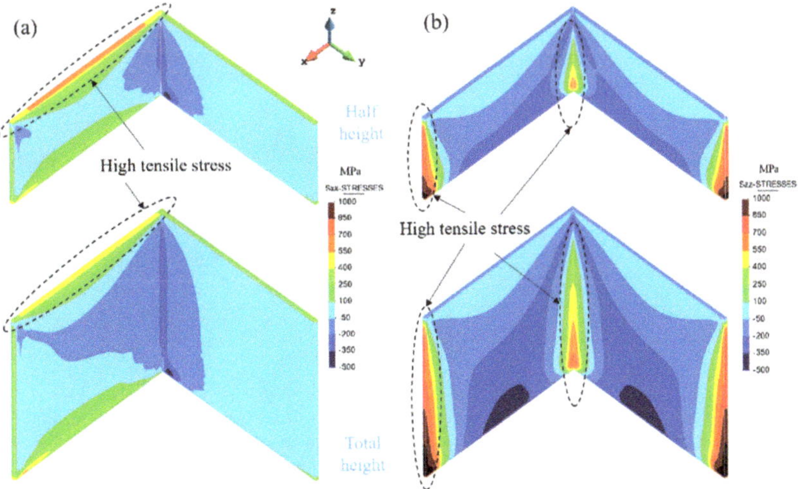

Figure 13. L-shape component: evolutions of (**a**) longitudinal stresses and (**b**) vertical stresses.

Yang et al. [41] also carried out a distortion analysis of thin-walled L-shaped and cross structures; their results showed that the side with longitudinal scan suffers larger distortion, whereas the side with transverse scan is distortion-free. Obviously, the mechanical response of thin-walled structures is closely related with the scan strategy.

5. Structural Optimization for Warpage Control

As above mentioned, the main difficulty for fabricating thin-walled AM components is their low structural stiffness and the high thermal stresses, especially at the sharp corners. In order to minimize the part warpage, a structural optimization strategy is proposed to enhance the overall stiffness of the component and particularly when open sections must be fabricated. The idea consists in adding vertical stiffeners able to resist the thermal stresses by locally increasing the structural stiffness and, at the same time, reducing the cooling rate at the edges of AM components as shown in Figure 14.

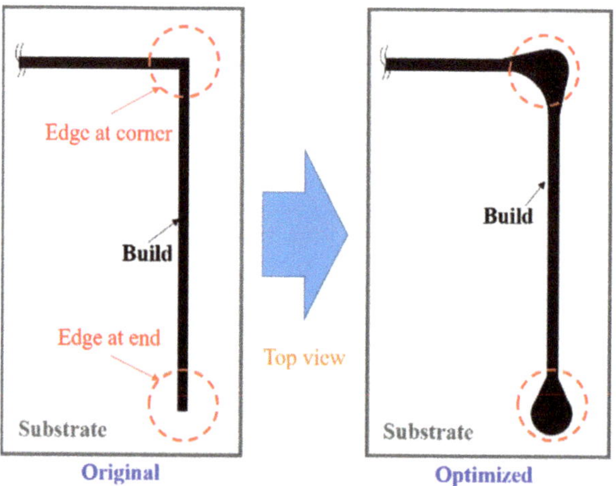

Figure 14. Proposed modification of the local stiffness of thin-walled structures.

Figure 15 shows the final warpage of both the open semi-cylindrical and L-shaped thin-walled parts with and without the vertical stiffeners. The warpage minimization is noticeably achieved.

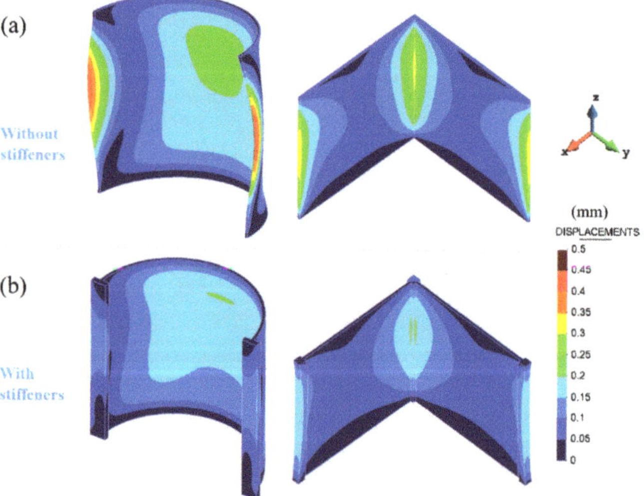

Figure 15. Warpage of the open (**a**) semi-cylindrical and (**b**) L-shaped thin-walled parts with and without vertical stiffeners.

Figure 16 shows the residual von Mises stresses of the open semi-cylindrical and L-shaped thin-walled structures. These residual stresses are mitigated by lowering the cooling rates at the edges of the two structures, thus increasing the local heat accumulation. Note that there still exist large residual stresses in the bottom area of both components. However, substrate preheating just before LPBF can successfully mitigate the residual stresses in this area to achieve high-quality fabrication of thin-walled components [5,7].

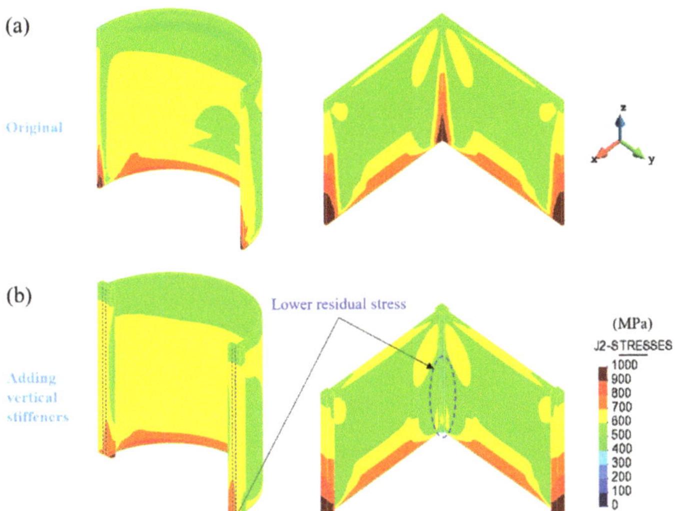

Figure 16. Residual von Mises stresses of the open (**a**) semi-cylindrical and (**b**) L-shaped parts with and without vertical stiffeners.

6. Conclusions

In this work, the warpage of several thin-walled structures fabricated by LPBF is investigated. The major conclusions of this study are:

1. The wall thickness plays a significant role on the warpage of the final part. Thicker walled structures present reduced warpage.
2. Increasing the build height as well as reducing the wall curvature causes larger warpage, as shown for the cylindrical structures.
3. Open sections (e.g., semi-cylinder, L-shape, etc.) are more prone to warpage than closed ones (e.g., cylinder and square section) because of their reduced structural stiffness.
4. The use of vertical stiffeners enables locally enhancing the structural stiffness of thin-walled structures, minimizing the residual warpage induced by the LPBF process.
5. FE analysis of LPBF processes is a useful tool to analyse different thin-walled structures in order to predict the actual warpage. The developed FE model has been calibrated with experimental 3D-scanning images.

Author Contributions: Conceptualization, X.L. (Xufei Lu) and M.C. (Michele Chiumenti); methodology and software, M.C. (Miguel Cervera) and H.T.; data curation, S.W. and X.L. (Xufei Lu); writing—original draft preparation, X.L. (Xufei Lu) and M.C. (Michele Chiumenti); writing—review and editing, M.C. (Miguel Cervera), and X.L. (Xin Lin); supervision, H.T. and X.L. (Xin Lin). All authors have read and agreed to the published version of the manuscript.

Funding: This work was funded by the National Key Technologies R & D Program (No. 2016YFB1100104), the European KYKLOS 4.0 project (An Advanced Circular and Agile Manufacturing Ecosystem based on rapid reconfigurable manufacturing process and individualized consumer preferences—Grant

Agreement No. 872570), the Severo Ochoa Programme for Centres of Excellence in R&D (CEX2018-000797-S) and the China Scholarship Council (No. 201906290011).

Data Availability Statement: Data are contained within the article.

Conflicts of Interest: The authors declare no conflict of interest.

References

1. Herzog, D.; Seyda, V.; Wycisk, E.; Emmelmann, C. Additive manufacturing of metals. *Acta Mater.* **2016**, *117*, 371–392. [CrossRef]
2. DebRoy, T.; Wei, H.L.; Zuback, J.S.; Mukherjee, T.; Elmer, J.W.; Milewski, J.O.; Beese, A.M.; Wilson-Heid, A.; De, A.; Zhang, W. Additive manufacturing of metallic components—Process, structure and properties. *Prog. Mater. Sci.* **2018**, *92*, 112–224. [CrossRef]
3. Lu, X.; Zhao, T.; Ji, X.; Hu, J.; Li, T.; Lin, X.; Huang, W. 3D printing well organized porous iron-nickel/ polyaniline nanocages multiscale supercapacitor. *J. Alloys Compd.* **2018**, *760*, 78–83. [CrossRef]
4. Zheng, M.; Wei, L.; Chen, J.; Zhang, Q.; Zhong, C.; Lin, X.; Huang, W. A novel method for the molten pool and porosity formation modelling in selective laser melting. *Int. J. Heat Mass Transf.* **2019**, *140*, 1091–1105. [CrossRef]
5. Lu, X.; Lin, X.; Chiumenti, M.; Cervera, M.; Li, J.; Ma, L.; Wei, L.; Hu, Y.; Huang, W. Finite element analysis and experimental validation of the thermomechanical behavior in Laser Laser Directed energy deposition of Ti-6Al-4V. *Addit. Manuf.* **2018**, *21*, 30–40.
6. Cao, J.; Gharghouri, M.A.; Nash, P. Finite-element analysis and experimental validation of thermal residual stress and distortion in electron beam additive manufactured Ti-6Al-4V build plates. *J. Mater. Process. Technol.* **2016**, *237*, 409–419. [CrossRef]
7. Lu, X.; Lin, X.; Chiumenti, M.; Cervera, M.; Hu, Y.; Ji, X.; Ma, L.; Yang, H.; Huang, W. Residual stress and distortion of rectangular and S-shaped Ti-6Al-4V parts by Directed Energy Deposition: Modelling and experimental calibration. *Addit. Manuf.* **2019**, *26*, 166–179. [CrossRef]
8. Lu, X.; Lin, X.; Chiumenti, M.; Cervera, M.; Hu, Y.; Ji, X.; Ma, L.; Huang, W. In situ Measurements and Thermo-mechanical Simulation of Ti-6Al-4V Laser Laser Directed energy deposition Processes. *Int. J. Mech. Sci.* **2019**, *153*, 119–130. [CrossRef]
9. Liu, Y.J.; Liu, Z.; Jiang, Y.; Wang, G.W.; Yang, Y.; Zhang, L.C. Gradient in microstructure and mechanical property of selective laser melted AlSi10Mg. *J. Alloys Compd.* **2018**, *735*, 1414–1421. [CrossRef]
10. Fang, Z.C.; Wu, Z.L.; Huang, C.G.; Wu, C.W. Review on residual stress in selective laser melting additive manufacturing of alloy parts. *Opt. Laser Technol.* **2020**, *129*, 106283. [CrossRef]
11. Lu, Y.; Wu, S.; Gan, Y.; Huang, T.; Yang, C.; Li, J.; Lin, J. Study on the microstructure, mechanical property and residual stress of SLM inconel-718 alloy manufactured by differing island scanning strategy. *Opt. Laser Technol.* **2015**, *75*, 197–206. [CrossRef]
12. Zaeh, M.F.; Branner, G. Investigations on residual stresses and deformations in selective laser melting. *Prod. Eng.* **2009**, *4*, 35–45. [CrossRef]
13. Vora, P.; Mumtaz, K.; Todd, I.; Hopkinson, N. AlSi12 in-situ alloy formation and residual stress reduction using anchorless selective laser melting. *Addit. Manuf.* **2015**, *7*, 12–19. [CrossRef]
14. Lu, X.; Cervera, M.; Chiumenti, M.; Li, J.; Ji, X.; Zhang, G.; Lin, X. Modeling of the Effect of the Building Strategy on the Thermomechancial Response of Ti-6Al-4V Rectangular Parts Manufactured by Laser Directed Energy Deposition. *Metals* **2020**, *10*, 1643. [CrossRef]
15. Yakout, M.; Elbestawi, M.; Veldhuis, S.; Nangle-Smith, S. Influence of thermal properties on residual stresses in SLM of aerospace alloys. *Rapid Prototyp. J.* **2020**, *26*, 213–222. [CrossRef]
16. Chiumenti, M.; Neiva, E.; Salsi, E.; Cervera, M.; Badia, S.; Moya, J.; Chen, Z.; Lee, C.; Davies, C. Numerical modelling and experimental validation in Selective Laser Melting. *Addit. Manuf.* **2017**, *18*, 171–185. [CrossRef]
17. Neiva, E.; Chiumenti, M.; Cervera, M.; Salsi, E.; Piscopo, G.; Badia, S.; Martín, A.F.; Chen, Z.; Lee, C.; Davies, C. Numerical modelling of heat transfer and experimental validation in powder-bed fusion with the virtual domain approximation. *Finite Elem. Anal. Des.* **2020**, *168*, 103343. [CrossRef]
18. Lindgren, L.-E.; Lundbäck, A.; Malmelöv, A. Thermal stresses and computational welding mechanics. *J. Therm. Stresses* **2019**, *42*, 107–121. [CrossRef]
19. Lundbäck, A.; Lindgren, L.-E. Modelling of metal deposition. *Finite Elem. Anal. Des.* **2011**, *47*, 1169–1177. [CrossRef]
20. Williams, R.; Catrin, J.; Davies, M.; Hooper, P.A. A pragmatic part scale model for residual stress and distortion prediction in powder bed fusion. *Addit. Manuf.* **2018**, *22*, 416–425. [CrossRef]
21. Baiges, J.; Chiumenti, M.; Moreira, C.A.; Cervera, M.; Codina, R. An Adaptive Finite Element strategy for the numerical simulation of Additive Manufacturing processes. *Addit. Manuf.* **2021**, *37*, 101650.
22. Lu, X.; Chiumenti, M.; Cervera, M.; Li, J.; Lin, X.; Ma, L.; Zhang, G.; Liang, E. Substrate design to minimize residual stresses in Directed Energy Deposition AM processes. *Mater. Des.* **2021**, *202*, 109525. [CrossRef]
23. Ramos, D.; Belblidia, F.; Sienz, J. New scanning strategy to reduce warpage in additive manufacturing. *Addit. Manuf.* **2019**, *28*, 554–564. [CrossRef]
24. Levkulich, N.; Semiatin, S.; Gockel, J.; Middendorf, J.; Dewald, A.; Klingbeil, N. The effect of process parameters on residual stress evolution and distortion in the laser powder bed fusion of Ti-6Al-4V. *Addit. Manuf.* **2019**, *28*, 475–484. [CrossRef]
25. Li, H.; Ramezani, M.; Chen, Z.; Singamneni, S. Effects of Process Parameters on Temperature and Stress Distributions During Selective Laser Melting of Ti-6Al-4V. *Trans. Indian Inst. Met.* **2019**, *72*, 3201–3214. [CrossRef]

26. Afazov, S.; Denmark, W.A.; Toralles, B.L.; Holloway, A.; Yaghi, A. Distortion prediction and compensation in selective laser melting. *Addit. Manuf.* **2017**, *17*, 15–22. [CrossRef]
27. Yaghi, A.; Ayvar-Soberanis, S.; Moturu, S.; Bilkhu, R.; Afazov, S. Design against distortion for additive manufacturing. *Addit. Manuf.* **2019**, *27*, 224–235. [CrossRef]
28. Biegler, M.; Elsner, B.A.; Graf, B.; Rethmeier, M. Geometric distortion-compensation via transient numerical simulation for directed energy deposition additive manufacturing. *Sci. Technol. Weld. Join.* **2020**, *25*, 468–475. [CrossRef]
29. Li, Z.; Xu, R.; Zhang, Z.; Kucukkoc, I. The influence of scan length on fabricating thin-walled components in selective laser melting. *Int. J. Mach. Tools Manuf.* **2018**, *126*, 1–12. [CrossRef]
30. Chen, C.; Xiao, Z.; Zhu, H.; Zeng, X. Deformation and control method of thin-walled part during laser powder bed fusion of Ti–6Al–4V alloy. *Int. J. Adv. Manuf. Technol.* **2020**, *110*, 3467–3478. [CrossRef]
31. Ahmed, A.; Majeed, A.; Atta, Z.; Jia, G. Dimensional Quality and Distortion Analysis of Thin-Walled Alloy Parts of AlSi10Mg Manufactured by Selective Laser Melting. *J. Manuf. Mater. Process.* **2019**, *3*, 51. [CrossRef]
32. Setien, I.; Chiumenti, M.; Van Der Veen, S.; Sebastian, M.S.; Garciandía, F.; Echeverría, A. Empirical methodology to determine inherent strains in additive manufacturing. *Comput. Math. Appl.* **2019**, *78*, 2282–2295. [CrossRef]
33. Chiumenti, M.; Lin, X.; Cervera, M.; Lei, W.; Zheng, Y.; Huang, W. Numerical simulation and experimental calibration of additive manufacturing by blown powder technology. Part I: Thermal analysis. *Rapid Prototyp. J.* **2017**, *23*, 448–463. [CrossRef]
34. Chiumenti, M.; Cervera, M.; Salmi, A.; De Saracibar, C.A.; Dialami, N.; Matsui, K. Finite element modeling of multi-pass welding and shaped metal deposition processes. *Comput. Methods Appl. Mech. Eng.* **2010**, *199*, 2343–2359. [CrossRef]
35. Chiumenti, M.; Cervera, M.; Dialami, N.; Wu, B.; Jinwei, L.; De Saracibar, C.A. Numerical modeling of the electron beam welding and its experimental validation. *Finite Elem. Anal. Des.* **2016**, *121*, 118–133. [CrossRef]
36. Cervera, M.; de Saracibar, C.A.; Chiumenti, M. *COMET: Coupled Mechanical and Thermal Analysis*; Data Input Manual, Version 5.0, Technical Report IT-308; CIMNE: Barcelona, Spain, 2002.
37. Dialami, N.; Cervera, M.; Chiumenti, M.; De Saracibar, C.A. A fast and accurate two-stage strategy to evaluate the effect of the pin tool profile on metal flow, torque and forces in friction stir welding. *Int. J. Mech. Sci.* **2017**, *122*, 215–227. [CrossRef]
38. Ribó, R.; Pasenau, M.; Escolano, E.; Pérez, J.; Coll, A.; Melendo, A. *GiD The Personal Pre and Postprocessor*; Reference Manual; CIMNE: Barcelona, Spain, 2006.
39. Zhang, W.; Tong, M.; Harrison, N.M. Resolution, energy and time dependency on layer scaling in finite element modelling of laser beam powder bed fusion additive manufacturing. *Addit. Manuf.* **2019**, *28*, 610–620. [CrossRef]
40. Dunbar, A.J.; Denlinger, E.R.; Gouge, M.F.; Michaleris, P. Experimental validation of finite element modeling for laser powder bed fusion deformation. *Addit. Manuf.* **2016**, *12*, 108–120. [CrossRef]
41. Yang, T.; Xie, D.; Yue, W.; Wang, S.; Rong, P.; Shen, L.; Zhao, J.; Wang, C. Distortion of Thin-Walled Structure Fabricated by Selective Laser Melting Based on Assumption of Constraining Force-Induced Distortion. *Metals* **2019**, *9*, 1281. [CrossRef]

Article

Deep Learning Sequence Methods in Multiphysics Modeling of Steel Solidification

Seid Koric [1,2,*] **and Diab W. Abueidda** [1,2]

[1] National Center for Supercomputing Applications, University of Illinois at Urbana Champaign, Urbana, IL 61801, USA; abueidd2@illinois.edu

[2] Department of Mechanical Science and Engineering, University of Illinois at Urbana Champaign, Urbana, IL 61801, USA

* Correspondence: koric@illinois.edu

Abstract: The solidifying steel follows highly nonlinear thermo-mechanical behavior depending on the loading history, temperature, and metallurgical phase fraction calculations (liquid, ferrite, and austenite). Numerical modeling with a computationally challenging multiphysics approach is used on high-performance computing to generate sufficient training and testing data for subsequent deep learning. We have demonstrated how the innovative sequence deep learning methods can learn from multiphysics modeling data of a solidifying slice traveling in a continuous caster and correctly and instantly capture the complex history and temperature-dependent phenomenon in test data samples never seen by the deep learning networks.

Keywords: sequence deep learning; neural networks; casting; steel; solidification; multiphysics

1. Introduction

While developments of advanced steel solidification processes, such as metal-based additive manufacturing, are making steady progress, the field is still dominated by more traditional techniques, such as ingot, foundry, and continuous casting. Continuous casting produces over 95% of the world's steel. Experiments are limited due to the molten steel's harsh environment and the many process variables, which affect their complex multiphysics phenomena. Advancement of these established manufacturing processes mostly relies on the improved quantitative understanding gained from sophisticated multiphysics numerical models. Remarkable advances in computing technology and numerical algorithms in the last 30 years have enabled more realistic and accurate multiphysics modeling of steel solidification processes on high-performance computing systems.

Lee et al. [1], followed by Teskeredžić et al. [2], demonstrated multiphysics modeling of steel solidification by coupling a fluid flow of molten steel thermal-stress models of solidifying shell to predict solidification and stress formation on simplified casting geometries. Koric and Thomas [3] incorporated a bounded Newton-Raphson method from an in-house code [4] into Abaqus [5] commercial finite element code, which provided an order of magnitude performance increase in solving solidification thermo-mechanical problems, including the explicit finite element formulation [6]. Later Koric et al. [7] devised and validated a new enhanced latent heat method to convert spatial and temporal superheat flux data into the finite-element thermo-mechanical model of the solidifying shell and mold. The modeling approach was demonstrated via simulation of the multiphysics phenomena in continuous commercial casters of carbon steel grades [8], and lately, stainless steel grades [9]. Besides pure Lagrangian approaches, researchers have also lately used a combination of Eulerian approach in liquid, Lagrangian in solid, and Eulerian-Lagrangian in mushy (semi-solid) zones in treating multiphysics phenomena in steel solidification [10] as well as meshless numerical methods [11]. Application of steel solidification multiphysics models in crack formation mechanisms was also the subject of recent research works [12,13]. However,

Citation: Koric, S.; Abueidda, D.W. Deep Learning Sequence Methods in Multiphysics Modeling of Steel Solidification. *Metals* **2021**, *11*, 494. https://doi.org/10.3390/met11030494

Academic Editor: Roberto Montanari

Received: 15 February 2021
Accepted: 12 March 2021
Published: 17 March 2021

Publisher's Note: MDPI stays neutral with regard to jurisdictional claims in published maps and institutional affiliations.

finding efficient and accurate multiphysics computational approaches, which can be widely applied on commodity computers, remains a challenge.

Lately, machine learning techniques, particularly deep learning, which is inspired by the biological structure and functioning of a brain, have accomplished significant successes in wide areas of science and engineering, such as natural language processing, computer vision, voice recognition, autonomous vehicle driving, medical diagnosis, and financial service. Numerical modeling in mechanics and material science is not an exception. Various surrogate deep learning data-driven models have been trained to learn and quickly inference the thermal conductivity and advanced manufacturing of composites [14,15], topologically optimized materials and structures [16,17], the fatigue of materials [18], nonlinear material response such as in plasticity and viscoplasticity [19,20], and many other applications.

2. Thermo-Mechanical Model of Steel Solidification

We use the modeling results from the existing multiphysics (thermo-mechanical) model [3] with a solidifying slice domain traveling in the Lagrangian frame of reference down the continuous caster to generate training and testing data for the sequential deep learning methods. The governing equation for thermal behavior of the solidifying shell is given in Equation (1)

$$\rho\left(\frac{\partial H}{\partial t}\right) = \nabla \cdot (k \nabla T), \tag{1}$$

where k is thermal conductivity, ρ is density, and H is specific enthalpy including the latent heat during phase transformations, such as in solidification and transition from δ-ferrite to austenite. Inertial effects are negligible in solidification problems, so using the quasi-static mechanical equilibrium in Equation (2) as the governing equation is appropriate:

$$\nabla \cdot \sigma(x) + b = 0, \tag{2}$$

where σ is the Cauchy stress tensor and b is the body force density vector. The rate representation of total strain in this thermo-elastic-viscoplastic model is given by Equation (3):

$$\dot{\varepsilon} = \dot{\varepsilon}_{el} + \dot{\varepsilon}_{ie} + \dot{\varepsilon}_{th}, \tag{3}$$

where $\dot{\varepsilon}_{el}, \dot{\varepsilon}_{ie}, \dot{\varepsilon}_{th}$ are the elastic, inelastic, and thermal strain rate tensors, respectively.

At temperatures close to and above the solidification/melting point, steel alloys show significant time- and temperature-dependent plastic behavior, including phase transformations. Kozlowski et al. [21] created a visco-plastic constitutive equation to model the austenite phase of steel, relating inelastic strain to stress, strain rate, temperature, and steel grade via carbon content, Equation (4)

$$\dot{\bar{\varepsilon}}_{ie}[\sec^{-1}] = f_C\left(\bar{\sigma}[\text{MPa}] - f_1\bar{\varepsilon}_{ie}|\bar{\varepsilon}_{ie}|^{f2-1}\right)^{f3} \exp\left(-\frac{Q}{T[\text{K}]}\right)$$

$$\text{where :}$$
$$Q = 44,465$$
$$f_1 = 130.5 - 5.128 \times 10^{-3}T[\text{K}]$$
$$f_2 = -0.6289 + 1.114 \times 10^{-3}T[\text{K}]$$
$$f_3 = 8.132 - 1.54 \times 10^{-3}T[\text{K}]$$
$$f_C = 46,550 + 71,400(\%C) + 12,000(\%C)^2$$

$$\tag{4}$$

where Q is an activation energy constant, $\bar{\sigma}$ (MPa) is Von Mises effective stress, f_1, f_2, f_3, f_c are empirical functions that depend on absolute temperature (K), and $\%C$ is carbon content (weight percent) representing steel grade (composition).

Another constitutive model, so called Zhu power law model [22] in Equation (5), was devised to simulate delta ferrite phase with relatively higher creep rate and weaker than austenite phase.

$$\dot{\bar{\varepsilon}}_{ie}(1/\text{ sec.}) = 0.1 \left| \frac{\bar{\sigma}(\text{MPa})}{f_{\delta c}(\%C)\left(\frac{T(^\circ K)}{300}\right)^{-5.52}(1+1000\dot{\bar{\varepsilon}}_{ie})^m} \right|^n$$

where :

$$f_{\delta c}(\%C) = 1.3678 \times 10^4 (\%C)^{-5.56\times10^{-2}}$$
$$m = -9.4156 \times 10^{-5} T(^\circ K) + 0.3495$$
$$n = \frac{1}{1.617\times10^{-4}T(^\circ K)-0.06166}$$

(5)

The delta-phase model given in Equation (5) is applied in the solid whenever delta-ferrite's volume fraction is more than 10%, to approximate the dominating influence of the very high-creep rates in the delta-ferrite phase of mixed-phase structures on the net mechanical behavior. While there are sophisticated constitutive models applied in the mushy and liquid zones in the previous works [23,24], in this work, the elastic-perfectly-plastic constitutive model with small yield stress is applied above the solidus temperature Tsol, to enforce negligible strength in those volatile zones. The highly nonlinear constitutive visco-plastic models in Equations (4) and (5) are efficiently integrated at the integration points in UMAT subroutine [3] and linked with Abaqus, which, in turn, solves the coupled thermo-mechanical governing Equations (1) and (2) by the implicit nonlinear finite element solution methodology.

The low carbon steel grade was chosen for this work with 0.09 wt%C, T_{sol} = 1482.4 °C, and T_{liq} = 1520.5 °C, whose solidification path is shown in the pseudo-binary iron-carbon phase diagram shown in Figure 1.

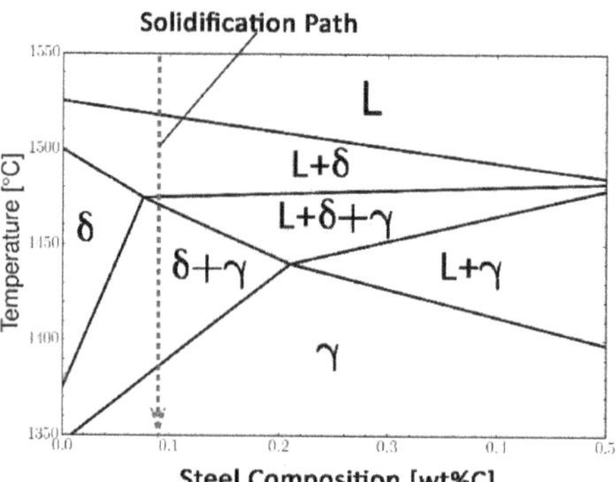

Figure 1. Pseudo-binary iron-carbon phase diagram with solidification path for 0.09 wt%C low carbon steel grade.

Highly temperature-dependent and nonlinear material properties for this steel grade are given in Figure 2a for elastic modulus and enthalpy and in Figure 2b for thermal conductivity and thermal expansion.

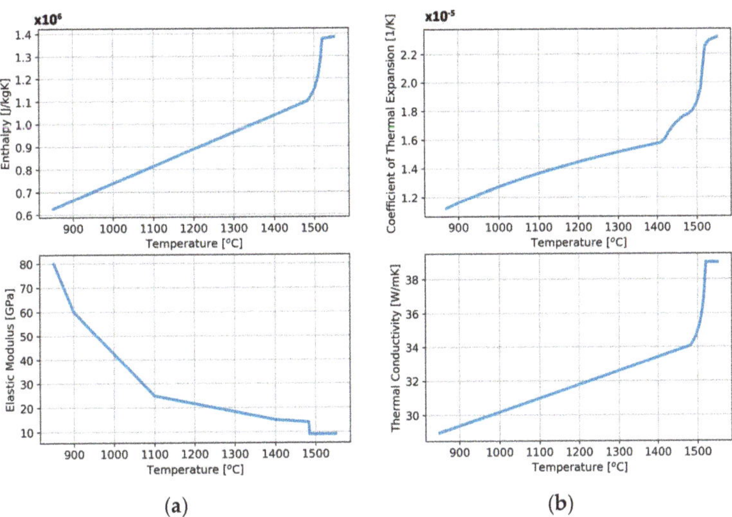

Figure 2. Temperature dependent material properties for 0.09 wt%C steel grade: (**a**) elastic modulus and enthalpy and (**b**) thermal conductivity and thermal expansion.

The phase fraction and temperature-dependent material properties in Figures 1 and 2 and other interfacial calculations are an integral part of UMAT and other user-defined subroutines, which are programmed and linked with Abaqus to provide a complete multiphysics model of steel solidification on the continuum level [3,8].

Generalized plane strain conditions can realistically recover a full 3D stress state in long objects under thermal loading [3,4], such as in continuous caster in Figure 3a with the considerable length and width. The slice domain in Figure 3b travels down the mold with casting velocity in a Lagrangian frame of reference. It consists of a single row of 300 coupled thermo-mechanical generalized plane strain elements with 602 nodes, which provide generalized plane strain condition in axial (z-direction). In addition, a second generalized plane strain condition was imposed in the y direction by coupling the vertical displacements of all nodes along the bottom edge of the domain. Time-dependent profiles of thermal fluxes leaving the chilled surface on the left side of the domain and their displacement due to mold taper and other interactions with the mold provide the solidifying slice domain's thermal and mechanical boundary conditions. These profiles' temporal variations are inputs for the sequence deep learning methods, while the calculated temperature and stress histories are their targets (labels).

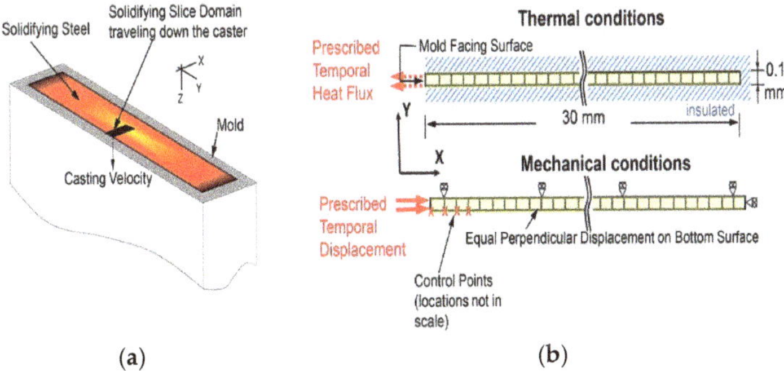

Figure 3. (**a**) Continuous caster with solidifying slice finite element domain and (**b**) thermal and mechanical boundary conditions.

3. Deep Learning Models

Deep learning is a subset of machine learning. Deep learning models consist of neural networks, a hierarchical organization of layers of neurons connected to other neurons' layers.

3.1. Dense Feedforward Neural Network

As shown in Figure 4, the most straightforward neural network, the so-called dense feedforward neural network, comprises linked layers of neurons that calculate predictions based on input data.

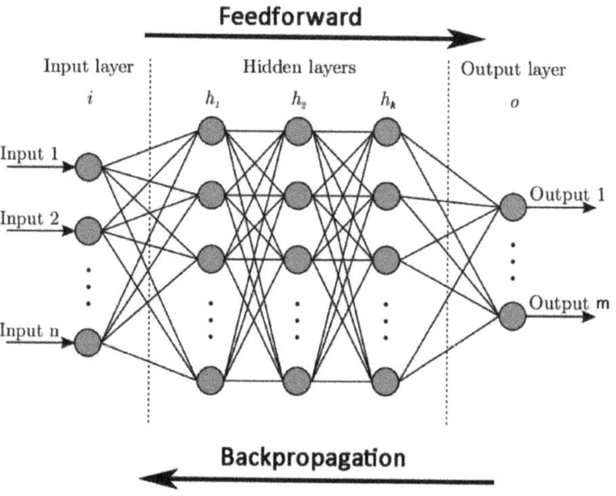

Figure 4. Dense feedforward deep neural network.

The layers with neurons feed information forward to other layers based on the received input, and this forms a complex network that learns with some feedback mechanism. The number of hidden layers, i.e., the layers in between input and output layers, defines a neural network's deepness. Neurons of successive layers are connected through associated weights and biases W and b. The output $\hat{Y}^{[l]}$ for a layer l is calculated as:

$$Z^{[l]} = W^{[l]}Z^{[l-1]} + b^{[l]}$$
$$\hat{Y}^{[l]} = f^{[l]}(Z^{[l]}) \qquad , \tag{6}$$

where $W^{[l]}(n_l \times n_{l-1})$ and $b^{[l]}$ ($n_{l-1} \times 1$) are matrix of weights and vector of biases, respectively which are updated after every training pass. $f^{[l]}$ is the activation function that transforms Z into output for every neuron in the layer l. Activation functions in neural networks are nonlinear functions such as Rectified Linear Unit (ReLu), Sigmoid, and Hyperbolic Tangent. They enable the neural network to learn almost any complex functional relationship between inputs and outputs. At the end of each feed-forward pass, the loss function L compares the predictions to the targets by calculating a loss value that measures how well the network's predictions match what was expected. Then, in a backpropagation process, the optimizer minimizes loss value iteratively with gradient descent in Equation (7) and other similar optimization techniques. The gradients of loss function L are calculated with respect to the weights in the last layer, and the weights are updated for each node before the same process is done for the previous layer and backward until all of the layers have had their weights adjusted. Then, a new k iteration with forward propagation starts again. After a reasonable amount of iterations, the series W^k should converge toward the minimum loss value location. The parameter γ is called the learning rate.

3.2. Recurrent Neural Networks

In a feedforward neural network, we assume that all inputs (and outputs) are independent of each other. However, we want to predict the next value in a sequence for many deep learning tasks, such as a word in a sentence, by knowing which words came before it. The most known sequence neural network is the family of recurrent neural networks (RNN). They are called recurrent because they perform the same task for every element of a sequence, with the output being dependent on the previous computations. Sequence learning predicts for a given input sequence $x_0, \ldots, x_T$ the matching outputs $\hat{y}_0, \ldots, \hat{y}_T$ at each time step while minimizing the loss function L between predictions $\hat{y}_0, \ldots, \hat{y}_T$ and actual targets $y_0, \ldots, y_T$. The diagram in Figure 5 shows RNN being unfolded in time into a full network for an entire sequence and is described by the propagation expressions in Equation (7).

$$s_t = f(Ux_t + Ws_{t-1} + b)$$
$$\hat{y}_t = Vs_t + c$$

(7)

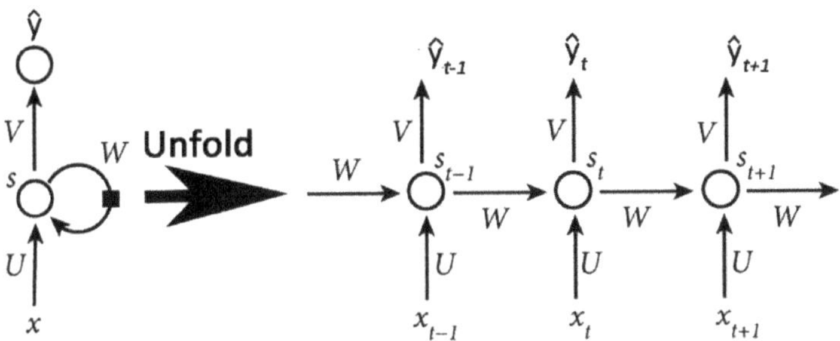

Figure 5. Recurrent neural network with its unfolding in time calculations.

s_t is the hidden state at time step t, also known as the "memory" of the recurrent network since it captures information about what happened in all the previous time steps. It is calculated from the previous hidden state and the input at the current step represented by W (hidden-to-hidden) and U (input-to-hidden) weight connections, respectively, f is the activation function, and b is a bias. Output prediction at time step t is calculated from s_t using V (hidden-to-output) weights and c bias. Similarly, to the dense feedforward network, all weights and biases are updated from their loss gradients by backpropagations. However, recurrent neural networks often consist of very deep computational graphs repeatedly (recurrently) applying the identical operation at each time step of a long-time sequence. This may cause the vanishing or exploding gradient problems during backpropagation and makes it challenging to train long sequence data with RNNs. To address the vanishing gradient problems of traditional RNN, the long short-term memory (LSTM) [25] and gated recurrent unit (GRU) [26] are devised. Hidden state (memory) cells in the LSTM and GRU advanced recurrent neural networks are designed to dynamically "forget" some old and unnecessary information via select gated units that control the flow of information inside a memory cell, thus avoiding multiplication of a long sequence of numbers during temporal backpropagation. GRU has a simpler gated unit architecture than LSTM and generally learns faster than LSTM. A recent work [19] has shown that GRU's formulation is less prone to overfitting in materially non-linear sequence learning and allows faster training due to the smaller number of trainable parameters. A comprehensive mathematical overview of LSTM and GRU networks can be found here [27].

3.3. Temporal Convolutional Neural Network

The temporal convolutional network (TCN) is a variant of Convolutional Neural Networks particularly adept for sequence learning tasks. The two-dimensional (standard) version of Convolutional Neural Networks (CNN) is generally considered the best deep learning model available today for computer vision tasks. Convolution is sliding a smaller matrix (known as a kernel or filter) over a larger matrix representing an image. At each run, we do an element-wise multiplication of the two matrix elements and add them. The kernel slides to another portion of the image matrix until the whole image is convoluted, and a feature map corresponding to a particular kernel is computed. The gradients of the loss function with respect to kernel weights are then calculated in a backpropagation process, and the weights are updated similar to dense feedforward artificial neural networks. There are many kernels in a typical computer vision CNN such as geometric shapes, edges, distribution of colors, etc. This makes these networks very robust in image classification and other similar data that contain 2D-spatial properties. TCN uses a one-dimensional version of CNN where the kernel slides along a single dimension, i.e., time. In these so-called casual convolutions, an output at time t is convoluted only with elements from time t and earlier in the previous layer, as shown in Figure 6. No information from the future is propagated into the past. To produce an output of the same length as the input layer, zero padding of length (kernel size $-$ 1), is added to keep subsequent layers the same size as previous ones. Multiple 1D convolutional layers are typically stacked on top of each other in order to learn from the previous time steps.

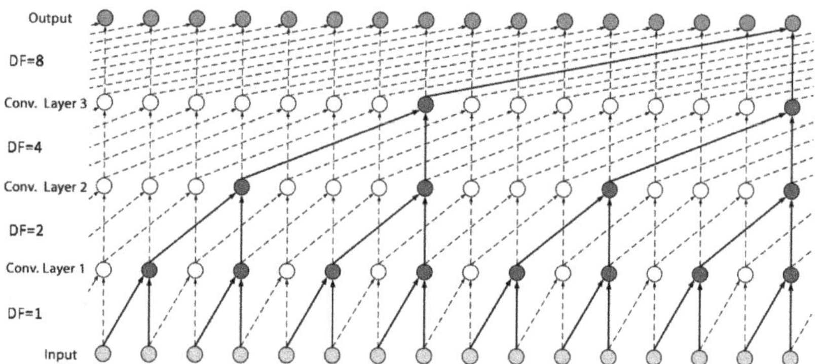

Figure 6. Temporal convolutional neural network.

The dilation factor (DF), or receptive field enlargement with gaps in entries of subsequent convolution layers, is increased exponentially to achieve large receptive field sizes. Since TCN does not perform recurring calculations, it is more robust with respect to vanishing and exploding gradients than any RNN network. It is more computationally and memory efficient. More detailed information about TCN can be found in [28].

4. Results and Discussion

Instantaneous heat flux profile on the chilled surface is essential for the multiphysics analysis that drives the solidifying shell's transient heat transfer. Similarly, the chilled solidifying shell surface's displacement profile due to the mechanical contact and interactions with the mold surface is an important boundary condition for mechanical analysis. Based on a variation of the time histories of heat flux and displacement boundary conditions, the multiphysics model of solidifying slice from Chapter 2 is run for 17 s of simulation time that the slice spends solidifying in the mold. Thus, we generate temperature and stress histories at the four spatial control points along the slice's bottom edge at the chilled surface and 1, 2, and 3 mm in the domain interior (see Figure 3), which are used as targets (labels) for the sequence deep learning methods. The points are chosen on those locations since casting quality depends on the solidification conditions on the chilled surface and its subsurface. It is known from the steel plant observations that the thermal flux has an overall decaying profile due to transient heat transfer, while the displacement profile has an increasing trend due to mold taper. After having several temporal points defined randomly within ranges of expected profile values, a radial basis interpolation with Gaussian function is used to connect (interpolate) the points and to emulate additional fluctuations and data noise observed in the actual flux and displacement profiles due to sudden changes in contact conditions and interfacial heat transfer between mold and steel surfaces. Many thousands of these thermal and displacement variations provided all possible scenarios that the solidifying shell encounters on its chilled surfaces while traveling down the caster, as well as a sufficient number of data samples generated for deep learning. One such test input data sample with displacement and heat flux profiles is given in Figure 7.

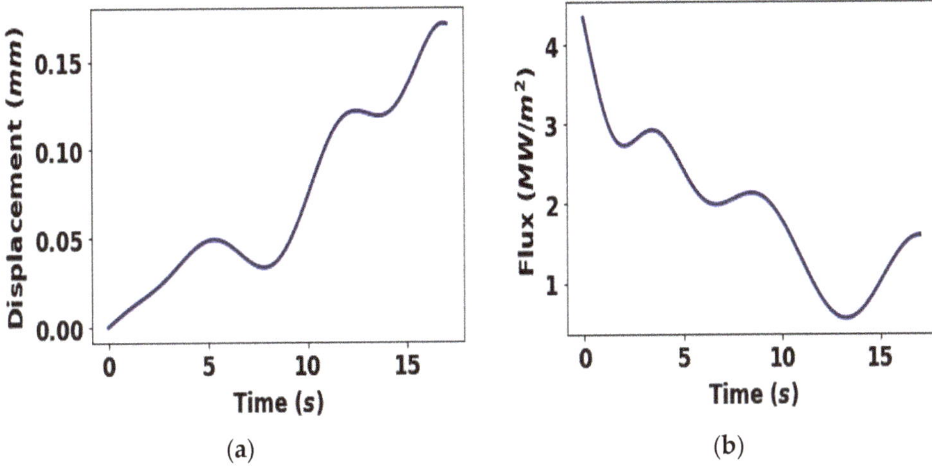

Figure 7. Example of a testing input data sample: (**a**) displacement profile and (**b**) flux profile.

Figure 8 shows the schematic diagram of a training process in sequence learning in this work. The modeling database provides inputs for sequence neural networks and temperature and stress sequences, i.e., labels (targets). The loss function compares predictions from the neural network to the labels and calculates a loss value. The optimizer minimizes the loss value and updates the weights of the network iteratively, as explained earlier.

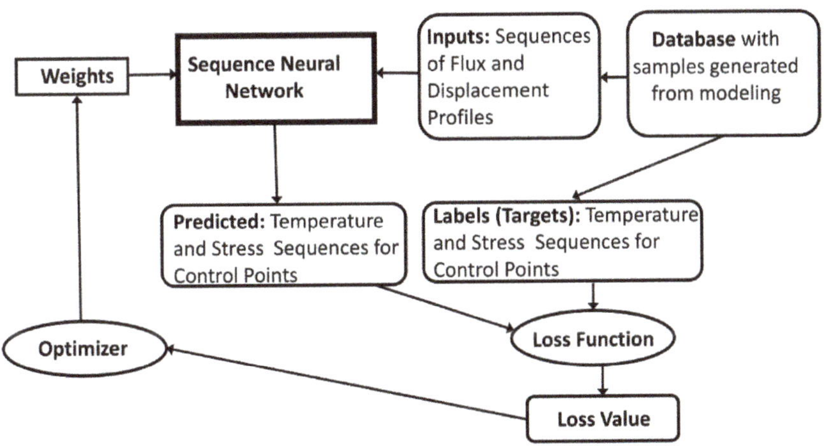

Figure 8. Training process in sequence deep learning.

The three deep learning models have a comparable number of approximately 3.5 million total trainable parameters, such as entries in W, U, V, b, and c weight and bias matrices in Figure 5 that connect a multitude of nodes and layers of the recurrent neural networks, and with each network consisting of 3 hidden layers. This provided an optimum between computational cost and minimization of errors. The input and target data for over 14,000 samples, each with 100 time steps, are generated using high-throughput computing capabilities of several nodes of the high-performance computing cluster called iForge at the National Center for Supercomputing Applications at the University of Illinois. Approximately, 80% of the generated data samples are randomly selected for training with the LSTM, GRU, and TCN sequence deep learning methods, discussed in Section 3, on an iForge's computing node equipped with GPUs using Keras [29] with TensorFlow backend. The remaining 20% of the data is set aside for testing and validation. Figure 9 depicts an example of each sequence learning model's convergence plots for 150 training epochs in terms of scaled mean absolute error (SMAE).

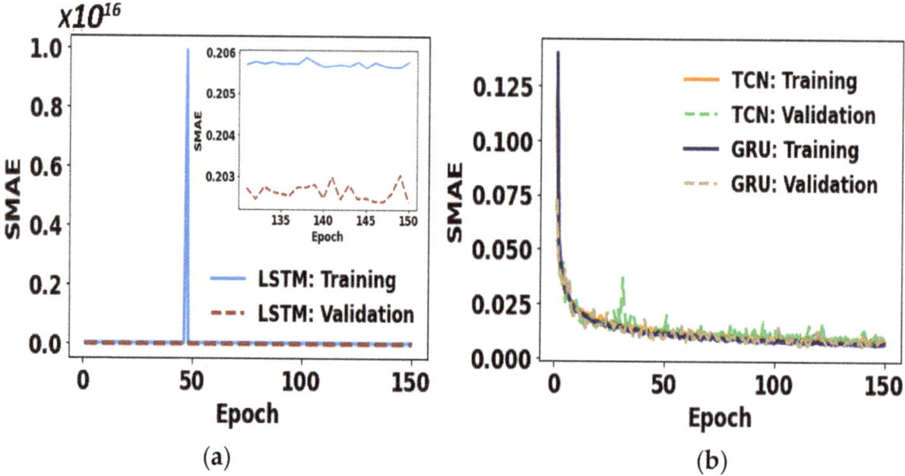

Figure 9. Convergence history: (**a**) LSTM and (**b**) GRU and TCN.

The training of the LSTM sequence learning model is generally extremely unstable, as can be seen in Figure 9. This is not the case with GRU and TCN. Both deep learning models provided accurate qualitative and quantitative results, with slight variations in the SMAE and steady converging. Figure 10 shows ground-truth stress and temperature results calculated for the four control points by the multiphysics model and the corresponding sequence learning predictions from the neural networks for the test input sample in Figure 7, which was never trained by any deep learning network.

The GRU and TCN deep learning models almost perfectly predicted the highly complex and coupled multiphysics thermo-mechanical phenomena of steel solidification with multiple visco-plastic constitutive laws, generating tensile and compressive residual stresses on the chilled surface and its subsurface. These stresses and strains are responsible for most cracks and other defects initiating in the continuous casting of steel [4]. The LSTM model, on the other hand, only roughly predicted the temperatures and stresses histories mainly due to its limited capabilities to cope with long sequences of data such as here with 100 time steps. We have inferred results from many other test samples, which again were never trained by the deep learning networks. Every time the GRU and TCN models were able to accurately predict the stress and temperature histories. This is summarized in Figure 11 in terms of average SMAE with its standard deviation for all the testing samples. The LSTM-based model has the highest error values among the models, while the GRU- and TCN-based models have similar error values. However, the TCN model was eight times faster in training than the GRU-based model due to its convolutional algorithm that avoids recurrent calculations and generally much better utilizes the GPU hardware.

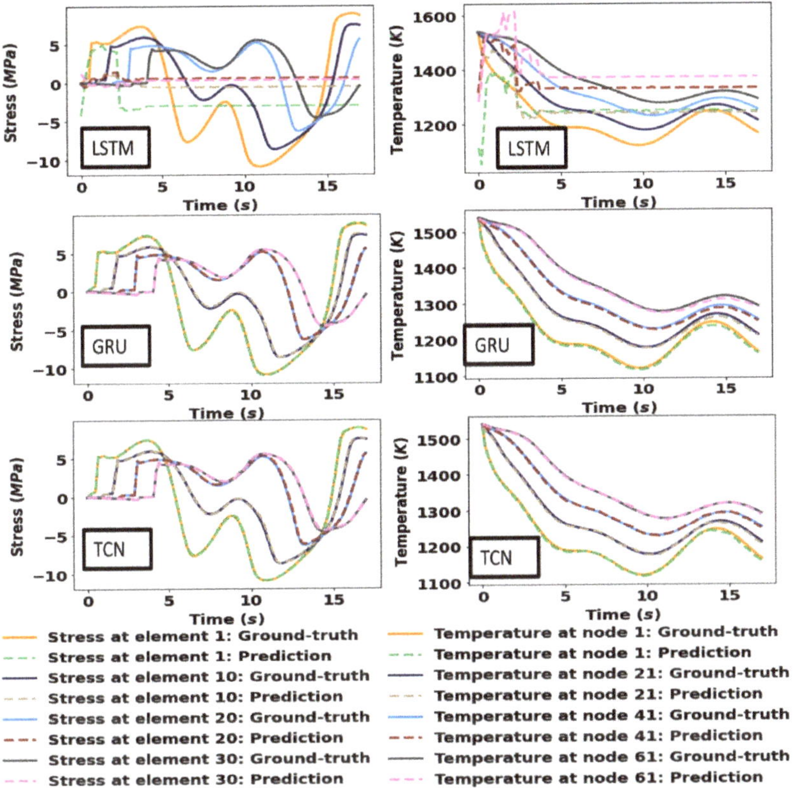

Figure 10. Output results, ground-truth, and deep learning predictions for the 4 control points, corresponding to the testing input sample in Figure 7.

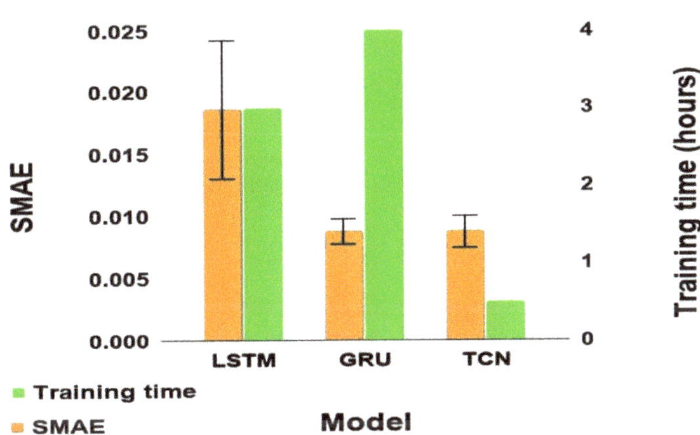

Figure 11. Training times and testing scaled mean absolute error (SMAE).

5. Conclusions

We have applied and compared the three types of deep sequence learning methods, short-term memory (LSTM), gated recurrent unit (GRU), and temporal convolutional network (TCN), to a multiphysics model of steel solidification. We have found that only the TCN- and GRU-based models could learn and gain enough artificial intelligence to accurately reproduce the extremely complex multiphysics behavior of a solidifying steel domain moving in a continuous caster. At the same time, the TCN-based model is also significantly more computationally efficient. Moreover, once appropriately trained on a high-end computing system with sufficient sequenced data generated from classical multiphysics finite element simulations, the model's trained weights and biases can be transferred to any low-end computer. Thus, the deep learning models developed in this work can almost instantly produce (inference) good quality results for unseen input data on laptops and without any modeling software. Finally, it is worth pointing that the deep learning methodology developed here is not confined to the slice domain or the constitutive models used in this work. Still, it is readily applicable to other solidification domains and constitutive models. These and other similar Artificial Intelligence (AI)-based surrogate data-driven methods will open the door for future highly accelerated multiphysics modeling, design, optimization, and process predictions in steel solidification and other complex manufacturing processes.

Author Contributions: Conceptualization, S.K. and D.W.A.; methodology, S.K. and D.W.A.; validation, S.K.; data curation, D.W.A.; writing—original draft preparation, S.K. and D.W.A.; writing—review and editing, S.K. and D.W.A.; visualization, S.K. and D.W.A. All authors have read and agreed to the published version of the manuscript.

Funding: This research received no external funding.

Institutional Review Board Statement: Not applicable.

Informed Consent Statement: Not applicable.

Data Availability Statement: The data that support the findings of this study are available from the authors upon reasonable request.

Acknowledgments: The authors would like to thank the National Center for Supercomputing Applications (NCSA) Industry Program and the Center for Artificial Intelligence Innovation for software and hardware support. The authors thank the Continuous Casting Center at the Colorado School of Mines for the support.

Conflicts of Interest: The authors declare no conflict of interest.

References

1. Lee, J.-E.; Yeo, T.-J.; Hwan, O.K.; Yoon, J.-K.; Yoon, U.-S. Prediction of cracks in continuously cast steel beam blank through fully coupled analysis of fluid flow, heat transfer, and deformation behavior of a solidifying shell. *Metall. Mater. Trans.* **2000**, *31*, 225–237. [CrossRef]
2. Teskeredžić, A.; Demirdžić, I.; Muzaferija, S. Numerical method for heat transfer, fluid flow, and stress analysis in phase-change problems. *Numer. Heat Transf. Part B Fundam.* **2002**, *42*, 437–459. [CrossRef]
3. Koric, S.; Thomas, B.G. Efficient thermo-mechanical model for solidification processes. *Int. J. Numer. Methods Eng.* **2006**, *66*, 1955–1989. [CrossRef]
4. Li, C.; Thomas, B.G. Thermomechanical finite-element model of shell behavior in continuous casting of steel. *Metall. Mater. Trans. B* **2004**, *35*, 1151–1172. [CrossRef]
5. *Abaqus/Standard User's Manual Version 2019*; Simulia Dassault Systèmes: Providence, RI, USA, 2019.
6. Koric, S.; Hibbeler, L.C.; Thomas, B.G. Explicit coupled thermo-mechanical finite element model of steel solidification. *Int. J. Numer. Methods Eng.* **2009**, *78*, 1–31. [CrossRef]
7. Koric, S.; Thomas, B.G.; Voller, V.R. Enhanced Latent Heat Method to Incorporate Superheat Effects into Fixed-Grid Multiphysics Simulations. *Numer. Heat Transf. Part B Fundam.* **2010**, *57*, 396–413. [CrossRef]
8. Koric, S.; Hibbeler, L.C.; Liu, R.; Thomas, B.G. Multiphysics Model of Metal Solidification on the Continuum Level. *Numer. Heat Transf. Part B Fundam.* **2010**, *58*, 371–392. [CrossRef]
9. Zappulla, M.L.; Cho, S.-M.; Koric, S.; Lee, H.-J.; Kim, S.-H.; Thomas, B.G. Multiphysics modeling of continuous casting of stainless steel. *J. Mater. Process. Technol.* **2020**, *278*, 116469. [CrossRef]

10. Zhang, S.; Guillemot, G.; Gandin, C.-A.; Bellet, M. A Partitioned Solution Algorithm for Concurrent Computation of Stress–Strain and Fluid Flow in Continuous Casting Process. *Met. Mater. Trans. A* **2021**, 1–18. [CrossRef]
11. Cai, L.; Wang, X.; Wei, J.; Yao, M.; Liu, Y. Element-Free Galerkin Method Modeling of Thermo-Elastic-Plastic Behavior for Continuous Casting Round Billet. *Met. Mater. Trans. B* **2021**, 1–11. [CrossRef]
12. Huitron, R.M.P.; Lopez, P.E.R.; Vuorinen, E.; Jentner, R.; Kärkkäinen, M.E. Converging criteria to characterize crack susceptibility in a micro-alloyed steel during continuous casting. *Mater. Sci. Eng. A* **2020**, *772*, 138691. [CrossRef]
13. Li, G.; Ji, C.; Zhu, M. Prediction of Internal Crack Initiation in Continuously Cast Blooms. *Met. Mater. Trans. B* **2021**, 1–15. [CrossRef]
14. Rong, Q.; Wei, H.; Huang, X.; Bao, H. Predicting the effective thermal conductivity of composites from cross sections images using deep learning methods. *Compos. Sci. Technol.* **2019**, *184*, 107861. [CrossRef]
15. Goli, E.; Vyas, S.; Koric, S.; Sobh, N.; Geubelle, P.H. ChemNet: A Deep Neural Network for Advanced Composites Manufacturing. *J. Phys. Chem. B* **2020**, *124*, 9428–9437. [CrossRef] [PubMed]
16. Abueidda, D.W.; Koric, S.; Sobh, N.A. Topology optimization of 2D structures with nonlinearities using deep learning. *Comput. Struct.* **2020**, *237*, 106283. [CrossRef]
17. Kollmann, H.T.; Abueidda, D.W.; Koric, S.; Guleryuz, E.; Sobh, N.A. Deep learning for topology optimization of 2D metamaterials. *Mater. Design* **2020**, *196*, 109098. [CrossRef]
18. Spear, A.D.; Kalidindi, S.R.; Meredig, B.K.; Le Graverend, A.J.-B. Data driven materials investigations: The next frontier in understanding and predicting fatigue behavior. *JOM* **2018**, *70*, 1143–1146. [CrossRef]
19. Mozafar, M.; Bostanabad, R.; Chen, W.; Ehmann, K.; Cao, J.; Bessa, M. Deep learning predicts path-dependent plasticity. *Proc. Natl. Acad. Sci. USA* **2019**, *116*, 26414–26420. [CrossRef]
20. Abueidda, D.W.; Koric, S.; Sobh, N.A.; Sehitoglu, H. Deep learning for plasticity and thermo-viscoplasticity. *Int. J. Plast.* **2021**, *136*, 102852. [CrossRef]
21. Kozlowski, P.F.; Thomas, B.G.; Azzi, J.A.; Wang, H. Simple constitutive equations for steel at high temperature. *Met. Mater. Trans. A* **1992**, *23*, 903–918. [CrossRef]
22. Zhu, H. Coupled Thermo-Mechanical Finite-Element Model with Application to Initial Solidification. Ph.D. Thesis, The University of Illinois at Urbana-Champaign, Urbana, IL, USA, May 1996.
23. Fachinotti, V.D.; Le Corre, S.; Triolet, N.; Bobadilla, M.; Bellet, M. Two-phase thermo-mechanical and macrosegregation modelling of binary alloys solidification with emphasis on the secondary cooling stage of steel slab continuous casting processes. *Int. J. Numer. Methods Eng.* **2006**, *67*, 1341–1384. [CrossRef]
24. Zhu, J.Z.; Guo, J.; Samonds, M.T. Numerical modeling of hot tearing formation in metal casting and its validations. *Int. J. Numer. Methods Eng.* **2011**, *87*, 289–308. [CrossRef]
25. Hochreiter, S.; Schmidhuber, J. Long Short-Term Memory. *Neural Comput.* **1997**, *9*, 1735–1780. [CrossRef]
26. Cho, K.; van Merrienboer, B.; Gulcehre, C.; Bahdanau, D.; Bougares, F.; Schwenk, H.; Bengio, Y. Learning Phrase Represen-tations using RNN Encoder-Decoder for Statistical Machine Translation. *arXiv* **2014**, arXiv:1406.1078.
27. Pattanayak, S. *Pro Deep Learning with TensorFlow, A Mathematical Approach to Advanced Artificial Intelligence in Python*, 1st ed.; Apress Media LLC, Springer Media: New York, NY, USA, 2017; pp. 262–277.
28. Alla, S.; Adari, S.K. *Beginning Anomaly Detection Using Python-Based Deep Learning*, 1st ed.; Apress Media LLC, Springer Media: New York, NY, USA, 2019; pp. 257–282.
29. Keras, Chollet, François. Available online: https://github.com/keras-team/keras (accessed on 8 February 2021).

 *metals*

Article

CFD Simulation Based Investigation of Cavitation Dynamics during High Intensity Ultrasonic Treatment of A356

Eric Riedel *, Niklas Bergedieck and Stefan Scharf

Institute of Manufacturing Technology and Quality Management, Otto-von-Guericke-University Magdeburg, Universitätsplatz 2, 39106 Magdeburg, Germany; niklas.bergedieck@ovgu.de (N.B.); stefan.scharf@ovgu.de (S.S.)
* Correspondence: eric.riedel@ovgu.de; Tel.: +49-391-67-57084

Received: 18 May 2020; Accepted: 16 November 2020; Published: 18 November 2020

Abstract: Ultrasonic treatment (UST) and its effects, primarily cavitation and acoustic streaming, are useful for a high range of industrial applications, e.g., welding, filtering, cleaning or emulsification. In the metallurgy and foundry industry, UST can be used to modify a material's microstructure by treating metal in the liquid or semi-solid state. Cavitation (formation, pulsating growth and implosion of tiny bubbles) and its shock waves, released during the implosion of the cavitation bubbles, are able to break forming structures and thus refine them. In this context, especially aluminium alloys are in the focus of the investigations. Aluminium alloys, e.g., A356, have a significantly wide range of industrial applications in automotive, aerospace and machine engineering, and UST is an effective and comparatively clean technology for its treatment. In recent years, the efforts for simulating the complex mechanisms of UST are increasing, and approaches for computing the complex cavitation dynamics below the radiator during high intensity ultrasonic treatment have come up. In this study, the capabilities of the established CFD simulation tool FLOW-3D to simulate the formation and dynamics of acoustic cavitation in aluminium A356 are investigated. The achieved results demonstrate the basic capability of the software to calculate the above-mentioned effects. Thus, the investigated software provides a solid basis for further development and integration of numerical models into an established software environment and could promote the integration of the simulation of UST in industry.

Keywords: aluminium; ultrasonic melt treatment; cavitation; CFD simulation; structure refinement

---

1. Introduction

Cavitation, caused by high-intensity ultrasonic treatment (UST), is used in a wide range of industrial applications and becomes more and more interesting for metallurgy and foundry processes. It can be used as an effective method for modifying a material's microstructure and to improve the material's mechanical properties [1,2]. Especially in the context of the treatment of aluminium alloys, many investigations were conducted in the last 20 years. Aluminium and its alloys are thereby of primary importance as they are used for the manufacturing of a wide variety of industrial components in automotive, aerospace, shipping and mechanical engineering. The UST of aluminium can be conducted in a liquid or semi-solid state [3–9]. In the process, extreme pressure amplitudes (higher than the material's cavitation threshold) are induced by an immersed sonotrode through high-frequency sinusoidal mechanical vibration, which causes the formation of cavitation bubbles during the phase of negative pressure amplitudes, pulsating growth and the collapse of cavities accompanied by energy release in the form of shock waves [1,10–12]. Since the pressure intensity decreases exponentially with growing distance, in most cases, cavitation activity is limited to a small cavitation cloud close to the sonotrodes tip [1,13]. The mentioned dynamics within the cavitation cloud, including temporal pressure and temperature changes, can evoke several refining metallurgical effects, mainly:

- Wetting: The aluminium melt inevitably contains low amounts of Al_2O_3, a by-product resulting from the hydrogen-absorbing reaction of aluminium and H_2O vapour. On the surfaces of these Al_2O_3 particles, hydrogen deposits prevent Al_2O_3 from wetting. Shock waves, resulting from the collapse of cavitation bubbles near the particle, remove the hydrogen deposits and make the Al_2O_3 particles available as nucleation sites for heterogeneous nucleation [9,14–16].
- Nucleation: The released shock waves result in a change of pressure ratios close to the collapsing bubbles. This leads to an increase of the alloy's solidification temperature, and thus to the formation of solid aluminium grains. Below liquidus temperature, a few of the grains are thermally stable and survive the drop to normal ambient pressure; they support microstructures' refinement as available nucleation sites [5,15–19].
- De-agglomeration: The releasing shock waves are able to separate agglomerated particles and, in this way, enhance the particle distribution and increase the amount of available nucleation sites within the scope of heterogeneous nucleation [1,16].
- Fragmentation: This effect takes place at temperatures below the liquidus temperature, when the alloy starts to solidify and dendrites form. Pulsating cavitation bubbles close to the dendrites and its roots bend the dendrite arms during pulsation regularly. Either during bending or through shock waves, dendrite arms break. The broken off dendrite arms henceforth are the basis for growing dendrite structures [1,16,20–26].

The knowledge of the development, size and dynamics of cavitation (zone) is of high importance for effective usage of the mentioned effects, e.g., for the process design of ultrasonic supported systems. In recent years, efforts for simulating ultrasonic treatment and its effects have increased. An overview of most of the published simulation studies so far was given in [27]. So far, much progress has been made, but looking at the efforts for simulating ultrasonic treatment in its entirety, it still is in the stage of development. While we presented a comprehensive approach for the simulation of UST in [27], the study lacked a detailed investigation of cavitation dynamics. Therefore, in this study, the capabilities of the established CFD (casting) simulation tool FLOW-3D, developed by Flow Science, Inc. (Santa Fe, NM, USA) to calculate and visualize cavitation in aluminium melt, here A356, were investigated.

2. Numerical Modelling

2.1. General

All modelling and simulation activities were performed using the CFD simulation software FLOW-3D v11.2, whereas for analysis purposes, the software FlowSight v11.2 was used. Both programs were developed by Flow Science, Inc. Within the scope of the investigations, a compromise had to be found between the resolution of the visual results (depending on the cell size) and the associated computing time (depending on the cell size, the time step definition and the CPU power). As a minimum requirement, the definition of the time steps should allow the most accurate reproduction of the sinusoidal curve on which the sonotrode movement is based (see Section 3.1). For this reason, the investigated time frame for the development of cavitation was set to 0.001 s, and the time step for calculation was set to 1.5625×10^{-6} s.

2.2. Meshing and Geometry

To minimize the required simulation time, only necessary and essential elements were modelled and simulated, without neglecting a real system. The fluid volume was $8 \times 8 \times 8$ mm (xyz), deducting the immersed sonotrode volume. The sonotrode used had a diameter d_s of 5 mm and an immersion depth of 4 mm, and its material was defined as ceramic. The selected dimensions allowed for a very small cell size with an acceptable amount of cells. In addition, particularly in the case of in situ experiments, sonotrodes with diameters of just several mm are used [19,25,28,29]. Since the investigations were conducted in isothermal conditions, heat transfer could be neglected. Thus, no sonotrode temperature was defined. The radiator movement was controlled by a sinusoidal translational velocity component in the z-direction, corresponding to an ultrasonic system with 20 kHz and a peak-to-peak amplitude of 35 μm, which leads to 20 calculated oscillations. The atmospheric pressure and temperature were set to 101,325 Pa and 293.15 K, respectively. Due to the reasons mentioned above, the cell size was 5×10^{-5} mm, and the primary three-dimensional system was discretised with 4,608,000 cells. That way, the corresponding visual results have a high resolution at acceptable computing times. To let acoustic waves propagate across the boundaries and avoid reflecting effects close to the radiator, multi-block meshing was defined. In this case, it means that the mesh block of the primary system (Mesh Block 1) is completely enclosed by another mesh block (Mesh Block 2) with a cell size of 1 mm. The boundaries of Mesh Block 1 were defined as inter-block mesh boundaries. Figure 1 displays the described modelling of Mesh Block 1.

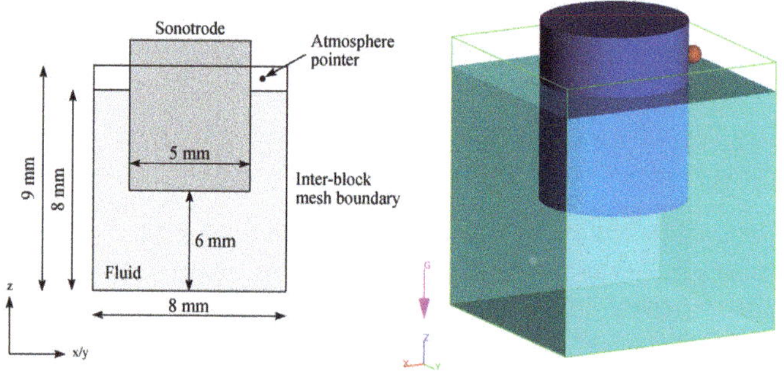

Figure 1. Geometric alignment of the simulation model setup.

2.3. Fluid

Fluid properties were set corresponding to aluminium silicon alloy A356 at 973.15 K, since the speed of sound within aluminium is known for this temperature from [30]. The main parameters used are presented in Table 1. To account for acoustic waves in fluids that would otherwise be treated as incompressible, the (limited) compressibility model was activated. That way, it is possible to simulate sharp pressure changes, like those occurring during UST. The compressibility coefficient β is defined as:

$$\beta = \frac{1}{\rho c^2} \tag{1}$$

where ρ and c are the respective fluid density and adiabatic speed of sound [31]. Therefore, the material's density and speed of sound can be considered for the calculation of acoustic pressure.

Table 1. Physical fluid parameters of A356 [30,31].

Parameter	A356	Unit
Density	2437	kg/m^3
Viscosity	0.0019	$kg/m/s$
Specific heat	1074	$J/kg/K$
Thermal conductivity	86.9	$W/m/K$
Liquidus temperature	881.15	K
Solidus temperature	825.55	K
Speed of sound	4600	m/s
Compressibility	1.94	$1/Pa$
Surface tension	0.871	kg/s^2

2.4. Cavitation Physics

Within the cavitation model, cavitation is measured by a transport equation, which calculates the advection, production and dissipation of the cavitation volume fraction, according to the following equations, which are the default equations for FLOW-3D v11.2 [31]:

$$\frac{DV_{cav}}{D_t} = Cav_{production} - Cav_{dissipation,} \tag{2}$$

$$C_p = C_e \frac{E_{turb}}{\sigma} \rho_l \rho_v \sqrt{\left[\frac{2}{3}\frac{p_{cav} - p}{\rho_l}\right]}(1 - f_{cav}) \tag{3}$$

$$C_d = C_c \frac{E_{turb}}{\sigma} \rho_l^2 \sqrt{\left[\frac{2}{3}\frac{p - p_{cav}}{\rho_l}\right]} f_{cav} \tag{4}$$

V_{cav} is the computed cavitation volume fraction; C_p is the cavitation production coefficient; C_d is the cavitation dissipation coefficient; C_e the evaporation coefficient; C_c the condensation coefficient; E_{turb} is the turbulent kinetic energy (alternatively 10% of the total kinetic energy if no turbulence model is selected); p_{cav} is the specified cavitation pressure; p is the local fluid pressure; σ is the material surface tension; and f_{cav} is the mass fraction of cavitation within the cell, with ρ_l and ρ_v as the densities of liquid and vapour. The coefficients used for cavitation production and dissipation are default values (0.02 and 0.01, respectively) that can be adjusted. The newly opened void region was treated as a fixed pressure bubble with the pressure equal to the cavitation pressure, which is a fixed parameter that currently does not depend on temperature [31]. Hydrogen gas at 973.15 K was assumed to be the gas species inside the cavitation bubbles. The respective gas density was calculated by the general gas equation:

$$pV = nR_ST \longrightarrow \frac{n}{V} = \frac{p}{R_ST,} \tag{5}$$

where p is the gas pressure, V the gas volume, n the amount of gas, R_m the universal gas constant and T the thermodynamic temperature of the gas [32]. Using the molar mass of H_2 and the data in Table 2, the gas density of cavitation bubbles was 0.025 kg/m^3. While in real cases, it is expected that the pressure within the bubbles changes with positive and negative pressure amplitudes resulting from the radiator, the average value still is the atmospheric pressure. Due to simplification, the average value of hydrogen density was used within a so-called one fluid simulation to reduce the complexity of the simulation model in which one density within the bubbles can be defined. Within the scope of the bubble and phase change model, the relationship between pressure, volume and temperature followed an adiabatic law. The pressure in each bubble was inversely proportional to its volume to the power of γ. Since the bubble and phase change model allows tracking collapsed bubbles with void particles, this feature was activated to analyse the distribution of cavitation activity within the analysed volume. For an accurate calculation and visualization, opening bubbles had to be resolved by a minimum of three cells across the diameter [31].

Table 2. Parameters and values used for the calculation of hydrogen density at 973 K [32].

Parameter	Value	Unit
Gas pressure p	101,325	Pa
Universal gas constant R_m	8314.41	J/(kmol K)
Temperature T	973.15	K
Molar mass M (H_2)	2.016	kg/kmol

The so-called general moving object (GMO) model calculates the radiator movement by a sinusoidal translational velocity component in the z-direction, corresponding to an ultrasonic system with 20 kHz and a peak-to-peak amplitude of 35 μm. The sinusoidal translational velocity is calculated by:

$$v(t) = \omega \cdot s_0 \cdot cos(\omega t + \phi_0), \tag{6}$$

with:

$$\omega = 2\pi f, \tag{7}$$

where $v(t)$ corresponds to the angular frequency, ω is the angular frequency with f as the frequency, s_0 is the amplitude (17.5 μm) and ϕ_0 is the phase angle. The GMO model calculates the kinetic energy, which is caused by the radiator movement, and its transfer on the fluid by momentum and mass conservation. For moving object/fluid coupling, the explicit numerical method was chosen, in which the fluid and GMO motions, i.e., the radiator, of each time step are calculated using force and velocity data from the previous time step. Besides gravitation and the activation of the GMO model for the movement of the radiator, the surface tension model, as well as the viscosity and turbulence model were of primary importance. All specific parameters and values used are listed in Table 3.

Table 3. Summary of the chosen models with the corresponding parameters.

Model	Parameter	A356	Unit
Bubble and phase change with adiabatic bubble and dynamic nucleation	Gamma	1.4	Without unit
	Pressure	101,325	Pa
Cavitation with empirical model for cavitation control active model for voids and activated cavitation potential model	Cavitation pressure (Cavitation threshold)	0	Pa
	Surface tension coeff.	0.871	kg/s²
	Density of cav.bubbles	0.025	kg/m³
	Cav. production coeff.	0.02	Without unit
	Cav. dissipation coeff.	0.01	Without unit
Surface tension model with explicit numerical approximation for surface tension pressure	Surface tension coeff.	0.871	kg/s²
	Temperature dependence	0	kg/s²/K
	Contact angle	90	Degrees

3. Results

3.1. Radiator Movement

First, the software's capabilities to follow the sinusoidal curve shape and to calculate the entered amplitude were analysed. The result is demonstrated in Figure 2. The chosen time step to calculate the oscillation can be considered sufficient to correctly generate the targeted amplitude of 35 μm. Both calculated amplitude and frequency correspond to the entered parameters and are not impaired by the chosen time steps. Therefore, the chosen time step can be considered a solid basis for the following investigations.

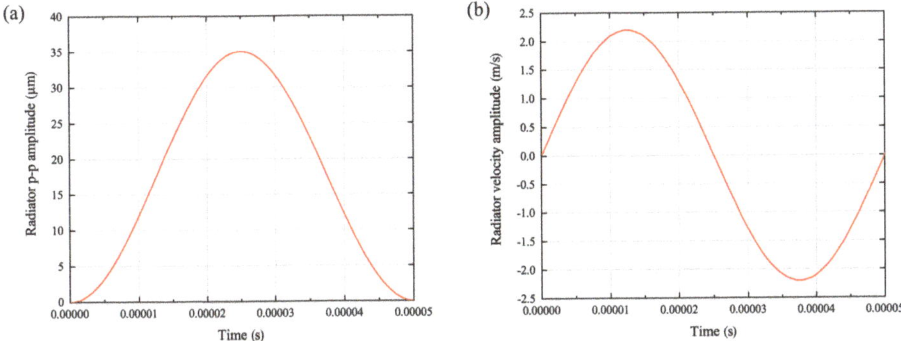

Figure 2. Analysis of calculated radiator movement: (**a**) actually calculated radiator amplitude and (**b**) velocity amplitude and frequency.

3.2. Cavitation

Figure 3 shows the juxtaposition of the cavitation volume fraction, pressure analysis and fluid fraction for five different times in the cells right below the radiator surface. The analysis of the cavitation volume fraction (Figure 3a) reveals a fast increase in cavitation activity within the first 0.0003 s. From there, the cavitation zone starts expanding in all directions in the form of little branches. The distribution of cavitation activity seems mostly homogeneous; only in some sectors, slightly stronger activities are measureable. The pressure analysis in Figure 3b) allows gaining better insight into the pressure conditions within the cavitation zone. It reveals the permanently varying pressure distribution within the cavitation zone and, in addition to it, the formation of locally occurring positive pressure hotspots, the values of which are much higher than the usual pressure amplitudes. Furthermore, the boundaries of the active cavitation zone towards the rest of the fluid are partially clearly recognisable. Volume fraction rendering (Figure 3c) highlights sectors with volume fraction values between 1.0 (100% fluid) and 0.0 (100% void). This way, e.g., the occurring bubbles or cavities are made visible, and better insight into the cavitation activity is possible. At t = 0.0001 s, a conspicuous ring structure is visible. Considering the radiator movement (Figure 2b), it becomes evident that the radiator reached its negative location peak at t = 0.0001 s. Therefore, it can be assumed that the ring structure arises from the fluid displacement caused by the radiator.

In addition to the cavitation volume fraction results shown in Figure 3a, the cavitation area dimensions in yz-layer in the cross-section of the sonotrode after t = 0.001 s are demonstrated in Figure 4. Starting from the radiator surface, the cavitation area also expands downwards. The cavitation area can thereby be divided in two zones: a primary zone (zone 1), where the cavitation activity is highest and a transition zone (zone 2) between zone 1 and the rest of the fluid, where the cavitation activity is continuingly decreasing to zero. Since the pressure amplitude decreases exponentially with growing distance to the radiator and cavitation dynamics are directly dependent on the pressure conditions, cavitation activity decreases as well.

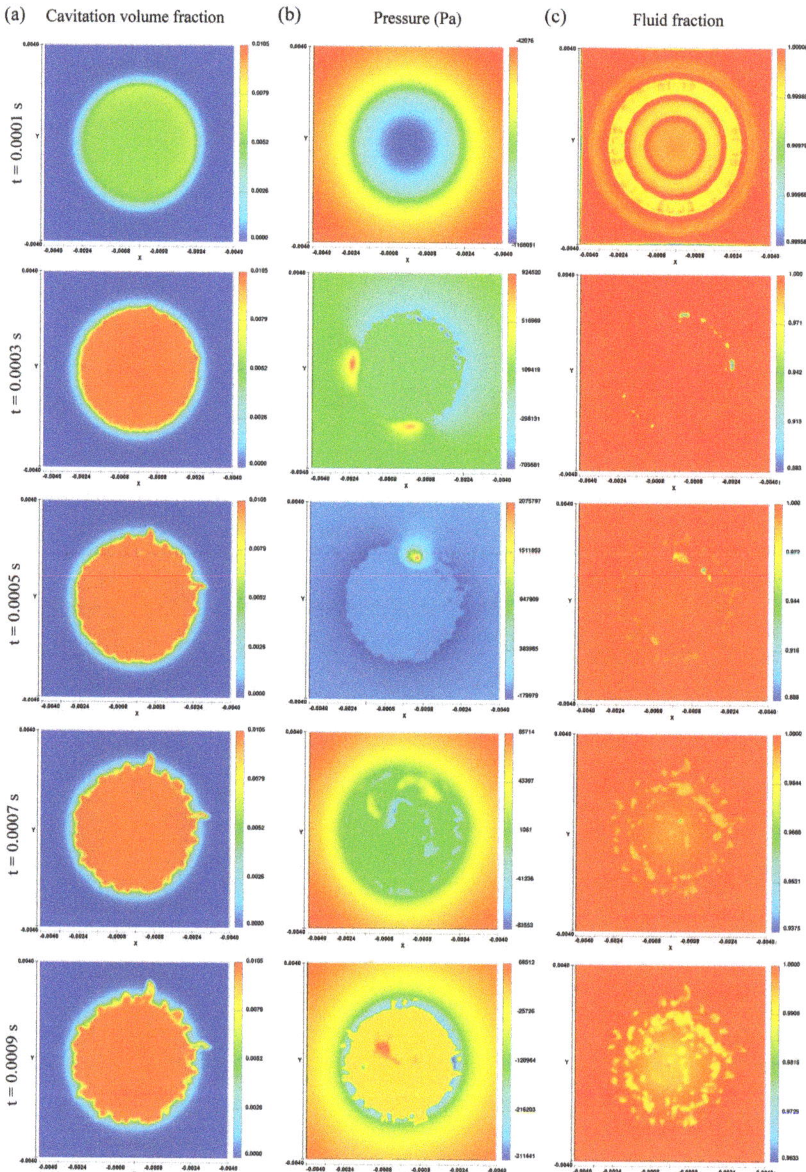

Figure 3. Development of the cavitation zone on the radiator's surface at different times: (**a**) cavitation volume fraction, (**b**) pressure and (**c**) fluid fraction.

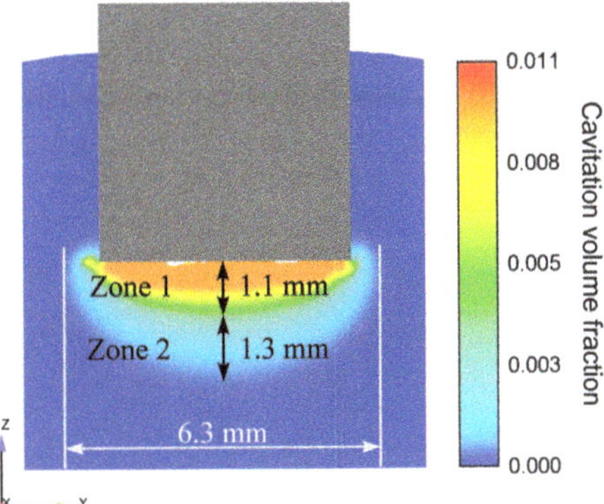

Figure 4. Dimensions of the cavitation zone in the central yz-layer after t = 0.001 s.

In addition to the results illustrated in Figure 3, the volume fraction distribution in and around emerging bubbles is calculated, as demonstrated in Figure 5. The displayed bubble in Figure 5a has a maximum size of four mesh cells, which corresponds to roughly 200 μm, and forms and collapses within 30 μs. Figure 5b shows the corresponding pressure values during bubbles' expansion and compression and after bubble collapse. Until the final collapse, the bubble is surrounded by a negative pressure field, which adjusts to the pressure values of the rest of the fluid with increasing distance to the bubble. After the bubble vanishes or collapses (t = 0.000240 s), a short pressure hotspot occurs.

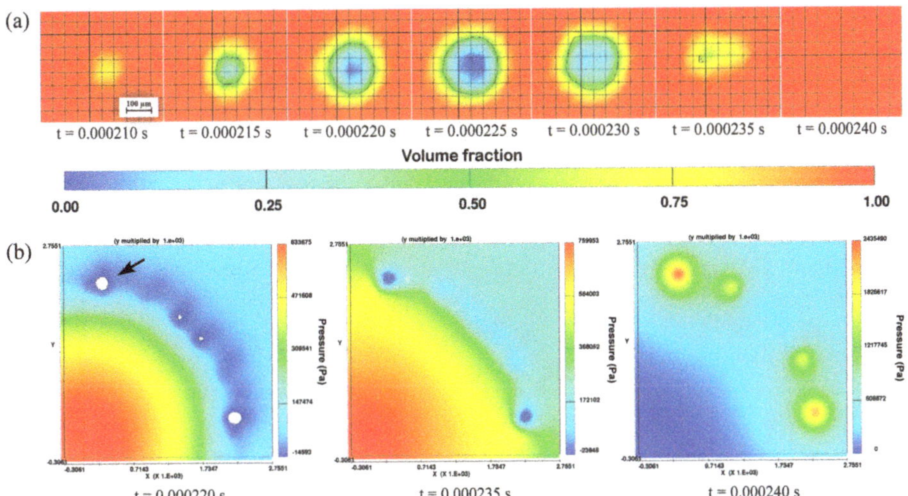

Figure 5. Bubble dynamics: (**a**) fluid fraction in and around the bubble and (**b**) pressure conditions in the area around the bubble and after collapse.

Apart from this, the volume fraction rendering shows that proceeding from the spherical radiator boundaries, more and more areas with volume fractions lower than 100% arise. From there, these areas increasingly appear right up to the radiator's surface centre with every oscillation and cover the whole radiator surface. Figures 6 and 7 support the observation that the occurring cavities seem to expand from the radiator surface edge to its centre. With every oscillation, the area without volume contact spreads in the direction of the surface centre. This occurs when the radiator is moving upwards, and a negative acoustic pressure builds up and disappears or collapses when the sonotrode changes its direction and a positive pressure amplitude builds up.

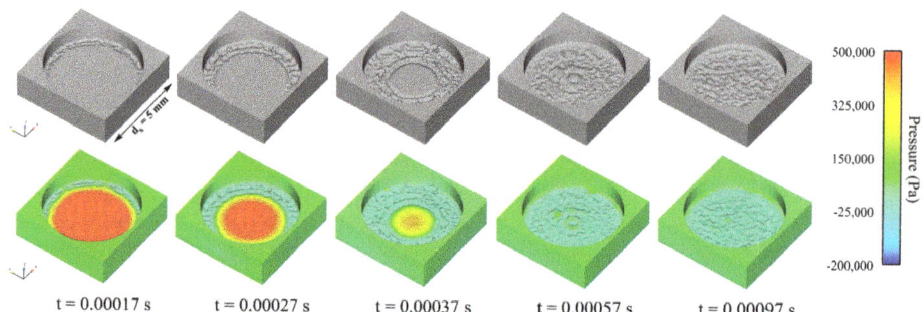

Figure 6. Expanding cavitation activity and respective pressure build-up.

3.2.1. Shielding Effect

Since it is known that cavitation influences the acoustic pressure propagation and that it is an important factor for cavitation dynamics, the pressure oscillation with and without the activated cavitation model was analysed. Figure 8 shows the comparison of the pressure developments in the centre line 4 mm below the radiator. Without the activated cavitation model, pressure oscillates sinusoidally around the ambient pressure of almost 100,000 Pa regularly, and no abnormalities are registered; whereas with the activated cavitation model, considerable fluctuations in pressure values are perceptible. It can be concluded that the cavitation activity has a damping influence on acoustic pressure development and propagation, even if pressure eruptions higher than the normal pressure oscillation (without cavitation) occur, due to collapsing bubbles, were measured.

3.2.2. Collapsing Bubbles

As already mentioned, the software is able to track collapsed bubbles via void particles as a quantitative measurement that fulfil the function of markers. Within the investigated time period (1000 μs), one-thousand seven-hundred fifty-two collapsed bubbles were registered. Figure 9 demonstrates the overlay of collapsing bubbles with the corresponding pressure values at that exact moment. It reveals that the collapse activity increases and decreases the more the positive pressure amplitude nears and leaves its local peak. Nearly all collapses take place close to the radiator surface; collapsing activity decreases with increasing distance to the radiator.

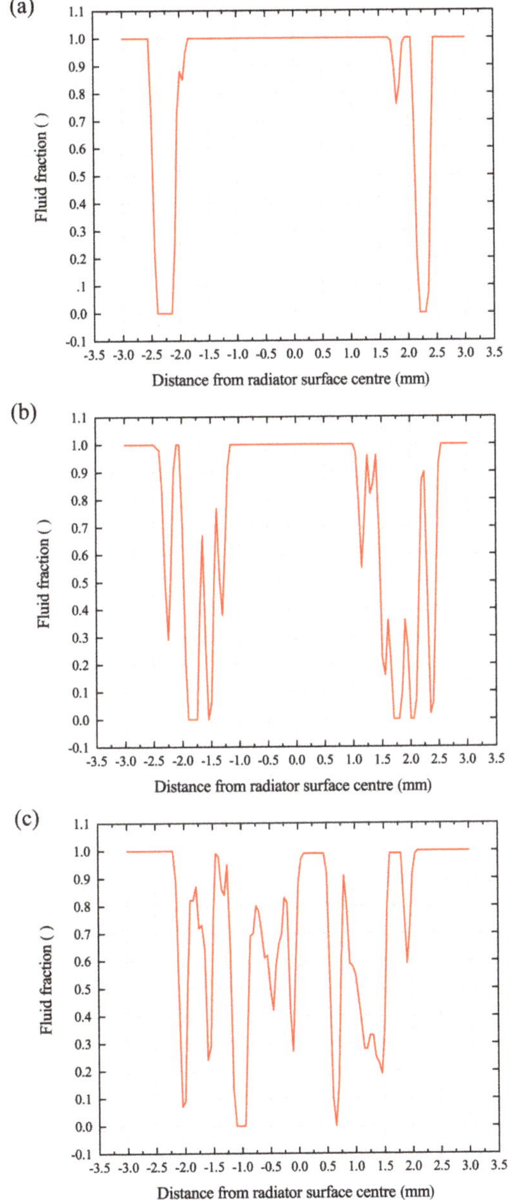

Figure 7. Fluid fraction analysis on the radiator surface: (**a**) t = 0.00027 s, (**b**) t = 0.00037 s and (**c**) t = 0.00097 s.

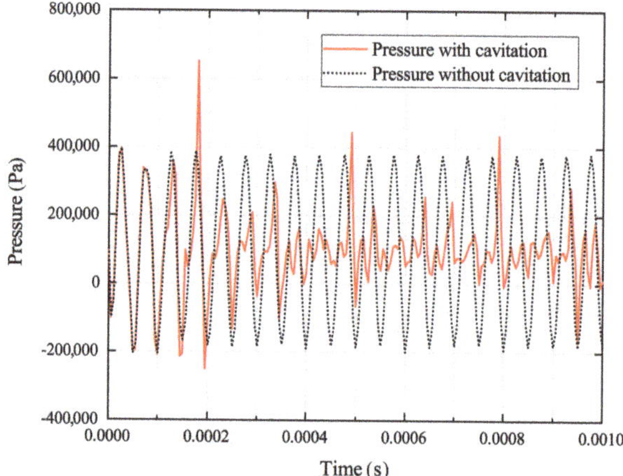

Figure 8. Comparison of acoustic pressure amplitudes 4 mm below the radiator with and without the activated cavitation model.

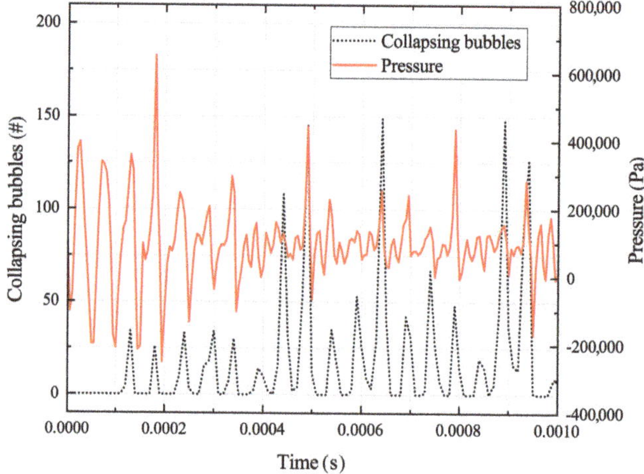

Figure 9. Bubble collapse distribution compared to pressure amplitudes.

4. Discussion

Based on the results obtained, it can be stated that the software is able to qualitatively simulate the acoustic cavitation effects. The accurate calculation of radiator movement, depending on a sufficient time step setup, was verified, as can be taken from Figure 2. The emergence and propagation of active cavitation areas, including cavitation volume fraction per cell, was calculated. The visualization of cavitation opening and closing depends on and is limited by the chosen cell size. For this reason, not all or just sufficiently sized cavitation bubbles are shown. A smaller cell size would result in a greater amount of opening and closing activity, but would require higher computing times. However, even if the cell size is too coarse and neglects opening and closing of every single bubble, the cavitation volume fraction indicated all areas of cavitation activity and collapsing bubbles were counted. In this context, the fluid fraction is another useful indicator for characterizing cavitation areas. Analysing the visualized cavitation bubbles, the qualitative accordance of bubble dynamics is recognizable: cavitation bubbles grow for a while and then suddenly contract. Since the pressure rendering allows for the pressure conditions around the bubbles and after their collapse to be analysed, it becomes clear that the pressure close to the bubble areas differs from the rest of the fluid. After the bubble collapse, positive pressure hotspots occur, adequate for the behaviour of shock wave formation after collapse. The amount of collapsed bubbles could therefore function as a quantitative benchmark for different ultrasonic systems and dimensions. The fact that the collapsing activity decreases with increasing distance to the radiator's surface matches the exponential decrease in oscillation energy and qualitative cavitation dynamic according to:

$$A = A_0 e^{-\alpha x} \tag{8}$$

$$I = I_0 e^{-\alpha x} \tag{9}$$

where A and I are the amplitude and intensity of a plane ultrasonic wave, α is the attenuation factor and x is the propagation distance [1].

Next to the fundamental simulation of cavitation dynamics, the influence of cavitation on the propagation of acoustic pressure is also taken into account. Even for a short investigated time frame, the correlation between cavitation development and the damping of acoustic pressures is simulated. Taking the expanding cavities on the radiator's surface into account, it stands to reason that they are one (major) factor of the shielding effect, as already described in literature. If open or void volumes exist on the surface, the radiator is not able to build up pressure in the same way it does at the areas where the radiator is in full contact with the fluid. This is logical, as the acoustic pressure P_A directly depends on the material's density and the speed of sound:

$$P_A = A_0 \rho c \omega \tag{10}$$

where $\omega = 2\pi f$ is the angular frequency [1]. A lower pressure built up at the void contact areas therefore will automatically lead to a lower acoustic pressure distribution.

From a holistic point of view, the software FLOW-3D (accurately) simulates acoustic cavitation and thus could be a helpful tool for the setup of ultrasonic systems and basic investigations (e.g., erosion analytics). Even if there is still a need for further development of the numerical models for the special application of ultrasonic treatment, FLOW-3D still represents a good basis for advanced numerical developments of acoustic cavitation models, e.g., detailed acoustic behaviour in different liquids. Due to the easy adjustment of fluid and system parameters like viscosity, surface tension, as well as frequency and amplitude, cavitation activity for different fluid states and ultrasonic systems can be investigated. Furthermore, the scale and cell size are largely unlimited within the software, thus allowing for very small-scale investigations with corresponding small cell sizes. Another interesting approach could also be so-called two fluid simulations, which are more complex, but allow a more comprehensive definition of another fluid or gas, i.e., hydrogen.

5. Conclusions

In this study, the capability of the CFD software FLOW-3D to simulate acoustic cavitation during ultrasonic treatment with a frequency of 20 kHz and a peak-to-peak amplitude of 35 μm within a time frame of 0.001 s was investigated. The results obtained can be summarized as follows:

- The occurrence, propagation and dynamics of cavitation can be simulated.
- The software allows the analysis of pressure conditions in and around the cavitation zone during bubble lifetime, as well as during and after collapse.
- Volume fraction along the bubble-fluid interface can be evaluated.
- Tracking of collapsed bubbles is possible.
- The influence of cavitation on pressure propagation (shielding effect) is taken into account.
- Further investigations should also take the possibility of so-called two fluid simulations into account.

In summary, it could be shown that some of the essential ultrasonic and cavitation effects can be calculated with the software used. The software thus provides a good basis for the further development of numerical, UST-specific models and thus could be a helpful tool for further developments of UST towards industrial application.

Author Contributions: Conceptualization, E.R.; data curation, E.R.; formal analysis, E.R.; funding acquisition, E.R. and S.S.; investigation, E.R.; Methodology, E.R.; project administration, E.R. and S.S.; resources, E.R.; software, E.R.; supervision, E.R.; validation, E.R.; visualization, E.R.; writing, original draft, E.R. and N.B.; writing, review and editing, E.R., N.B., and S.S. All authors read and agreed to the published version of the manuscript.

Funding: This research was funded by the Investment Bank Saxony Anhalt/European Regional Development Fund project ZS/2018/08/94143.

Conflicts of Interest: The authors declare no conflict of interest.

Abbreviations

The following abbreviation is used in this manuscript:

UST   Ultrasonic treatment

References

1. Eskin, G.; Eskin, D. *Ultrasonic Treatment of Light Alloy Melts*, 2nd ed.; CRC Press: Boca Raton, FL, USA, 2014.
2. Meek, T.; Han, Q. *Ultrasonic Processing of Materials*; Technical Report; U.S. Department of Energy—Energy Efficiency and Renewable Energy: Washington, DC, USA, 2006.
3. Eskin, G. Influence of cavitation treatment of melts on the processes of nucleation and growth of crystals during solidification of ingots and castings from light alloys. *Ultrason. Sonochem.* **1994**, *1*, S59–S63. [CrossRef]
4. Eskin, G. Broad prospects for commercial application of the ultrasonic (cavitation) melt treatment of light alloys. *Ultrason. Sonochem.* **2001**, *8*, 319–325. [CrossRef]
5. Wang, G.; Dargusch, M.; Qian, M.; Eskin, D.; StJohn, D. The role of ultrasonic treatment in refining the as-cast grain structure during the solidification of an Al–2Cu alloy. *J. Cryst. Growth* **2014**, *408*, 119–124. [CrossRef]
6. Zhang, X.; Kang, J.; Wang, S.; Ma, J.; Huang, T. The effect of ultrasonic processing on solidification microstructure and heat transfer in stainless steel melt. *Ultrason. Sonochem.* **2015**, *27*, 307–315. [CrossRef] [PubMed]
7. Eskin, D. Ultrasonic melt processing: Achievements and challenges. *Mater. Sci. Forum* **2015**, *828*, 112–118. [CrossRef]
8. Khalifa, W.; Tsunekawa, Y. Production of grain-refined AC7A Al–Mg alloy via solidification in ultrasonic field. *Trans. Nonferr. Met. Soc. China* **2016**, *26*, 930–937. [CrossRef]
9. Eskin, D. Ultrasonic processing of molten and solidifying aluminium alloys: Overview and outlook. *Mater. Sci. Technol.* **2017**, *33*, 636–645. [CrossRef]
10. Brennen, C. *Cavitation and Bubble Dynamics*; Oxford University Press: Oxford, UK, 1995.
11. Feng, H.; Yu, S.; Li, Y.; Gong, L. Effect of ultrasonic treatment on microstructures of hypereutectic AlSi alloy. *J. Mater. Process. Technol.* **2008**, *208*, 330–335. [CrossRef]
12. Leong, T.; Ashokkumar, M.; Kentish, S. The fundamentals of power ultrasound—A review. *Acoust. Aust.* **2011**, *39*, 54–63.

13. Tzanakis, I.; Lebon, G.; Eskin, D.; Pericleous, K. Investigation of the factors influencing cavitation intensity during the ultrasonic treatment of molten aluminium. *Mater. Des.* **2016**, *90*, 979–983. [CrossRef]

14. Huang, H.; Xu, Y.; Da, S.; Han, Y.; Jun, W.; Sun, B. Effect of ultrasonic melt treatment on structure refinement of solidified high purity aluminum. *Trans. Nonferr. Met. Soc. China* **2014**, *24*, 2414–2419. [CrossRef]

15. Tuan, N.; Puga, H.; Barbosa, J.; Pinto, A. Grain refinement of Al-Mg-Sc alloy by ultrasonic treatment. *Met. Mater. Int.* **2015**, *21*, 72–78. [CrossRef]

16. Eskin, D.; Tzanakis, I.; Wang, F.; Lebon, G.; Subroto, T.; Pericleous, K.; Mi, J. Fundamental studies of ultrasonic melt processing. *Ultrason. Sonochem.* **2019**, *52*, 455–467. [CrossRef] [PubMed]

17. Zhang, Y.; Kateryna, S.; Li, T. Effect of ultrasonic treatment on formation of iron-containing intermetallic compounds in Al–Si alloys. *China Foundry* **2016**, *13*, 316–321. [CrossRef]

18. Youn, J.; Kim, Y. Nucleation enhancement of Al alloys by high intensity ultrasound. *Jpn. J. Appl. Phys.* **2009**, *48*, 07GM14. [CrossRef]

19. Huang, H.; Shu, D.; Zeng, J.; Bian, F.; Fu, Y.; Wang, J.; Sun, B. In situ small angle X-ray scattering investigation of ultrasound induced nucleation in a metallic alloy melt. *Scr. Mater.* **2015**, *106*, 21–25. [CrossRef]

20. Wang, F.; Eskin, D.; Connolley, T.; Mi, J. Effect of ultrasonic melt treatment on the refinement of primary Al_3Ti intermetallic in an Al–0.4Ti alloy. *J. Cryst. Growth* **2016**, *435*, 24–30. [CrossRef]

21. Atamanenko, T.; Eskin, D.; Zhang, L.; Katgerman, L. Criteria of grain refinement induced by ultrasonic melt treatment of aluminum alloys containing Zr and Ti. *Metall. Mater. Trans. A* **2010**, *41*, 2056–2066. [CrossRef]

22. Shu, D.; Sun, B.; Mi, J.; Grant, P. A high-speed imaging and modeling study of dendrite fragmentation caused by ultrasonic cavitation. *Metall. Mater. Trans. A* **2012**, *43*, 3755–3766. [CrossRef]

23. Chow, R.; Blindt, R.; Chivers, R.; Povey, M. The sonocrystallisation of ice in sucrose solutions: Primary and secondary nucleation. *Ultrasonics* **2003**, *41*, 595–604. [CrossRef]

24. Wang, S.; Kang, J.; Zhang, X.; Guo, Z. Dendrites fragmentation induced by oscillating cavitation bubbles in ultrasound field. *Ultrasonics* **2017**, *83*, 26–32. [CrossRef] [PubMed]

25. Wang, F.; Tzanakis, I.; Eskin, D.; Mi, J.; Connolley, T. In situ observation of ultrasonic cavitation-induced fragmentation of the primary crystals formed in Al alloys. *Ultrason. Sonochem.* **2017**, *39*, 66–76. [CrossRef] [PubMed]

26. Wang, S.; Kang, J.; Guo, Z.; Lee, T.; Zhang, X.; Wang, Q.; Deng, C.; Mi, J. In situ high speed imaging study and modelling of the fatigue fragmentation of dendritic structures in ultrasonic fields. *Acta Mater.* **2019**, *165*, 388–397. [CrossRef]

27. Riedel, E.; Liepe, M.; Scharf, S. Simulation of ultrasonic induced cavitation and acoustic streaming in liquid and solidifying aluminum. *Metals* **2020**, *10*, 476. [CrossRef]

28. Tzanakis, I.; Xu, W.; Lebon, G.; Eskin, D.; Pericleous, K.; Lee, P. In situ synchrotron radiography and spectrum analysis of transient cavitation bubbles in molten aluminium alloy. *Phys. Procedia* **2015**, *70*, 841–845. [CrossRef]

29. Tzanakis, I.; Xu, W.; Eskin, D.; Lee, P.; Kotsovinos, N. In situ observation and analysis of ultrasonic capillary effect in molten aluminium. *Ultrason. Sonochem.* **2015**, *27*, 72–80. [CrossRef]

30. Tzanakis, I.; Lebon, G.; Eskin, D.; Pericleous, K. Characterizing the cavitation development and acoustic spectrum in various liquids. *Ultrason. Sonochem.* **2017**, *34*, 651–662. [CrossRef]

31. *FLOW-3D v11.2 Users Manual*; Flow Science, Inc.: Santa Fe, NM, USA, 2017.

32. von Böckh, P.; Stripf, M. *Technische Thermodynamik*; Springer: Berlin/Heidelberg, Germany, 2015.

Publisher's Note: MDPI stays neutral with regard to jurisdictional claims in published maps and institutional affiliations.

MDPI

St. Alban-Anlage 66

4052 Basel

Switzerland

www.mdpi.com

Metals Editorial Office

E-mail: metals@mdpi.com

www.mdpi.com/journal/metals

9 783725 810857